DAM PROJECTS AND THE GROWTH OF AMERICAN ARCHAEOLOGY

Dedication

This book is dedicated to those men and women of the IASP, RBS, and the cooperating institutions who set the path for those of us who followed.

DAM PROJECTS AND THE GROWTH OF AMERICAN ARCHAEOLOGY

The River Basin Surveys and the Interagency Archeological Salvage Program

Editors

Kimball M. Banks

Jon S. Czaplicki

Routledge
Taylor & Francis Group
LONDON AND NEW YORK

First published 2014 by Left Coast Press, Inc.

Published 2016 by Routledge
2 Park Square, Milton Park, Abingdon, Oxon OX14 4RN
711 Third Avenue, New York, NY 10017, USA

Routledge is an imprint of the Taylor & Francis Group, an informa business

Library of Congress Cataloging-in-Publication Data:

Dam projects and the growth of American archaeology: the River Basin Surveys and the Interagency Archeological Salvage Program/edited by Kimball M. Banks and Jon S. Czaplicki.
pages cm
Summary: "The Smithsonian Institution's River Basin Surveys and the Interagency Archeological Salvage Program were the most ambitious archaeological projects ever undertaken in the United States. Administered by the National Park Service from 1945–1969, the programs had profound effects—methodological, theoretical, and historical—on American archaeology, many of which are still being felt today. They stimulated the public's interest in heritage preservation, led to the passage of the National Historic Preservation Act, served as the model for rescue archaeology in other countries, and helped launch the 'New Archaeology.' This book examines the impacts of these two programs on the development of American archaeology"—Provided by publisher.
Includes bibliographical references and index.
ISBN 978-1-61132-174-6 (hardback) — ISBN 978-1-61132-176-0 (institutional eBook) — ISBN 978-1-61132-766-3 (consumer eBook)
1. Archaeology—United States—History—20th century. 2. Salvage archaeology—United States—History—20th century. 3. Archaeological surveying—United States—History—20th century. 4. Archaeology and state—United States—History—20th century. 5. Historic preservation—United States—History—20th century. 6. River Basin Surveys—History. 7. Interagency Archeological Salvage Program (U.S.)—History. 8. Dams—United States—Design and construction—History—20th century. 9. Dams—Environmental aspects—United States—History—20th century. I. Banks, Kimball M. II. Czaplicki, Jon S.
CC101.U6D36 2014
973—dc23

2013048333

ISBN 978-1-61132-174-6 hardback

Contents

Illustrations

Figures

Tables

Acknowledgments

This volume is dedicated to the men and women of the RBS. The objective is to call attention to the activities of the IASP, the RBS, and the other cooperating institutions that, more than any, affected the state of the discipline today. The authors wish to thank all those, including the contributors herein, who expressed interest in and encouraged the production of this volume. The authors also want to thank Fern Swenson and Amy Munson of the State Historical Society of North Dakota for their support, as well as Mark Lynott (retired) and Karin Roberts of the Midwest Archeological Center of the National Park Service for their assistance in research associated with the preparation of this volume. Mark Rudo of the National Park Service's San Francisco Regional Office tracked down the location of several RBS manuscripts. Special thanks go to Tom Thiessen (retired) of the Midwest Archeological Center of National Park Service, and Deb Hull-Walski of the Smithsonian Institution for their enthusiasm for the project, assistance in the research, lively discussions, and review and comments on the contributions. James Krakker also of the Smithsonian Institution assisted the authors in researching information during their visit to the Smithsonian. An especial thanks to Will Husted for his support and insights and James Richardson for allowing us to use his letters and photographs.

Preparation of this book was funded in part by the North Dakota Humanities Council, a nonprofit, independent state partner of the National Endowment for the Humanities. Any views, findings, conclusions, or recommendations expressed in this publication do not necessarily represent those of the North Dakota Humanities Council or the National Endowment for the Humanities.

CHAPTER 1

Introduction: The *Flood Control Act of 1944* and the Growth of American Archaeology

Kimball M. Banks and Jon S. Czaplicki

The year is 1944. The allied forces have turned the tide, and the possibility of an end to World War II can be envisioned. Congress and the president begin to contemplate a post-war America that would preclude a return to pre-war, depression-era conditions. The question that both faced was how to stimulate and shape a post-war economy. At the same time, America was undergoing a demographic shift that began during the depression. Americans had discovered the West. Population west of the Mississippi River was growing at an ever-faster pace, and the expectation was that this trend would continue when the vets—soldiers, sailors, marines, and airmen—returned. The problem was that much of the West was arid and could not support large, urban populations without adequate sources of water. Coupled with that, access to electricity was becoming increasingly common, especially in urban areas. Yet, precisely these urban areas—Los Angeles, Denver, Phoenix, Salt Lake City, Las Vegas—were growing the fastest.

Further, rivers in the West, such as the Missouri, Columbia, Sacramento, American, and Colorado, were subject to rapid fluctuations in levels, often going from low, slow flows during the summer and fall to severe, violent floods in the spring. The effect of this pattern was most evident in the Missouri River Basin (Basin). As recently as 1942, a rapid snowmelt resulted in the worst flood in the Basin in the previous 60 years. The river ravaged Omaha, Council Bluffs, Kansas City, and St. Louis. The result was millions of dollars in damage. The river encompasses 529,350 square miles, the largest drainage basin in the United States. In 1942, though,

---

Dam Projects and the Growth of American Archaeology: The River Basin Surveys and the Interagency Archeological Salvage Program, edited by Kimball M. Banks and Jon Czaplicki, 11–22.

the river was unregulated. Other than Fort Peck Dam built in the 1930s, the flow from Montana to the confluence with the Mississippi River was unchecked. Congress addressed these issues—employment for returning veterans, demographic changes, and flood control—in one package: the *Flood Control Act of 1944*. The program would employ veterans, help provide water to growing urban centers and agriculture, and attempt to manage rivers that previously had not been controlled. In doing so, with this act Congress inadvertently and accidentally changed the face of American archaeology and, ultimately, how we practice the discipline today.

The *Flood Control Act of 1944*

The objective of the act was to exercise "jurisdiction over the rivers of the Nation through the construction of works of improvement, for navigation or flood control" (§ 1, P. L. 78–534). The act created one of the largest, if not the largest civil works program in American history. Initially the act authorized the expenditure of over $1 billion for the construction of more than 275 dams and reservoirs in 29 states. Almost every major drainage would be affected; although the greatest effort and expenditure of funds was for projects in the West. Almost two-thirds of the authorized projects were in basins west of the Mississippi, especially in the Basin and basins in Texas and California. Projects involved river channelization; the construction of dams and reservoirs, levees, and dykes; irrigation projects; and the construction of facilities for generating and transmitting hydropower. The U.S. Army Corps of Engineers (Corps) and the Bureau of Reclamation (Reclamation) were the primary federal agencies charged with overseeing and administering the projects.

The biggest impact was in the Basin. The Pick-Sloan Plan authorized 316 separate projects that included the construction of 112 dams, hundreds of miles of levees and other flood protection structures, the irrigation of 4.3 million acres, and the generation of 2.6 million kilowatts of hydroelectric power. The plan called for the inundation of 1,400 square miles along the mainstem (that is, the upper Missouri River). Except for Canyon Ferry Dam, which was located towards the headwaters of the river and constructed by Reclamation the Corps would build and operate five dams and reservoirs on the mainstem, primarily for flood control, hydropower generation, and lower basin navigation. The Corps constructed Fort Peck, the sixth dam on the mainstem, in the late 1930s, and they continue to operate it today. In addition to Canyon Ferry, Reclamation would construct and operate the dams on the tributaries primarily for irrigation and hydropower.

Most of the construction in the Basin occurred between the late 1940s and early 1960s but continued on into the early 1970s. By the time construction ceased within the Basin, six dams on the mainstem and 60 dams on the tributaries had been built. Seven dams now regulate the main stem.

Starting at the upstream end, these are Canyon Ferry, Fort Peck, Garrison, Oahe, Big Bend, Fort Randall, and Gavins Point. The dams on the tributaries are distributed throughout the Basin and are much smaller.

The five reservoirs in North and South Dakota inundated over 81 percent of the 756 miles of river in the two states (Lehmer 1971). The reservoirs have a combined storage capacity of 76 million acre feet and a surface area exceeding one million acres, and they have transformed one-third of the Missouri River ecosystem into lake environments. Three of the five dams—Garrison, Oahe, and Fort Randall—back up waters within the Fort Berthold, Standing Rock, Cheyenne River, Crow Creek, and Lower Brule Indian reservations. The construction of these dams and reservoirs resulted in the last major taking of Indian lands by the federal government. This taking adversely affected the tribes and necessitated the displacement of many tribal members, as the lands were their most productive agriculturally (Lawson 2012).

The *Flood Control Act* and the Archaeological Community

The magnitude of the program and its potential impact to archaeological sites awakened the archaeological community. Government-sponsored, archaeology salvage projects in conjunction with civil works projects came into vogue during the Depression. The most noticeable were those associated with the Tennessee Valley Authority. These salvage projects were funded as a means of employment. Concurrently with these government-sponsored projects, Congress passed the *Historic Sites Act of 1935*.

Armed with this history, the archaeological community, through the Committee for the Recovery of Archaeological Remains (CRAR), lobbied Congress and federal agencies to address the impact of dam and reservoir construction on archaeological sites (Wendorf and Thompson 2002). The result was arguably the most massive and extensive archaeological program ever undertaken in the United States (Figure 1.1). This program was the Interagency Archeological Salvage Program (IASP), which the National Park Service (NPS) administered. The IASP oversaw a number of components dedicated to salvage archaeology. Components included CRAR, the constructing agencies, and cooperating museums, historical societies, and universities. Each had a role in the IASP. The CRAR provided "quality control," review, and scientific oversight. The constructing agencies shared project information and helped the archaeological activities in limited ways. The NPS funded, administered, and coordinated the program. Cooperating organizations did the archaeological salvage activities. These organizations included museums, state historical societies, universities, and most prominently, the River Basin Surveys (RBS), which the Smithsonian Institution (SI) administered.

Figure 1.1 RBS field crew off to work. Photograph courtesy of the National Anthropological Archives, Smithsonian Institution (NAA 39SL00-50).

Congress funded the IASP through the NPS; the IASP, in turn, transferred funds to the RBS and to cooperating museums, state historical societies, and universities. Between 1946 and 1967, Congress appropriated $10,861,297 for the IASP. The NPS transferred approximately one-third of these appropriations to the SI for the RBS program and another two-thirds to contracts with cooperating organizations; the remainder the NPS spent directly (Thiessen 1999, Table 1). The RBS conducted far more extensive archaeological activity than any of the other cooperating institutions, and as such, it became the icon of the IASP effort (Figure 1.2).

Over one-third of the appropriated funds, $4,256,860, were expended on archaeological projects in the Basin, and the rest was spent on investigations elsewhere in the United States. The relatively greater amount of funds expended on the archaeology in the Basin reflected the number

Figure 1.2 An RBS camp. Photograph courtesy of the National Anthropological Archives, Smithsonian Institution (NAA 395T1-108).

of projects authorized there under the Pick-Sloan Plan; more projects were authorized for the Basin than elsewhere in the country. Adjusting for inflation and averaged from 1955, the total amount Congress spent on archaeological projects in today's dollars equaled approximately \$88,139,425. In the Basin, it was \$32,460,000—the price of a small dam today (Banks et al. 2012)! However, the amount spent on the RBS in the Basin pales in comparison to the costs of constructing the dams here. The construction of the six dams on the mainstem alone cost \$1,027,500,000; in today's dollars that equals \$7,478,218,769. The budget for the archaeology in the Basin represents only 0.004 percent of the cost of just the six dams. If the costs for the other 60 dams in the Basin are considered, the percentage would even be smaller.

The RBS operated from 1945 to 1969 when it was terminated and its facilities and some of its personnel were transferred to the NPS, which continued to conduct IASP work through research contracts with cooperators and by undertaking limited research itself (mostly in the Oahe Reservoir). The IASP continued to operate until 1975 (Thiessen 1999). For the first 19 years, Frank H. H. Roberts, Jr. directed the RBS from his office at the SI. He set the direction and focus and provided overall coordination for the RBS program. The RBS and IASP undertook or sponsored projects throughout the United States. To coordinate RBS activities, Roberts established five field offices hosted by various universities. These were located at the University of Nebraska in Lincoln (the Missouri Basin Project 1946–1962) (Figure 1.3); the University of Oregon

Figure 1.3 The Missouri Basin Project office in Lincoln, Nebraska. Photograph courtesy of the National Anthropological Archives, Smithsonian Institution (NAA 00L177).

Figure 1.4 The distribution of RBS project offices and the number of RBS projects in the various states.

in Eugene (Columbia Plateau 1946–1952); the University of Texas in Austin (Texas and Oklahoma 1946–1952); the University of California, Berkeley (California 1946–1952); and the University of Georgia in Athens (Georgia and surrounding states 1950–1952) (Figure 1.4). In addition, Roberts administered projects in the middle Atlantic and northeastern states that were run out of the Washington office. Located just down the mall from Robert's office at the SI was the Department of the Interior and the NPS headquarters. Coordination between the SI and the IASP over the ensuing years was crucial for the success of the RBS. The organization that Roberts set in motion lasted through the life of the RBS.

The 24 years that the RBS operated makes it one of the most enduring, if not the longest-lasting archaeology program in the United States (Figure 1.4). Although archaeological projects were undertaken throughout the country, the RBS and IASP focused their activities in the western states in tandem with the majority of the construction projects. Of the 22 states west of the Mississippi, the RBS and NPS-funded cooperators were active in all but Nevada, with the largest amount of funds and effort expended in the Basin. One of the most important legacies of the RBS and IASP was the impact these programs had on the legislative framework that now guides historic preservation in the United States. In 1966, the *National Historic Preservation Act* (NHPA) was enacted. The NHPA passed into law three years before the *National Environmental Policy Act*, the most comprehensive piece of environmental legislation and which enfolds historic preservation. Congress passed NHPA in part because of the public interest in historic preservation that the RBS and IASP spurred (Figure 1.5) (cf. Baldwin 1966; Silverberg 1967). The NHPA recognized that the federal government had an inherent interest in and a responsibility to preserve our nation's heritage. Amendments to the act also recognized that historic preservation activities were allowable

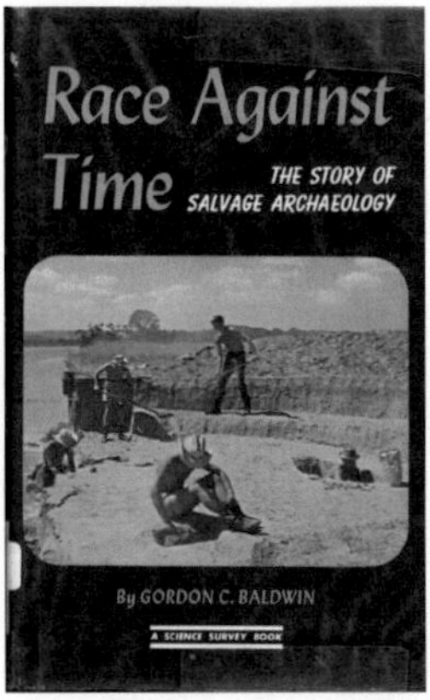

Figure 1.5 One of two books for the general public on salvage archaeology.

project costs and that Native Americans were direct participants in the preservation of sites important to their heritage or traditional practices. In 1974, Congress passed the *Archaeological and Historic Preservation Act* which further legitimized and strengthened the concept that historic preservation activities were a legitimate expenditure.

Because of their geographical reach and duration, the RBS and IASP had major impacts on American archaeology. Unknown to the practitioners at the time, the RBS represented a major turning point in the growth and development of American archaeology in many ways (Banks 1994).

- The IASP, with the RBS as its major component, ushered in the era of cultural resource management (CRM)—then called "salvage archaeology"—that dominates American archaeology today (Hester 1963; Jennings 1985).
- The RBS regularly employed a multidisciplinary approach to site investigation (Jennings 1985; Wood 2006, 2011).
- The RBS implemented laboratory and field techniques and approaches to doing archaeology that are still employed today.
- The RBS and other cooperating institutions employed and trained a large number of the next generation of archaeologists, both men and women, who in turn, changed American archaeology and are still influencing it today.
- The IASP, along with the RBS and other cooperating institutions, helped reawaken public interest in America's archaeological and historic heritage and helped spur passage of major cultural resource and historic preservation legislation in the 1960s and 1970s. The most important is the *National Historic Preservation Act*, which continues to protect our cultural heritage.
- The IASP inexorably linked federal construction projects and cultural resource preservation; funding of these programs led the federal government to recognize that it had a responsibility to preserve and protect our national heritage and that the expenditure of funds on historic preservation is a legitimate project cost.

- The projects that the RBS and cooperating institutions undertook expanded our understanding and interpretation of archaeology and constituted the foundation of current research in many parts of the United States. Many of the reports, papers, and documents remain the baseline data in numerous regions (Petsche 1968).

In short, RBS archaeology was not academic archaeology. The RBS and IASP represented a new way of doing archaeology, and as such, it transformed American archaeology (Jennings 1985). Many of the leaders in the field today might have entered other professions if it were not for their employment in the RBS and IASP. These factors were the impetus for this volume.

About This Volume

The volume explores the history of the RBS and IASP and their contributions to American archaeology. Part one explores the relationship between the SI, the RBS, the NPS, and the IASP. The chapters in part two examine the activities of each of the individual offices and the projects in their respective regions. Part three separately considers the contributions of the programs to physical anthropology, historical archaeology, and paleontology. The contributions in part four examine the social impacts of the programs, specifically impact to Native Americans and the role of women in the RBS. The book closes with part five: reflections on life in the RBS and IASP and, finally, how the two programs and earlier federally funded archaeological projects impacted the way we practice American archaeology today.

The National Park Service and the Smithsonian Institution

Thomas D. Thiessen, Deborah Hull-Walski, and Lynn M. Snyder's chapter **The National Park Service and the Smithsonian Institution: Partners in Salvage Archaeology** focuses on the roles the SI and the NPS played in the development, funding, administration, and success of the RBS and IASP and the relationship between the two. They discuss the role of the IASP in the NPS and the crucial part it played in funding the RBS and cooperating institutions throughout the programs' existence and the SI's role in the development and administration the RBS.

The RBS and IASP Programs

W. Raymond Wood's chapter **The Lincoln Office and the Upper Missouri River Basin** is a personal reflection on his experience as a graduate student working in the Missouri Basin Projects' office of the RBS. The Missouri Basin Projects' office was the most active and longest-enduring of all the field offices. His experience with the RBS was the stimulus for and the beginning of over 40 years of professional work focusing on the prehistory and history of the upper middle Missouri region. It is through the lens of this experience that he looks back on his years with the RBS.

California hosted a number of RBS projects primarily on the American and Sacramento rivers, which are the major rivers in the state, and their tributaries. Michael J. Moratto and Francis A. Riddell address these projects in their chapter **Archaeological "Appraisals" of Twenty Stream Basins in California: Investigations by the Smithsonian Institution River Basin Surveys, 1947–1951**. This chapter incorporates one of the last papers of Francis Riddell, one of many archaeologists who came to the fore in the California projects.

R. Lee Lyman addresses the projects administered in the Eugene, Oregon field office in the chapter **The River Basin Surveys in the Columbia Plateau**. The Eugene field office oversaw projects in the Pacific Northwest, especially in the Columbia River Basin. These projects put the Northwest on the archaeological map.

The field office at the University of Texas, Austin, administered and managed projects in Texas and Oklahoma. Edward B. Jelks examines the activities in Texas in **Archaeological Salvage at Texas Reservoir Construction Projects: 1945–1969**. Jelks directed the office during the period when the office closed, then reopened under the NPS, and was finally transferred to the University of Texas. It was through his involvement in these activities that he became not only a major figure in Texas archaeology but the ideal person to review the accomplishments of the RBS/IASP in Texas.

The work of the RBS in Georgia, Alabama, Tennessee, and other southeastern states has not often been recognized or acknowledged. Richard A. Krause corrects this situation in **Federal Archaeology in the Southeast**. Krause participated in IASP-funded projects as a graduate student. He puts the activities of the RBS in an historical context by looking for the roots of the projects in federal archaeological projects during the 1930s and earlier.

Although the Washington, DC office administered the entire RBS program, the office also managed and administered specific projects in the several states in the Northeast and middle Atlantic regions. Susan C. Prezzano's **Contributions of the Washington Office to the Archaeology of the Northeast and middle Atlantic** regions looks at these projects. Similarly, Prezzano looks for the roots of some of the projects in the federal archaeological projects undertaken in the 1930s.

Don D. Fowler reviews IASP projects in **Upper Colorado River Archaeological Salvage Projects**. Although late in the history of the IASP, the Glen Canyon Project was one of the biggest and most publically visible in the Southwest during this period. As a graduate student, Fowler participated on the Glen Canyon Project under Jesse Jennings. Fowler's long-time interest in the history of archaeology in the Southwest and intimate knowledge of the Glen Canyon Project make him the ideal person to take a retrospective look at this major IASP program overseen by the National Park Service.

The RBS and IASP Impacts on American Archaeology

Several American physical anthropologists cut their teeth as students on RBS and IASP projects and became leaders in the field. Most of the physical anthropological activities were in the Missouri River Basin, where many sites yielded large collections of human remains. Vanessa A. Mirro's **Contributions to Physical Anthropology by the Smithsonian Institution's River Basin Surveys** reviews the sites' histories with the projects and later contributions.

Similarly, other graduate students who began their careers on RBS projects, especially those in the Missouri River Basin, were instrumental in the development of historical archaeology in the United States. To a large extent, projects that the RBS and cooperating institutions undertook led to the development of historic archaeology in the western United States. William B. Lees' **Missouri Basin Projects and the Emergence of Historical Archaeology on the Great Plains** reviews these people and their contributions.

Paleontological investigations were included in projects during the early years of the Missouri Basin Projects office. Theodore White directed most of these investigations; he also assisted archaeologists in identifying and interpreting faunal remains, thereby laying the foundation for archaeozoological investigation. John W. Hoganson's chapter **Paleontological Significance of the Missouri River Basin Surveys** looks at this little recognized aspect of the RBS.

Some facets of the RBS and IASP have received little or no attention during the intervening years since the end of these programs; several of these are the subject of the last four chapters. The success of the RBS and IASP is a tribute to the dedicated men and women who labored long hours in the field under conditions seldom encountered by archaeologists today. Two aspects of the RBS have yet to be appreciated: the role of women and the participation of and effects on Native Americans. Ruthann Knudson has enthusiastically addressed the role of women and the RBS and IASP in **Women in Reservoir Salvage Archaeology**. She enlightens readers with attitudes and prejudices that confronted women during most of the life of the RBS.

Native Americans living along the Upper Missouri River lost tribal lands under the *Flood Control Act* (Lawson 2012). As noted above, this loss represents the last major taking of Indian lands by the federal government. These lands were rich farmland where their ancestors had lived and farmed for centuries and, as such, were important historically and spiritually to the tribes. Ironically, some tribal members worked on RBS projects as laborers excavating sites where their ancestors or those of other tribes had lived. Gerard Baker's **The Missouri River: The Backbone to Survival** examines this sad but important facet of American history associated with the *Flood Control Act* and the RBS/IASP.

The next to the last chapter is a personal reflection on daily life in the RBS. Field seasons were often short, especially in the Plains. Field camps were usually isolated. Living conditions were primitive. Through his personal lens, David Mayer Gradwohl's **Flapping Tents, Outhouses, Hail Storms, and Running for Water: An Insider's View of the Life in Interagency Archeological Salvage Program Field Camps, 1950s–1970s** focuses on life in the field and a personal perspective that is both entertaining and educational. Field conditions encountered by RBS archaeologists are probably unfamiliar to many archaeologists today, but unforgettable to those who experienced them.

The final contribution is Francis P. McManamon's chapter **From RBS to CRM: Late Twentieth-Century Developments in American Archaeology.** McManamon examines the role the RBS, IASP, and the effect that the earlier federally funded archaeological programs had on the development of how American archaeology is practiced today. He also examines the history of the laws that today guide and govern the practice of CRM. Today, most American archaeologists work in the arena of CRM. McManamon, who served as the Departmental Consulting Archeologist for the Department of the Interior for 26 years, assesses the critical but under-appreciated role the RBS and IASP had in implementing and stimulating the changes to American archaeology that were just beginning as the RBS program ended.

Because research for this volume was exhaustive, all citations have been placed in the list of references at the end of this volume. In essence, this section represents an update of Petsche's 1968 compendium as citations subsequent to his publication are included. Petsche's volume and this section provide an estimation of the number of publications that were generated during or relate to the RBS and IASP.

The RBS and IASP: The Final Chapter

The RBS and IASP are not well known today to many archaeologists, and they are usually just a footnote, if that, in most textbooks on American archaeology. Yet, their impact on American archaeology has been significant (cf. Banks 1994). We hope this volume awakens archaeologists—professionals, students, and avocational enthusiasts, as well as the interested public—to the many contributions this massive federal undertaking made to the profession. Much was learned and much remains to be learned from the vast collections, notes, reports, and publications that accumulated from 30 years of survey, excavation, and analyses by a dedicated group of professional archaeologists, eager students, and interested volunteers.

We also hope this volume sparks an interest among professionals and students to take another look at these collections and archival data. Our hope is that this volume will help archaeologists today and tomorrow understand how the discipline got to where it is today. The words of James Richardson in a letter to his parents while serving on an RBS crew, sums it up best: "I'm having the time of my life here and I'm almost sure archaeology is my meat" (Richardson 1957, Letter dated August 9, 1975).

PART I

THE NATIONAL PARK SERVICE AND THE SMITHSONIAN INSTITUTION

CHAPTER 2

The National Park Service and the Smithsonian Institution: Partners in Salvage Archaeology

Thomas D. Thiessen, Deborah Hull-Walski, and Lynn M. Snyder

Introduction

For three decades, the U.S. government conducted a massive program to salvage archaeological resources threatened by dam construction throughout most of the 48 contiguous states. These dams were planned and constructed by the U.S. Army Corps of Engineers (Corps) and the Bureau of Reclamation (Reclamation). The archaeological salvage program was conducted through the coordinated efforts of scores of colleges and universities, museums, and historical and archaeological societies in conjunction with the National Park Service (NPS) and the Smithsonian Institution (SI). The work of most of the participants focused on specific reservoir projects and lasted for only one or a few seasons of fieldwork. However, the NPS and the SI participated on a continuous basis from the program's inception in 1945 until its end (in the case of the NPS) or nearly so (in the instance of the SI). Throughout the 30 years of the program's existence (1945–1975), the NPS served as the program's coordinator and interface between the two construction agencies and the research institutions.

The River Basin Surveys (RBS), a unit formed within the SI exclusively for salvage program research, existed from 1945 to the end of June, 1969. Several histories of the national salvage effort—focusing largely on the RBS—are available. The most comprehensive are Jennings (1985), Glenn (1994), and Wendorf and Thompson (2002). There are even two popular books that tell the story of the salvage program to young readers

Dam Projects and the Growth of American Archaeology: The River Basin Surveys and the Interagency Archeological Salvage Program, edited by Kimball M. Banks and Jon Czaplicki, 25–38.

(Baldwin 1966; Silverberg 1967). In addition, a summary article about the salvage program provides a brief, but broad, perspective on the program's history (Snyder et al. 2000).

Historical Beginnings

As World War II drew to a close, several prominent archaeologists, backed by some of the nation's leading scientific societies, lobbied for the federal government to take action to recover archaeological information threatened by a massive program of planned dam building throughout the United States. These men, led principally by Frederick Johnson of the Robert S. Peabody Foundation for Archaeology at Phillips Academy, formed a small advisory body of four members in early 1945. This group was called the Committee for the Recovery of Archaeological Remains (CRAR), and its purpose was to "stimulate government action for recovery of data threatened by construction of new reservoirs and to ensure that the resulting salvage archaeology met the highest professional standards" (Wendorf and Thompson 2002:318). The CRAR also had two liaison members representing the National Research Council and the SI. Frank H. H. Roberts, Jr., a highly respected archaeologist and the assistant chief of the Bureau of American Ethnology (BAE), was the SI's representative (Stephenson 1967). Throughout most of the CRAR's existence, the group served as a lobbying and advisory committee independent of the federal government.

In April, 1945, two CRAR members, Frederick Johnson and John Otis Brew, through the offices of Waldo Gifford Leland, Director of the American Council of Learned Societies, met with Arthur E. Demaray, then Associate Director (later Director) of the NPS (Wendorf and Thompson 2002:321–322). They found Demaray to be enthusiastic about a national program of archaeological data recovery and together they envisioned the program as a cooperative venture between the NPS and the SI. Under the *Historic Sites Act of 1935*, the NPS had broad legislative authority to conduct archaeological investigations for the ultimate purpose of preserving historic and archaeological resources of national importance. The *Park, Parkway, and Recreational Area Study Act of 1936* also authorized the NPS to provide planning assistance for recreational facilities to be administered by other agencies. The NPS thus viewed archaeological surveys as within its legislative authority (Kahler 1947:216; Wendorf and Thompson 2002:322). However, the NPS did not view research as part of its mission at the time and was not adequately staffed to undertake a national program of salvage archaeology (Glenn 1994:11; Kelly 1940; Wendorf and Thompson 2002:322). The SI, of course, was the traditional scientific research arm of the federal government and was staffed and equipped to undertake the necessary research for that purpose. Consequently, the NPS role was to administer the program and assist the dam-building agencies with the planning of museum facilities, and the SI would conduct the actual archaeological research.

Johnson and Brew then met with Alexander Wetmore, the Secretary of the SI, who was also enthusiastic about the potential relationship and proposed that the BAE carry out SI's research role under the leadership of Roberts (Wendorf and Thompson 2002:323). An agreement between the NPS and the SI was drawn up and signed by Wetmore for the SI and by Director Newton B. Drury for the NPS. The Secretary of the Interior approved the agreement on October 9, 1945, and thus established the Interagency Archeological and Paleontological Salvage Program (usually shortened to the Interagency Archeological Salvage Program [IASP]). Under the generally vague terms of the agreement, the IASP represented a loose relationship between the NPS and SI. The NPS provided recreational planning assistance to the Corps and Reclamation for their projects, and the SI, through the RBS, advised on the number, location, and importance of archaeological and paleontological resources in the reservoir areas. In time, the primary role of NPS in the IASP was as the liaison between the construction agencies and the RBS (which involved the exchange of information both ways). As such, the NPS facilitated field research by cooperating institutions and organizations and developed and justified cost estimates for IASP work (Snyder et al. 2000:18).

The River Basin Surveys and the Smithsonian Institution

To carry out its role in this cooperative relationship, the SI created the RBS as a branch within the BAE (Wendorf and Thompson 2002:324). From 1946, when fieldwork first began, until 1969 when the SI turned the program over to the NPS, the RBS was a major research program of the SI. Roberts was designated as director of the RBS in October, 1945, and remained in that capacity until 1963. Robert L. Stephenson became acting director in 1963 and Warren W. Caldwell became director in 1966 (Thiessen 1999:21).

Roberts, on staff with the BAE since 1926, was a well-respected and renowned archaeologist for his work at sites in the Southwest and at the Lindenmeier Site in northern Colorado. From the time of his appointment to the RBS until 1963, Roberts devoted most of his time administering the archaeology program rather than in actual fieldwork. As he put it, "I must now content myself with doing my archaeology by proxy" (Stephenson 1967:89). Roberts, however, was not an armchair participant in any aspect of the RBS. He established several field offices in regions of the United States where reservoir construction would be especially concentrated. The RBS projects were staged from these offices and augmented by researchers from the BAE office in Washington. Staff were hired for these offices, which generally were closely affiliated with institutions of higher learning. The Washington office oversaw projects associated with reservoirs in the middle Atlantic and northeastern states (Prezzano, this volume). For the vast Missouri River Basin (Basin), where more than 100 reservoirs were planned, an office called the Missouri Basin Project (MBP; Wood, this volume) was established in July, 1946, at the University of Nebraska in Lincoln. Other field offices were soon established in Eugene, Oregon (see Lyman, this volume); Berkeley, California (Moratto and Riddell, this volume); Austin, Texas (Jelks, this volume); and briefly at Athens, Georgia (Krause, this volume). All of these offices, except the one in Lincoln, were closed or turned over to the NPS during the early 1950s.

Roberts oversaw the development of the procedures by which the field offices were to operate, secured equipment and supplies for them, determined the senior staffing requirements, developed the budgets, outlined procedures for fieldwork, and with few exceptions visited every major excavation that the SI crews were working on during his annual inspection trips (Stephenson 1967:89). Although Roberts and the assistant secretary of the SI recognized that Roberts needed a well-qualified and experienced professional to work with him. Such an assistant did not materialize until 1963 when Stephenson became assistant director. Until then, Roberts made do with a secretary and a few other archaeologists, who, when they were not in the field, served as acting director in Roberts' absences.

During its 24 years of operation, the RBS employed hundreds of individuals in the field and offices, many on a seasonal or temporary basis (Gradwohl, this volume). In staffing the RBS, members of the SI's various anthropological staff were used whenever possible. Dr. Waldo R. Wedel, associate curator of archaeology at the U.S. National Museum (USNM) and a pioneer in Plains archaeology, was detailed in 1946 to the RBS to establish field headquarters at Lincoln and assemble personnel. Also from the USNM were Joseph R. Caldwell and Ralph Solecki. Caldwell was detailed and later transferred to RBS to conduct work in Georgia at the Allatoona Reservoir; Solecki was transferred to the RBS to work on sites in West Virginia. Dr. Philip Drucker, anthropologist on the regular staff of the BAE, was sent to direct work on the Columbia-Snake River Basin and to set up the field office in Eugene, Oregon. Dr. Gordon R. Willey, also of the BAE, was detailed to the RBS in 1947 for work in Tennessee.

NPS and IASP

On a national scope, the chief archeologist of the NPS, based in Washington, DC, oversaw NPS involvement in the IASP. Archaeologists based in each of the NPS regional offices oversaw regional coordination and contracting. NPS regional archaeologists administered contracts for IASP field research from the time that the NPS began to award such contracts (in 1950) through the end of the IASP program. However, after the closure of all the RBS field offices except the Lincoln office, archaeologists stationed in the NPS regional office in Omaha, Nebraska, coordinated NPS involvement with the RBS. From 1947 through 1975, these regional NPS representatives to the IASP successively included Jesse D. Jennings, Gordon C. Baldwin (later a writer of novels and popular books on anthropological subjects), Paul L. Beaubien, Wilfred D. Logan, and Carl R. Falk.

Of the NPS chief archeologists who were deeply involved in IASP coordination, John M. Corbett was a monumental, though under-appreciated, figure (Figure 2.1). A veteran of the Office of Strategic Services in World War II (a forerunner of today's Central Intelligence Agency), he served from 1948 until his retirement in 1972. His importance to the program and effectiveness in coordinating NPS involvement have been lauded in retrospect by contemporaries who survived him (Jennings 1994:159; McGimsey 2004:5; Wood 2011:44). Corbett is remembered for advocating innovative approaches to salvage research, especially remote sensing (McGimsey 2004:150; Wood et al. 1984:29; Wood, this volume).

Figure 2.1 John M. Corbett (center), Chief Archaeologist of the National Park Service, and Frank H. H. Roberts, Jr., Director, River Basin Surveys (right), standing by a River Basin Surveys vehicle parked in front of the Missouri Basin Project office at 1517 O Street, Lincoln, Nebraska. Lawrence Tomsyck, administrative officer of the MBP, stands at left. Photograph courtesy of the National Anthropological Archives, Smithsonian Institution (NAA00-L526), and the Midwest Archeological Center.

Funding the Programs

Initially, the program was funded by transferring money to NPS from the construction agencies, the Corps, and Reclamation. The NPS retained a small percentage for its program expenses and passed the remainder to the SI for operation of the RBS. Such funds were, in fact, transferred in 1946 and 1947, but in November, 1947, the Bureau of the Budget ruled that neither agency had legislative authority to fund archaeological research and pointed out that only the NPS possessed such authority (Thiessen 1999:15). Henceforth, the NPS requested annual appropriations for salvage work outside the Basin. Because Reclamation was designated the lead agency to integrate and coordinate the water resource development program in the Basin (through the Pick-Sloan Plan), funds for IASP work in the Basin—through estimates developed by the NPS—were part of Reclamation's annual budget request to Congress. Consequently, funds for Basin work were tracked separately from funds for work in other river basins. Funding levels for the IASP and the RBS fluctuated substantially, leading at times to RBS staff reductions and closure of field offices. In times of more ample funding, the RBS fielded up to 25 crews in a single fiscal year as in fiscal year 1963 (Roberts 1964).

Cooperation from Other Institutions

It was recognized early that the scale of the field research would be far too large for the SI to undertake alone. The CRAR, the RBS, and the NPS encouraged state and local universities, colleges, museums, and historical and archaeological societies to supplement the efforts of the SI by undertaking field survey and excavation projects in reservoir areas and thus contribute to the overall goals of the IASP (Wendorf and Thompson 2002:323, 324–325). During the first year of IASP fieldwork (fiscal year 1947), 23 institutions or organizations conducted field investigations or provided support for RBS operations. By the end of the RBS (1969), the number had grown to more than 77. The cooperators that worked under contract to the NPS after 1969 have not been tallied, so the actual number during the entire thirty-year span of the IASP is undoubtedly larger. In the early years, the cooperators' work was supported solely by the resources of each institution or organization; federal funds were not provided. However, this changed in 1950 when the NPS began to contract for salvage investigations on a cost-sharing basis (Roberts 1951:66). The contracts were meant to defray the cooperators' costs to participate in the IASP, with NPS contributing funds and the cooperators contributing money and "soft" resources (personnel and equipment) toward the work.

Paleontology and the IASP

Paleontology was included in the purview of the salvage program through explicit language in the 1945 memorandum of understanding (Thiessen 1999:87–88; Hoganson, this volume). The RBS staff included Theodore E. White, a paleontologist from Harvard University, who was typically assisted by one or two student helpers. From April, 1947, through mid-1953, these men searched for fossils of extinct animals in many of the same reservoirs where archaeological research was taking place. By fiscal year 1954, White appears to have transferred to the NPS, and regular RBS paleontological field research no longer took place. White was loaned to the RBS on a couple of occasions later to complete analysis and reports on paleontological data collected earlier. His work for the RBS included studies of the fossil remains of extinct animals, as well as analysis of faunal remains recovered from archaeological sites. He is recognized as one of the founders of zooarchaeology (Wood 2011:79–80 and this volume). Hoganson, this volume, summarizes White's contributions to paleontology through his activities with the RBS. Later on, the IASP again conducted paleontological research but only on a small scale. In 1972 and 1973 the Midwest Archeological Center—the NPS successor to the RBS Lincoln office—contracted with the University of Iowa for limited paleontological research that focused on recovery of fossil remains of large and small mammals in the Soil Conservation Service's Waubonsie Creek Watershed project area in Mills and Fremont counties, Iowa. Holmes Semken of the Department of Geology directed this research.

Reviewing RBS and IASP Activities

Each summer, NPS and RBS representatives, sometimes accompanied by archaeologists from cooperating institutions and members of the CRAR, traveled by automobile to visit RBS and cooperating institution field crews working in the Basin. John Corbett and Frank Roberts often led these inspection visits, along with the current head of the Lincoln RBS office and an NPS archaeologist from the Omaha regional office. See Wood, this volume, for a photograph of such an inspection party posing with members of a cooperator's excavation crew in the 1950s.

NPS and RBS activities in the IASP were reviewed at annual meetings with the CRAR, and they were usually hosted by the NPS in Washington, DC These two-day affairs were also attended by representatives of other federal agencies, who, along with NPS and RBS representatives, presented reports on archaeological activities conducted in association with projects of their respective agencies. Following passage of the *Federal Advisory Committee Act* in 1972, the CRAR lost a degree of its independence from the federal government and was only convened at the invitation of the

NPS as a formal advisory body to that agency. As the NPS phased out the IASP after passage of the *Archeological and Historic Preservation Act* (AHPA) in 1974, the NPS declined to convene meetings with the CRAR. The last CRAR meeting was held in 1976 and not at NPS expense as was customary (McGimsey 2004:132; Wendorf and Thompson 2002:327).

The End of the RBS and IASP

The RBS administration moved from the BAE in 1965 when the BAE and the NMNH Department of Anthropology merged to become the SI Office of Anthropology. At the same time, the USNM was formally split into the National Museum of Natural History (NMNH) and the National Museum of American History; the Office of Anthropology was placed under the NMNH. This office administered the RBS until 1967 when dissatisfaction with its progress in correcting deficiencies and complaints about administrative problems led to the placement of the RBS under the director of the Museum. Scientific leadership and direction for the RBS was to be provided through a "Scientific Advisory Committee" consisting of four SI Office of Anthropology staff members, including Waldo Wedel (Thiessen 1999:38).

Two subsequent memoranda of agreement between the NPS and the SI in 1961 and 1965 reaffirmed the cooperative relationship between the NPS and the RBS (Thiessen 1999:91–92, 95–96). These documents contained language requiring that the agreement be reviewed every three years so that it "can either be revised to fit new conditions or be terminated if it is no longer desirable" (Thiessen 1999:92, 96). The review that followed the 1965 agreement, which was conducted on January 30 and 31, 1968, led to the termination of the RBS the following year. By this point, the work of the RBS was largely confined to the Basin, and the NPS indicated a willingness to assume the transfer of the RBS responsibilities (Glenn 1994:12).

The NPS Midwest Archeological Center was established on July 1, 1969, from the facilities and staff of the RBS Lincoln office. The report of the ad hoc committee that conducted the program review in 1968 expressed concern about the delay in producing RBS publications that detailed research results. In addition, the ambiguity of the administrative relationships of the RBS to both the SI, of which the RBS was a part, and the NPS, which funded the RBS operation, was viewed with apprehension (Thiessen 1999:40). These concerns, plus the fact that all the Missouri River mainstem reservoirs were operational—which were the primary focus of RBS operations in the 1960s—led to the decision to terminate the RBS program. The NPS continued to justify and request funds for national IASP research, which was primarily accomplished through contracts with cooperating institutions though the Midwest Archeological Center conducted salvage investigations in Missouri River mainstem reservoirs for several more years (Snyder et al. 2000:19). The NPS's "external" archaeological program changed drastically in 1975 when offices of the newly formed Interagency Archeological Services program were opened in San Francisco, Denver, and Atlanta. That year marked the end of the IASP.

Several years after the RBS program ended, SI spokesman Clifford Evans offered a candid explanation to the CRAR for closing the program:

> The Smithsonian does not have the staff; the Smithsonian does not have the space; other organizations already charged with salvage, such as State archeologists, seem more logical to support such activities; if a national program of salvage archeology (or public archeology) is contemplated, it must have local backing and State support even if the funds are Federal; Congress requested the Smithsonian Institution to end some of its programs originally proposed for short-term, and from past experiences it would appear as if centralized coordination in a Federal agency, such as the Smithsonian Institution, is not the best solution when other Federal bodies are charged with the recording of sites of historical and scientific significance, etc. (Evans 1973:132–133).

Despite the Missouri River mainstem reservoirs having achieved operational status, ending the RBS in 1969 may have been premature in light of continuing resource destruction in these and other reservoir areas. About the time the RBS was terminated, NPS personnel at the Midwest Archeological Center recognized that archaeological resources were being lost in operational reservoirs because of erosion by fluctuations in reservoir pool levels and wave action. Consequently, shoreline surveys and excavations at eroding sites in some of the mainstem reservoirs were initiated in 1969 and continued through 1973 by means of a combination of contracts with cooperating institutions and crews fielded directly by NPS personnel from the Midwest Archeological Center (Thiessen 1994). Most of this work centered in the Oahe and Garrison reservoirs in South and North Dakota respectively, but because available funding was very limited, the overall effort was miniscule compared to the magnitude of the resource loss.

The IASP did not last long after the RBS program ended. The AHPA authorized federal agencies to spend money for archaeological investigations necessitated by their activities. This authorization led to the diminishment and eventual disappearance of NPS appropriations to assist other agencies with archaeological surveys and excavations. Many other agencies, among them the Corps, Reclamation, Bureau of Land Management, and U.S. Forest Service, began to add archaeologists to their staff and took responsibility for addressing their own archaeological resource needs. Nevertheless, these agencies are indebted to the IASP researchers who laid much of the groundwork for the eventual development of the subfield of archaeology that has come to be widely known as cultural resource management.

The RBS and Advancements in Field and Laboratory Procedures

The RBS, in particular, was responsible for methodological advances in the field of archaeology, including the careful use of earthmoving machinery in excavations, consultation of aerial imagery for the detection and interpretation of archaeological sites, and analysis and interpretation of zooarchaeological remains and dendrochronological samples as a routine aspect of investigations. The RBS also adopted the trinomial site numbering system, a standardized approach to site numbering originally adapted from earlier Works Project Administration research in Nebraska and now utilized in most states (Wood, this volume).

Keeping track of all the documentation and artifactual material generated by the RBS was an issue of concern from the beginning. As a result, a consistent set of field methods and recording procedures and formats was developed and instituted for field and laboratory operations. Standardized field and laboratory forms insured adequate documentation and context for recovered survey and excavation data, as well as the vast body of artifactual materials produced by the program. The completed forms were copied in triplicate and filed in both RBS field offices and the SI in Washington as a hedge against information loss.

A positive and practical aspect of the program was the creation of central laboratories. The laboratory procedures devised in the MBP Lincoln office were adopted by other RBS field offices and became a model for many of the cooperating institutions as well (Figure 2.2). In the laboratories, specimens were identified and processed; photography

Figure 2.2 Specimen processing desks. Records of the Lincoln, Nebraska, office. Photo by Wayne L. Nelson, December, 1964. Left to right: Lee Madison, Donald Blakeslee, Gaillard Jackson, Clarence Johnson, and John Ritch. Photograph courtesy of the National Anthropological Archives, Smithsonian Institution (NAA 00-L1288).

and drafting were carried out; maps, field notes, and other records were filed; and analytical work was carried out for the preparation of full technical reports for publication (Jennings 1985:293). The *Handbook for Basic Archeological Specimen Processing*, developed by Dean Clark in the MBP laboratory, provided further standardization. The standard methods that RBS staff used in documenting the collections ensured their continued availability and importance to future researchers (Thiessen and Roberts 2009).

Collections

The quantity and geographic scope of the RBS and IASP materials are unparalleled in the history of American archaeology. Surveys in more than 500 reservoir areas in 43 states led to the recording of an estimated 20,000 sites (Brew 1968:3; Snyder et al. 2000:16). Over 500 major excavations were conducted to further document significant prehistoric and historic sites prior to inundation. From California to Virginia, RBS and IASP fieldworkers collected over three million artifacts and specimens. In addition to artifactual materials, zooarchaeological assemblages were recovered along with carefully collected and documented paleobotanical specimens. An extensive series of soil samples also exists for many regions, and wood and charcoal samples suitable for radiocarbon dating and dendrochronological study were collected throughout the program. The collection of such scientific samples, however, was not done on a consistent basis and depended on the training and inclination of the investigators. Finally, extensive collections of human skeletal remains also were recovered (see Mirro, this volume).

The archival and archaeological collections produced by the RBS and IASP contain evidence that is critical to providing a detailed picture of centuries of prehistoric life and early contact with European traders and settlers. These collections also are crucially important for documenting the intellectual history of the discipline for a period in which American archaeology experienced significant change. During this post-war period, archaeology expanded in an unprecedented manner institutionally and organizationally, as the discipline redefined its primary objectives.

Publications

Despite chronic problems with analysis and reporting subsequent to fieldwork, the RBS and other IASP participants achieved a prodigious publication record. The SI published 39 River Basin Surveys Papers in its BAE Bulletin series and the Publications in Salvage Archeology series issued from the Lincoln office consists of 13 numbered volumes. A 1968 bibliography of works resulting directly from this program lists more than 2,600 published and unpublished reports associated with IASP projects (Petsche 1968). The IASP also produced a number of major scholarly syntheses, several of which Jennings (1985) noted in his review of the RBS. These syntheses included Wedel's (1961) masterful summary of Plains prehistory from Paleoindian to historic time; Lehmer's (1971) ordering of data from scores of newly dug sites to document and describe the Middle Missouri and Coalescent traditions in North and South Dakota; and Wood's (1967) monograph on the Huff Site and the prehistoric Mandan.

Indeed, the amount of scholarship resulting from the materials of the RBS and IASP is staggering—and yet the potential of the collections and data is far from exhausted. Many of the collections have never been studied or published. As contemporary archaeological philosophy and practices concentrate increasingly on conservation and preservation of archaeological resources, the RBS and IASP collections take on greater significance and research value. Also, as changing research interests prompt new questions and methods of materials analysis become increasingly sophisticated, anthropologists are eager to use this material (Department of Anthropology 2002).

Accessing the RBS and IASP Collections

The greatest strength of the salvage program collections for scholarly studies—their sheer size and comprehensive scope—does, however, pose challenges to effective information management and ease of access. Today, the archaeological collections and records generated by RBS and other IASP research are housed and curated in many repositories across the nation. The SI is a major repository for these materials. The NMNH, Department of Anthropology, for example, cares for over one million specimens from more than 2,500 RBS sites and localities; and, the National Anthropological Archives (NAA) serves as a major repository for RBS archival and documentary materials. The NAA houses more than 2,000 original site files, approximately 800 maps, field drawings, and illustrations, and nearly 20,000 field and laboratory photographs and 35mm color slides of RBS excavations. The NAA also retains extensive administrative records and correspondence of the Washington and regional offices, as well as the personal records and correspondence of many of the SI personnel associated with the RBS. These collections are located at the Smithsonian Museum Support Center in Suitland, Maryland. In addition, the SI Archives in Washington, DC contain RBS housekeeping files, memoranda of telephone conversations, photographic material, and limited audiovisual materials. Online access to these collections is available through the SI Collections Search Center.

The scattering of collections and the associated potential loss of data is an important challenge faced by contemporary and future archaeologists and researchers. Despite the establishment of central laboratories for the RBS, the collections were never gathered into a single permanent repository. Much to the frustration of researchers interested in archaeological surveys and excavations done in a particular locale or region, no convenient or ready way exists to ascertain the existence, extent, or location of many individual collections that may be associated with that work. Modern web technology, however, offers the potential for institutional website listings of RBS and IASP archaeological collections and records; such listings would be an invaluable tool for present-day and future researchers (Snyder et al. 2000:20). Commenting on the importance of online RBS resources, Dr. Edward B. Jelks (1995), an RBS pioneer, stated:

> Archeological field and analytical methods in the 1940s, 1950s, and 1960s were primitive in some respects when compared to more advanced methods in common use today. There is much to be learned about past cultures by reinterpreting RBS data through the use of more modern techniques and methods. . . . The heavy investment of the federal government—and of many universities and museums—in collecting these specimens and data should be protected by the implementation of a program to ensure that the RBS specimens and records are preserved, inventoried, and made available for future study by students and scholars.

In addition to the collections being scattered in various repositories, many RBS and IASP collections have suffered from inadequate facilities and curation. Overcrowded storage, lack of inventory control, and poor curatorial oversight potentially have damaged the research value of many IASP artifacts and associated documents. Many federal agencies and professional societies, including the Department of the Interior, the Corps, the Society of American Archaeology, and the Society for Historical Archaeology, are working to generate policies, standards, and guidelines for the curation of such archaeological collections and are seeking ways to correct curatorial problems (e.g., Ferguson 2001a, 2001b; Snyder et al. 2000:20).

Conclusion

The RBS and the other IASP cooperating institutions and organizations contributed in innumerable ways to the advancement of knowledge about America's archaeological past, and to the growth of American archaeology as a discipline. Through their efforts, basic culture history sequences were established for many regions of the nation, providing foundations for subsequent research that continues today. The RBS and

IASP research also created a rich legacy of collections and associated records that continue to have value for researchers and will do so for future generations of archaeologists.

Cooperation between RBS and the other IASP researchers stimulated the development and growth of anthropological and archaeological departments and programs at many local institutions of higher learning, as well as the training of future generations of archaeologists. Dozens of students who worked on RBS and cooperators' research projects went on to fruitful careers in archaeology.

And last but not least, the philosophy which underlay the IASP—that is, the study and recovery of archaeological data before its destruction by the advance of economic and technological progress—created a foundation for the development of cultural resource management precepts as they are known and practiced today (McManamon, this volume). Truly, American archaeological practitioners of today owe a huge debt to their RBS and other IASP forerunners.

PART II

THE RBS AND IASP PROGRAMS

CHAPTER 3

The Lincoln Office and the Upper Missouri River Basin

W. Raymond Wood

American archaeology virtually ground to a halt during World War II. Many of the nation's archaeologists were in the armed services, and resources were sparse and hard to obtain for those who were not. During the war, the U.S. Army Corps of Engineers (Corps) was fully engaged in war-related tasks: building airfields, repairing demolished bridges, and other such activities supporting our advancing combat troops.

That role was to change dramatically with the war's end. The Corps, together with the Bureau of Reclamation (Reclamation), was planning a massive dam-building program on virtually every stream in the Missouri River Basin (Basin), including five massive ones on the mainstem of the Missouri itself in North and South Dakota, as well as plans for other river systems in the United States. Numerous prominent archaeologists quickly banded together to address this threat to the nation's archaeological resources (Wendorf and Thompson 2002). Thiessen, Hull-Walski, and Snyder (this volume and Thiessen 1999) have documented the resulting development of the Interagency Archeological Salvage Program (IASP) and its role in the Basin between 1946 and 1975. The IASP ranks as one of the largest archaeological salvage programs ever undertaken in the United States and, at the time, in the entire world. Nationwide, there were 273 planned reservoirs in 29 states; nearly half of them, 105, were in the Basin.

A number of institutions cooperated with the Missouri Basin Project (MBP) in the IASP. Over the years, 10 state universities and historical societies that ranged along the river from Kansas to Idaho, as well as the Science Museum of St. Paul and the University of Wisconsin, joined the MBP field parties on the upper Missouri. Their contribution to the overall program was significant indeed. These organizations conducted at

---

Dam Projects and the Growth of American Archaeology: The River Basin Surveys and the Interagency Archeological Salvage Program, edited by Kimball M. Banks and Jon Czaplicki, 41–52.

least one-third of the nearly 100 major excavations in sites in the upper Missouri River Basin (Lehmer 1971:193–195, 200).

Among the several area offices organized under the Smithsonian Institution (SI) and the IASP was an MBP office in Lincoln, Nebraska, established to cope with the dams on the drawing boards of the Corps and Reclamation under what was known as the Pick-Sloan Plan. Lincoln was likely chosen because of its somewhat central location; it had been the headquarters for the Nebraska State Archaeological Survey; and the RBS had been offered space on the campus of the University of Nebraska. The Nebraska Survey had conducted numerous digs in the state, and a core of professionals was in residence. The first director of the RBS was Frank H. H. Roberts, Jr., a respected Southwestern archaeologist at the Smithsonian Institution. Under his direction, Waldo R. Wedel became overall director of the MBP, which began operations beginning in July, 1946. A succession of chiefs directed the field office in Lincoln, beginning with Paul L. Cooper and later Ralph D. Brown, Robert L. Stephenson, Warren W. Caldwell, and a few temporary chiefs.

The Lincoln office began operations in the basement of Love Library on the University of Nebraska campus and later moved to the basement of the newly constructed Burnett Hall, where it shared space with John L. Champe's Laboratory of Anthropology. This space soon became crowded and the office moved to 1517 O Street, then to 1835 P Street and, following its absorption by the National Park Service in mid-1969, to a building on North 27th Street. For years, museum specialist Dean E. Clark received the incoming specimens and field documents and prepared them for curation (Figure 3.1). Administrative clerk Lawrence L. Tomsyck served as administrative officer. Secretaries were available to type up and proof manuscripts; and a series of cartographers (including myself) prepared finished drawings of site and feature maps for publication.

Figure 3.1 Dean Clark cataloging specimens at the MRBS field office in Lincoln, Nebraska (National Anthropological Archives, MRBS photograph).

Archaeologists hired for the Lincoln office immediately began surveying the banks of streams slated for inundation. Following the fieldwork, the archaeologist in charge of the individual reservoir survey produced a document listing the sites that were recorded, their presumed function, legal description, their relationship to the reservoir boundaries, a map of their location within the reservoir, and a few photographs. The document concluded with recommendations for excavation and estimates for the number of excavation units (EUs) required to do so. (Between 1948 and the conclusion of the work, an EU was estimated to cost \$20,000 to \$26,000 each; see Moratto and Lyman, this volume.) Such "preliminary appraisals" eventually were prepared for most or all of the 123 reservoirs proposed under the Pick-Sloan Plan. Theodore E. White prepared a few appraisals on the paleontological resources of a reservoir, but most such commentary was restricted to a paragraph in these reports. The survey teams that began work the first season in 1946 were principally led by J. Joe Bauxar, Wesley R. Bliss, Paul L. Cooper, Jack T. Hughes, and Marvin F. (Gus) Kivett. Other major surveys were carried out by Franklin Fenenga and George Metcalf. In the end, of course, only 60 of the smaller impoundments were built, but the five mainstem reservoirs on the Missouri River in North and South Dakota were completed.

Mainstem reservoirs received the bulk of attention by archaeologists, clearly overriding the smaller impoundments. This emphasis was criticized at the time, but there was justification for this focus. The mainstem reservoirs housed the unique histories of the Plains Village Tradition, whereas those on tributaries contained information primarily on hunters and gatherers that could be found elsewhere, scattered along the many secondary streams in the northern and northwestern Plains.

The actual losses to the archaeological record by flooding were far more appalling than we ever dreamed at the time. When Donald J. Lehmer and I began working on the Missouri River in North Dakota in 1966, we knew of only five earthlodge village sites along a seven-mile stretch of the terraces lining the valley on the property of the Cross Ranch, all of them Plains village sites. That is, we knew of only 0.7 sites per linear river mile. Since our focus was precisely on those villages, we neither looked for, nor did we find by chance, any further sites. But the University of North Dakota surveyed that same seven-mile reach of the river in 1979, and inspected each of the local landforms from the floodplain to the upland margin (Weston et al. 1980). The team recorded 133 sites—nearly 20 sites per linear river mile. This number, of course, included remains, such as lithic scatters and historic sites, which were not customarily recorded in earlier surveys. Nonetheless, on the basis of the Cross Ranch survey results, we did not lose the 300 plus sites approximately one site per river mile that the MBP appraisals cited for the three hundred-mile-long Oahe

Reservoir. Taking a conservative approach, at nearly 20 sites per mile, no less than 3,000 sites were inundated and lost or destroyed in the Oahe Reservoir and perhaps more.

The archaeologists stationed in Lincoln had little data to build on as they proceeded to choose sites for investigation, for the existing archaeological information was slim indeed. After all, only a few years had elapsed since Duncan Strong demolished the idea that the Great Plains had been essentially uninhabited before the arrival of horses (Strong 1933). George F. Will and Thad. C. Hecker (1944) had produced a rudimentary overview of the village cultures in North Dakota; the state of South Dakota had issued a few archaeological studies; and the work of Duncan Strong (1935) and his avocational allies had outlined the prehistory of the state of Nebraska. In short, substantive reports scarcely existed for the areas most under threat, a culture-historical synthesis existed only for Nebraska, and a geologic-climatic framework for the west was still being formulated. It was not until 1958 that an outline for the prehistoric northwestern Plains emerged (Mulloy 1958).

Investigation Results

The geologic-climatic units proposed by Ernst Antevs (Anathermal, Altithermal, Medithermal), and now universally accepted (with substantial revisions) were published in 1948. The units rapidly found favor in the western United States for they provided a model for the environments occupied by Paleoindian and Archaic cultures (Antevs 1948). Antevs' model aroused great interest in Great Plains archaeologists at the time, and Richard Page Wheeler's work in three reservoirs in Wyoming and South Dakota began to help fill in the Archaic Period on the High Plains (Wheeler 1995–1997) as did the work of E. Mott Davis on Paleoindians in the Medicine Creek Reservoir of western Nebraska. Nationally, there was less precedent for what today we call geoarchaeology, but on the Plains geoarchaeology remained an important component in the investigations of early Holocene sites. Under Robert Stephenson, the Lincoln office began a program for geochronology of the Basin, though ultimately it did not produce significant results.

The MBP archaeologists faced the same problems that confront cultural resource management archaeologists today: a long period of laboratory time after the field season is over, which restricts their investigations to the areas to be impacted by construction projects and the need to return to the field the following season. Little wonder they had so little time for a real synthesis, one of the criticisms leveled at the MBP. The only significant synthesis that derives solely from that work is that of Donald J. Lehmer (1954, 1971). Lehmer devised a brilliant synthesis for the

Figure 3.2 The Sully Site (39SL4), the largest of the earthlodge villages on the upper Missouri River; it was fully inundated by the Oahe Reservoir (National Anthropological Archives, MRBS photograph 39SL4-11).

Plains Village cultures that still is used (in highly modified form), though today it is not applicable for the entire area it was meant to encompass. Nonetheless, few such syntheses have driven research for as long.

Work carried out in the five major reservoirs along the mainstem of the Missouri River focused on the villages of ITS earthlodge-dwelling residents. Work carried out in the five major reservoirs along the main stem of the Missouri River focused on the villages of their earthlodge and dwelling and gardening residents. Major sites of the ancestral Mandan (Lehmer's Middle Missouri Tradition) and numerous sites of the ancestral and historic Arikaras were investigated (Lehmer's Coalescent Tradition). The striking nature of these earthlodge sites, conspicuous on the ground and even visible on aerial photographs, made them overly tempting targets for investigation (Figure 3.2). Woodland, Archaic, and earlier sites were buried and left no trace on the surface, so unless they were exposed in cut banks or by other erosion, they escaped discovery; though some were detected by investigations in village sites.

Suspected Woodland burial mounds were a clear exception to this generalization and were reported in reservoir appraisals along the major streams. Excavation of the Bald Hill Mounds in the Jamestown Reservoir by Gordon Hewes comprised the principal reported investigations of this site type along a tributary stream of the Missouri (Hewes 1948). This work, as well as later investigations by Wood and Neuman at various mound sites along the Missouri River, led to the definition of the Sonota complex, the first Woodland burial complex and its occupational sites to be defined in the upper Missouri Basin (Neuman 1975; Wood 1960).

The Plains Conference

The annual Plains Archaeological (later Anthropological) Conference, then held for the most part in Lincoln, Nebraska, was a forum for distributing new information derived from the salvage work. The conference, initiated in 1931 but suspended during World War II, was resumed in 1947, the year following the opening of the MBP office in Lincoln. The meeting permitted IASP archaeologists and other investigators to consult informally with colleagues concerning the interpretation and significance of their past summer's digs, and provided the principal critical input on MBP archaeological investigations and interpretations from sources outside the Lincoln office. That office also was a major contributor to the management of the conference, and hosted an open house yearly that showcased its research to the participants and often led to spirited discussions of its displays.

Field crews often visited one another over weekends during the summer field season to socialize and exchange ideas. In April, 1953, a group of Smithsonian Institution and state representatives made a brief field trip to the Fort Randall and Oahe reservoirs in South Dakota, and on their return to Lincoln convened what became known as "The Accidental (or 10½th) Plains Conference." The session produced a new taxonomy and chronology for the Central Plains and Middle Missouri areas (Stephenson 1954:15–21). Field-based "Half Plains Conferences" continued to be held for several years, often at the airport in Pierre, South Dakota, to share information and review mutual problems. These "mini" conferences were well attended and exceedingly useful. The MBP fieldwork provided training for hundreds of students, many of whom, like myself, never had a formal field school in archaeology. Some of those students remained in the Plains after they obtained advanced degrees and continued to contribute to the prehistory of the Great Plains.

In the Field

The MBP archaeologists were recruited from a wide range of sources, many of them lacking a Ph.D. at the time. Donald J. Lehmer, for example, had been working in the American Southwest, but prepared his dissertation at Harvard on the work he did in the Oahe Reservoir in 1950–1951. J. J. Hoffman left a doctoral program at the University of Pittsburgh. Douglas D. Enger was a doctoral student at the University of Utah. Doctorates in the profession were rare at the time, and most of those with a Ph.D. had jobs they were not willing to leave, creating a job market for those at the M.A. level. The new recruits brought with them a wide range of experience, and coffee breaks in the Lincoln office covered a remarkable range of subjects. Recruits were thrown into new settings, for which there was a steep learning curve. They had to absorb quickly what little archaeological information existed of the region in which they were to work and adapt to circumstances there all the while knowing that they might be assigned the following year to another project in a different region. Many field parties operated from rented or abandoned houses along or near their sites; others were housed in tent camps of varying formality (those of Robert Stephenson, a former officer in the Marine Corps, were meticulously organized). At least one of the tent camps was "fortified" against marauding cattle by erecting a corral. In 1955 the State Historical Society of North Dakota crew bunked in the girl's dormitory of the school at Fort Yates on the Standing Rock Indian Reservation (there was no summer school) (Figure 3.3).

Varying members of the Smithsonian Institution, the Committee for the Recovery of Archaeological Remains (CRAR), and/or the Omaha office of the National Park Service visited crews along the river every summer to inspect the work in progress and to consult with the supervising archaeologists. Paul L. Beaubien and, later, Wilfred D. Logan represented the Omaha office on these tours, though other archaeologists sometimes accompanied them, often brought watermelons or intoxicating treats for the crew (Figure 3.4).

Figure 3.3 Field camp of Robert L. Stephenson at the Potts Site (39CO19), Oahe Reservoir, Corson County, South Dakota, July, 1961 (National Anthropological Archives, MRBS photograph 39CO00-45).

Figure 3.4 Supervisors and visitors at the excavation of Fort Berthold I, Garrison Reservoir, North Dakota, summer, 1954. Left to right: unknown, Raymond S. Price, W. Raymond Wood, Carlyle S. Smith, Alan R. Woolworth, Paul L. Beaubien, G. Hubert Smith, and Lee G. Madison.

Innovations

Site Recording

What became known as the "Smithsonian trinomial site designation" was an innovation of the Lincoln office. Paul Cooper, when he was with the Nebraska State Archaeological Survey, had devised a system whereby each county in Nebraska was given a two-letter abbreviation followed by a numeral designating the number of that site within the county. With the advent of the MBP, the system was expanded to include a prefix that denoted the number of each state (in alphabetical order before the admission of Hawaii and Alaska; Champe 1949:46). This simple but effective system enabled rapid filing and recovery of site data, a fact that has led to its adoption in all but a handful of states. Standardized burial, feature, and other forms were adopted that led to consistent data recording and recovery (Thiessen and Roberts 2009).

Zooarchaeology

The RBS retained Theodore E. White, formerly of Harvard University, to conduct paleontological surveys and appraisals of the reservoirs. His knowledge of the vertebrate skeleton, however, soon led him to identify the bones being returned to the laboratory by his archaeological

Figure 3.5 The beginnings of "mechanized archaeology." Marvin F. Kivett's excavation of Upper Republican houses in the Medicine Creek Reservoir, 1948 (Bureau of Reclamation photo 25FT17-124).

colleagues, a task that quickly led him to query the significance of the remains he was studying. In 1952, he began publishing a series of papers in *American Antiquity* that heralded the beginning of the field of zooarchaeology (e.g., White 1952b). Among his groundbreaking studies, seminal in every sense of the word, were those that detailed butchering techniques and a method of calculating the dietary percentage of food animals consumed by site inhabitants; he also introduced the concept of the minimum number of individuals (MNI) represented in a sample of bone (a concept he borrowed from paleontology). White is today considered one of the founding fathers of zooarchaeology, a lasting legacy of the River Basin Surveys.

"Mechanized Archaeology"

Marvin Kivett of the Nebraska State Historical Society pioneered the use of mechanical equipment to expose house sites of the Upper Republican culture in the Medicine Creek Reservoir in Nebraska (Figure 3.5). Such practices had been eschewed in the past, and initially were resisted by some archaeologists. However, because of the unprecedented demands of salvage work, others soon mimicked the use of heavy equipment, and its use rapidly became standard procedure for deeply buried sites and even for those at modest depths. Buried over 50 feet deep, the Scottsbluff/Milnesand Lime Creek Site in western Nebraska clearly illustrated the necessity for the use of such equipment, though various MBP crews and state agencies also used them to expose village features in sites along the Middle Missouri River, as Alan R. Woolworth and I did at the Paul Brave and Demery village sites.

Aerial Photography

In 1952, Ralph Solecki flew photographer Nathaniel Dewell more than 8,000 kilometers (5,000 miles) to obtain aerial views of sites in Missouri Basin reservoirs. Though aerial photography was nothing new, no such coordinated survey of this type had previously been attempted in the United States. Solecki focused on the main stem of the Missouri River where the project obtained dramatic results. His flight not only documented sites but also provided graphic evidence of the level of destruction that would occur when the reservoirs filled. Chief Archeologist of the National Park Service John M. Corbett later contracted with the ITEK Corporation in 1965 to obtain false-color and infrared photographs of village sites along the mainstem of the Missouri. These photographs illuminated many aspects of the sites, but they were rarely used or the technique pursued because the MBP was nearing its end. And the very high quality of the 1938 U.S. Department of Agriculture aerial coverage had been sufficient for most purposes, certainly in locating and mapping earthlodge villages. The latter images were widely used in surveys and evaluations, being the most readily available form of remote sensing.

Remote Sensing

The utility of the ITEK photographs led John Corbett to be a major figure in establishing the Remote Sensing Division of the National Park Service in Albuquerque, New Mexico. That division was to become instrumental in pioneering and advocating these techniques of finding, mapping, and interpreting sites across the United States until its operations were suspended during the Reagan administration. Under the direction of Thomas Lyons and Thomas Avery, the division produced *A Handbook for Archeologists and Cultural Resource Managers*, followed by nine supplements devoted to the problems attending different geographical regions of the United States, including one for the American Great Plains (Lyons and Avery 1977; Wood et al. 1984).

Publication

The data recovered were published in a wide variety of government, state, and other outlets, including bulletins published by the Bureau of American Ethnology, and *Publications in Salvage Archeology*, a special series released by the Lincoln office. The range of other publications can be judged by a state-by-state bibliography issued in 1968 (Petsche 1968). Excavation results also were used in innumerable master's theses and doctoral dissertations. One of the most significant of them was James Deetz's *The Dynamics of Stylistic Change in Arikara Ceramics* (Deetz 1965). Although the conclusions in Deetz's study are no longer accepted, it was an important work that stimulated innovative approaches in what was to become known as the "New Archaeology."

Historical Archaeology

G. Hubert Smith was assigned to the Lincoln office in 1950 to conduct investigations in historic-era archaeological sites in the Missouri Basin. He excavated and reported a number of military and fur-trading posts along the Missouri in North and South Dakota (e.g., Smith 1960a, 1960b), and the State Historical Society of North Dakota likewise excavated two trading posts in that state. No thought was given at the time to the investigation of sites such as farmsteads or ranches. Instead focus was on those that were deemed "historically important." Smith later became one of the founders of the Society for Historical Archaeology.

Conclusions

Culture-historical considerations reigned as the paradigm of the time that the MBP was in operation, and most honest critics would also agree that the data the River Basin Surveys recovered generally were published more promptly than are many investigations today. But some reports were shelved or delayed because dams continued to be built, waters continued to rise, more surveys were required, and more sites were to be dug. Salvage from the ground usually took priority over other considerations. The MBP publication record was better than its contemporary critics alleged, though a more serious criticism was that they were rarely formally refereed out-of-house. Data from numerous important excavated sites have remained in archives since they were dug, but manuscripts and data continue to be rescued from those archives. A number of them have been published, including a series produced by Richard P. Wheeler involving Hypsithermal-age Archaic sites in reservoirs in Wyoming and South Dakota (Wheeler 1995–1997). State cooperators also did very well at publication, and numerous reports—like my own dissertation on the Huff village site—appeared in Smithsonian Institution outlets. However limited the data in these sources may be by current standards, the MBP databank remains a vibrant source of information.

In a follow-up of a reservoir that MBP personnel had surveyed in Wyoming in 1947 but was never built, Julie Francis found that the recommendations for work in that reservoir were remarkably insightful and consistent with those made 40 years later (Francis 1994:187). Yet there have been sometimes savage criticisms of the work done under this program. Most of these criticisms were unjustified and historically naive. Professional historians learn even as mere novices that past actions are to be judged in terms of the *prevailing contemporary paradigms*, that is, the standards of the actors of the past, not those of today. Burnishing one's reputation at the expense of the past is unforgivable.

It would be an interesting exercise to speculate what the participants in conceiving, developing, and running the MBP would be capable of contributing under the sway of present-day method and theory. Had they not had to continually fight for funding, carry out administrative duties, and swing from fieldwork to laboratory work and back again with little respite, I have no doubt they would be as spectacular and groundbreaking as those we presently see today in *American Antiquity*.

CHAPTER 4

Archaeological "Appraisals" of Twenty Stream Basins in California: Investigations by the Smithsonian Institution's River Basin Surveys, 1947–1951

Michael J. Moratto and Francis A. Riddell

Introduction

This chapter is in three parts, of which the first presents an overview of the Smithsonian Institution's (SI) River Basin Surveys (RBS) archaeological work in California between 1947 and 1951. In Part II, Francis ("Fritz") Riddell, who passed away in 2002, reminisces about his experiences with RBS field projects and personnel in California as well as other western states. In the final part, benefiting from more than 60 years of hindsight, we assess contributions of the RBS program to the development of California archaeology. Part II was presented originally in the session, Interagency Archeological Salvage Program and the River Basin Surveys, at the annual meeting of the Society for American Archaeology in 2001. This was perhaps the last archaeological paper that Fritz Riddell wrote, as he left us soon thereafter.

Part I: The RBS in California

Notable among the public works begun in California during the Great Depression were the Friant Dam (constructed 1937–1942) and the Shasta Dam (constructed 1937–1945) on the San Joaquin, and Sacramento rivers, respectively. These federal projects set the stage for the many California reservoirs that the U.S. Army Corps of Engineers (Corps) and

Dam Projects and the Growth of American Archaeology: The River Basin Surveys and the Interagency Archeological Salvage Program, edited by Kimball M. Banks and Jon Czaplicki, 53–65.

the U.S. Bureau of Reclamation (Reclamation) proposed as part of their nationwide plans to stimulate the economy, create jobs, and enhance America's infrastructure—especially for flood control, irrigation, and hydroelectric power generation—after WW II. Hundreds of dams were to be built throughout the United States, not only along major rivers such as the Columbia, Missouri, and Colorado, but also on countless smaller watercourses. In California, federal agencies envisioned dams on such streams as Littlejohns and Hoods creeks in the Central Valley; the American and Kaweah rivers in the Sierra Nevada; Dry and Putah creeks in the North Coast Ranges; and the Santa Ynez River in the South Coast Ranges. Subsequently, most of the proposed reservoirs were constructed during the first few decades of the post-War era.

By the mid-1940s, scholars had become aware that the dam and reservoir projects would destroy or flood thousands of historic and prehistoric sites and that a program of salvage archaeology would be needed to mitigate some of the foreseeable impacts. The Committee for the Recovery of Archaeological Remains (CRAR) was formed in 1945 to help meet the challenge and immediately began to advise and coordinate with the SI and the National Park Service (NPS) to develop a salvage program. To further this goal,

> [i]n conjunction with the construction agencies, a scheme based on interagency agreements was devised and financed under which archaeologists and historians made surveys of dam sites, the pool areas, the sites of sometimes very large construction towns, and borrow pits. . . . Thus, the Inter-Agency Archaeological Salvage Program came into being (Brew 1968:3).

Pursuant to a 1945 agreement (revised in 1961 and 1965), the NPS became responsible for "over-all programming, funding, and administration" while the SI functioned as an "advisor to the NPS in the planning and programming stages" and as "the major cooperating agency in salvage operations and the publication of archaeological research" (Petsche 1968:iii). In California, fieldwork was performed under the local direction of the University of California, Berkeley (UCB), Department of Anthropology and, until June 30, 1949, UCB's California Archaeological Survey (CAS). From July 1, 1949, until June 30, 1961, the survey's official title was the University of California Archaeological Survey (UCAS); later it was known as the University of California Archaeological Research Facility (UCARF). Professor Robert F. Heizer established the CAS in 1948 and thereafter directed it and its successors until his death in 1979 (Heizer 1972:37–38). During the years of RBS operations in California, the CAS/UCAS was staffed mainly by UCB students:

- Franklin Fenenga, Archaeologist, and Francis A. Riddell, Assistant Archaeologist, 1948–1949;

- Franklin Fenenga, Archaeologist, and Arnold Pilling, Assistant Archaeologist, 1949–1950; and
- William Wallace, Archaeologist (July 1, 1950–January 31, 1951),Clement Meighan, Archaeologist (February 1, 1951–July 31, 1951), and Donald Lathrap, Assistant Archaeologist (Heizer 1972:37).

The RBS opened a field office at UCB in 1948 with Franklin Fenenga in charge. The Berkeley office was closed in 1950 due to intermittent operation. Fenenga then transferred to the RBS's Lincoln, Nebraska, office and remained there until 1960 where he focused on Plains archaeology (Butler 1995:224). For its California projects, the RBS often "borrowed" archaeologists from CAS/UCAS. For example, CAS "loaned" Fritz Riddell for the Farmington Reservoir survey in the Central Valley (Riddell 1949:1). The CAS also provided equipment, film and processing, and mileage reimbursement. The Corps, Reclamation, and NPS supplied maps, and the UCB Museum of Anthropology (now the Phoebe A. Hearst Museum of Anthropology) gave access to records of previous fieldwork in the region.

The rationale for RBS field surveys in California was straightforward. For example, at Folsom Reservoir,

> [t]he purpose of the survey was the customary aim of the River Basin Survey's preliminary studies: To determine the number and extent of archaeological remains in the area to be inundated by the proposed dam, in order to learn whether or not aboriginal sites of major significance, from the cultural and scientific point of view, occur there which should be excavated to salvage materials and data from them, before filling of the reservoir makes such work impossible (Drucker 1948h:1–2).

By today's standards, the methods of the initial surveys or "appraisals" were rather perfunctory.

> Frank [Fenenga] once told me that while the appraisals were fun to do, they were also frustrating in that he could not really investigate the most interesting resources found. The usual procedure was that he was furnished with a car of some sort, given a few dollars for gas and food, and a map to get him and an accomplice or two on their way. These were not rigorous transect surveys, but rather "go look where you think the sites ought to be" sorts of projects (Butler 1995:274).

Survey reports were written promptly, mostly by students or staff affiliated with UCB. Reports were submitted by the author or by Philip Drucker (who sometimes also was the author), Field Director of the Pacific Coast area, and approved for distribution by Frank H. H. Roberts, Jr., Associate Director of the Bureau of American Ethnology and Director of the RBS. Just as the field surveys were far from thorough, the appraisal reports were often cursory. For example, in the double-spaced 1.5-page report on the surveys of Indian Valley, Sly Park, and Dry Creek reservoirs, it was concluded that,

> None of these reservoir areas appears to have been suitable for intensive occupation by Indians in aboriginal times, for no archaeological remains of any importance were found in them. They were probably visited sporadically for gathering wild plant foods and for hunting. 2 small, very superficial camp sites suggesting such use were found in the Dry Creek Reservoir area, but these remains are too insignificant to warrant further investigation (Drucker 1948f:2).

Later archaeological investigations at Dry Creek Reservoir (Lake Sonoma) in the North Coast Ranges would reveal more than 30 archaeological sites representing thousands of years of human activity (e.g., see Basgall 1993; Baumhoff and Orlins 1979). Similarly, at New Melones Reservoir in the central Sierra Nevada, later studies resulted in the discovery of nearly 700 cultural sites attesting to 10,000 years of culture history (Moratto 2002) in the area where the RBS field team had recorded only four sites. This is not to fault the RBS investigators, however, because they were severely constrained with respect to time and funds, and they were pursuing very limited objectives. Typically one or two archaeologists were expected to drive to the study location, conduct the survey, and document sites in an area of hundreds or thousands of acres, with just a few days allowed for fieldwork. Moreover, at the time of the RBS program in California, almost no attention was paid to historical sites, such as nineteenth-century mines, ranches, or town sites, and consequently these were seldom recorded.

Table 4.1 summarizes the RBS surveys of 20 stream basins in California. Of this number: seven projects are located in the North Coast Ranges (i.e., Black Butte, Coyote Valley, Dry Creek, Indian Valley, Kelsey Creek, Monticello, and Wilson Valley reservoirs); one is in the South Coast Ranges (Cachuma Reservoir); and 12 lie in the Sierra Nevada foothills or at the eastern margin of the Central Valley (Bear, Burns, Farmington, Folsom, Isabella, Mariposa, New Melones, Pine Flat, Owens, Sly Park, Success, and Terminus reservoirs). The surface areas of the proposed reservoirs varied from a few hundred to more than 10,000 acres; but since none of the field surveys was intensive and much of the searching for sites was intuitive, the RBS field teams actually examined far smaller areas. The number of discovered archaeological sites ranged from none at Sly Park to 50 at Monticello. Some of the more "promising" sites—typically larger midden deposits with house pits—were often recommended for excavation (Table 4.1).

It may be of interest to briefly discuss cost estimates for excavation. For budgetary purposes in the late 1940s, RBS employed a standard,

Table 4.1 River Basin Surveys projects in California (1947–1951)

Project name	County/ Counties	Year	Field archaeologists	Area surveyed	Sites investigated	References
Bear Reservoir, aka Bear Lake, on Bear Creek	Mariposa	1951 or before	Albert Mohr	No data	No data	Mohr (1951)
Black Butte Reservoir (now Black Butte Lake) on Stony, North Fork Stony, Burris and Grindstone Creeks	Glenn, Tehama	1948	David Fredrickson, Albert Mohr	~4,640 ac	26 sites recorded, salvage work recommended at 4-Gle-5, -10, and -11 and 4-Teh-6 and -10	Mohr (1949); Treganza and Heiksen (1969); Woolfenden (1969)
Burns Reservoir on Burns Creek	Merced	1951	Albert Mohr	~830 ac	6 sites recorded, including a petroglyph site, 4-Mer-52	Mohr (1951); Peak & Associates (1981)
Cachuma Reservoir, aka Bradbury Dam/Lake Cachuma, on Santa Ynez River	Santa Barbara	1951	Martin Baumhoff, Albert Mohr	~3,100 ac	19 sites recorded; 2 tested (4-SBr-477, -485)	Baumhoff (1951); Mohr (ca. 1950)
Coyote Valley Reservoir, aka Coyote Dam/Lake Mendocino, on East Fork Russian River	Mendocino	1947	Franklin Fenenga	~1,920 ac	No data	Drucker (1948g); Fenenga (1947a); Treganza (1957)
Dry Creek Reservoir, aka Lake Sonoma, Impounding Dry Creek and Warm Springs Creek behind Warm Springs Dam	Sonoma	1948	Clarence Smith	~2,700 ac	2 sites recorded	Basgall (1993); Baumhoff and Orlins (1979); Drucker (1948f)
Farmington Reservoir, on Littlejohns Creek	San Joaquin, Stanislaus	1949	David Fredrickson, Albert Mohr, Francis Riddell	~4,000 ac	24 sites recorded; excavation of 4-Sta-6 recommended; testing proposed at 5 sites; 4-Sta-6 excavated	Mohr (1948); Riddell (1949); Treganza (1952)

(*Continued*)

Table 4.1 (*Continued*)

Project name	County/ Counties	Year	Field archaeologists	Area surveyed	Sites investigated	References
Folsom Reservoir, on the American River at and above the confluence of North and South Forks	El Dorado, Placer, and Sacramento	1947	Franklin Fenenga assisted by graduate students from the UCB Department of Anthropology	~12,800 ac	1 site recorded; others presumed lost due to gold mining or perhaps obscured by dense brush	Drucker (1948h); Fenenga (ca. 1950a)
Indian Valley Reservoir, North Fork Cache Creek	Lake	1948	Clarence Smith	~4,000 ac	None reported	Drucker (1948f)
Isabella Reservoir, aka Lake Isabella on Kern River and South Fork Kern River	Kern	1947	Franklin Fenenga, Francis Riddell	~10,880 ac	14 recorded; 2 recommended for excavation	Drucker (1947b); Fenenga (1947d)
Kelsey Creek Reservoir, on Kelsey Creek	Lake	1947	Clarence Smith	~1,600 ac	13 sites recorded; excavations proposed at 3 sites (4-Lak-16, 4-Lak-18, and 4-Lak-23)	Drucker (1948a); C. Smith (1947)
Mariposa Reservoir on Mariposa Creek	Mariposa	1948	W. King, Albert Mohr	". . . a few hundred acres"	3 sites recorded; others presumed to have been destroyed by placer mining	Osborne (1948a)
Monticello Reservoir, aka Lake Berryessa, on Putah Creek	Napa	1947	Clarence Smith, assisted by graduate students from the UCB Department of Anthropology	~20,000 ac	50 sites recorded, some very large; 5 recommended for excavation: 4-Nap-9, 4-Nap-14, 4-Nap-19, 4-Nap-27, 4-Nap-47	Drucker (1948b); Elsasser (ca. late 1940s); Treganza (ca. 1950)
New Melones Reservoir, Stanislaus River, South Fork, Stanislaus River, and Angels Creek	Calaveras, Tuolumne	1948	David Fredrickson and Albert Mohr	~12,500 Ac	4 prehistoric sites documented; several Gold Rush-era sites also mentioned, but not recorded. Further archaeological work not recommended	Fredrickson (1949); Moratto (2002)

Pine Flat Reservoir, on Kings River	Fresno	1947	Franklin Fenenga, assisted by graduate students from the UCB Department of Anthropology	~7,680 ac	29 sites recorded; testing of 3 sites recommended; 3 sites (Fre-27, -29, and 30) subsequently were excavated	Drucker (1948c); Fenenga (1947b)
Owens Reservoir above Owens Dam	Mariposa		Albert Mohr			Mohr (1951)
Sly Park Reservoir, aka Jenkinson Lake, Above Sly Park Dam on Sly Park Creek	El Dorado	1948	Clarence Smith	~650 ac	None reported	Drucker (1948f)
Success Reservoir, aka Success Lake, on North and South Forks and mainstem Tule River	Tulare	1947	Franklin Fenenga	~4,200 ac	2 sites recorded; further work not recommended	Drucker (1948d); Fenenga (1947c); Meighan et al. (1988b)
Terminus Reservoir, aka Lake Kaweah, on Kaweah River	Tulare	1948	Franklin Fenenga	~2560 ac	10 sites recorded, including polychrome pictographs	Berryman and Elsasser (1966); Fenenga (ca. 1950b, 1952); Drucker (ca. 1948); Meighan (1958); Meighan et al. (1988a)
Wilson Valley Reservoir, on Cache Creek	Lake	1947	Clarence Smith assisted by graduate students from the UCB Department of Anthropology	~4,480 ac	14 sites recorded, 13 of these with house pits.	Drucker (1948e); C. Smith (ca. late 1940s)

all-inclusive "Excavation Unit" (EU) defined as follows (Drucker 1948c: Appendix):

> The Excavation Unit is an operational unit designed to provide labor, supervisory personnel, and equipment for the recovery of archaeological materials in the field; technical and clerical personnel, laboratory space, and laboratory equipment for processing these materials and preparing them for study; and finally for study of the materials and data recovered, and preparation and publication of a technical report (for which the archaeological supervisor will be responsible).

01	Personal services:	
1	Archaeologist-Supervisor for 1 year (P-3)	$4,460.00
1	Field Assistant, 4 months at $250 per month	$1,000.00
2	Laboratory Assistants, 1 year at $200 per month	$4,800.00
1	Clerk Typist (CAF-3)	$2,170.00
10	Laborers, 80 days at $8 per day	$6,400.00
02	Travel and per diem	$ 930.00
03	Transportation of things	$ 240.00
04	Communication services	$ 100.00
05	Rents and utility services	$ 700.00
08	Supplies and materials	$ 700.00
09	Equipment	$3,000.00*
Total cost per Excavation Unit		$24,500.00

*After the first year in the case of extended programs this item would be considerably reduced because most of the major equipment would have been purchased.

The costs of excavation at individual sites could then be calculated on the basis of a fraction of the EU total. To illustrate, Drucker (1947b) proposed to allocate 1/8 of an EU, or $3,062.50, to the investigation of two sites (i.e., approximately 1/16 EU, or $1,531.25 per site) at Isabella Reservoir, and he (Drucker 1948e) proposed to expend 3/16 EU, or $4,533.75 in the excavation of two sites at Wilson Valley Reservoir.

Part II: The Archaeological Salvage Program in California by Francis A. Riddell*

In preparation for this paper, I had to brush off a lot of old dust and cobwebs, both in my memory and on old correspondence and reports that I had not thought of, or in some cases seen, for more than 50 years. As a consequence, I cannot be too sure that what I say here will be entirely fact, or partly fancy; it will probably be a mixture of the two.

Interest in the prehistory of California was shown by the Smithsonian Institution in its early support of work done by Paul Schumacher in the 1870s at sites on the southern coast and by J. P. Harrington at the

Burton Mound near Santa Barbara in 1923 (Moratto 1984:121–123). However, not until the 1940s did its involvement become of greater significance after the passage of federal legislation establishing the River Basin Surveys program administered by the Smithsonian Institution.

There was, however, some federal involvement in California archaeology in the 1930s and into the 1940s by means of the Works Progress Administration—most notably in Orange County by Winterbourne (Moratto 1984:149) and at Buena Vista Lake in Kern County (Wedel 1941). As a personal note, possibly of historic interest is the fact that Franklin Fenenga, my brother Harry, and I prepared a request for a WPA project with the assistance of my father, Harry S. Riddell, Sr., an engineer with Reclamation for the Central Valley Project. This was in 1940 and the three of us were students at the Sacramento Junior College and under the guidance of Dr. Jeremiah B. Lillard whose interest in central California archaeology was rather widely acclaimed at that time for having demonstrated a three-step cultural sequence for the Sacramento Delta region (Towne 1997). In our field surveys with the Junior College, we had recorded several protohistoric sites at which house pits were still visible. Our project was accepted by the WPA but needed the o.k. of the Sacramento City School Board for their management of the project. Dr. Lillard's health was waning and our country was on the brink of war, so the school board shelved our project, much to our dismay. I cite this to indicate that there was a cadre of us who, after the war, were sufficiently trained and experienced to move quickly when the call for archaeologists came to fill the seasonal RBS jobs.

Federal legislation enacted toward the end of, or shortly after, WW II, had a profound effect on the pace of archaeological activities throughout the nation, at least where the impounding of rivers for flood control, power and water, or highway construction, were concerned. The former was particularly evident in the Missouri River Basin, but also a factor in the Columbia River drainage. But not only was archaeology undergoing a massive expansion in the Missouri Basin and the Pacific Northwest, but California was experiencing the beginning of a vast water development plan that utilized both state and federal funding. In California, Shasta and Friant dams had been completed in the early 1940s by the federal government. Some preliminary site surveys on the McCloud River within the Shasta Reservoir impound area was done by Franklin Fenenga and other Berkeley students. This survey and many that followed after WW II in California were summed up by Heizer in the following statement:

> No attempt is made here to trace the course of federally supported . . . or federally mandated . . . archeological salvage surveys and excavations. There has, since the 1930's, been continuous activity in this kind of work, but on the whole the published record is quite minimal if assessed in amount of time and money devoted to it. This is not to say that the various

> reasons why such rescue archeology was done in California at the time were wrong, but merely that the effort was relatively unproductive in terms of making generally available the results of the endeavor (Heizer 1978:13).

In this statement, Heizer was not too far off the mark, at least seen at that point in time. My first "hands-on" experience with the RBS program came in the summer of 1947 when Dick Daugherty and I made surveys of proposed reservoir areas in a number of western states, excluding California. Similarly, other two-man teams were going to other proposed reservoirs on the Columbia and Snake rivers. The teams included such men as Franklin Fenenga, Clarence Smith, Harry Riddell, Dave Cole, Daugherty, and me, as well as Phil Drucker and Douglas Osborne and others. For transportation in 1947 Daugherty and I used my Model A Ford with which we made surveys in Washington, Oregon, Idaho, and Montana, and often spent little more than a day or two at the reservoir sites.

A letter to me from Frank H. H. Roberts, Jr. the RBS Director, dated June 10, 1948 says, in part:

> We are in hopes that we can get things lined up satisfactorily soon so that we can get started on the surveys. If we are not able to get certification for you we probably can take you on the same status as you had last summer if your [*sic*] willing.

My status for both summers was as Expert Laborer, which paid $60.00 a week, but I did receive mileage reimbursement for the use of my Model A.

By July, 1948, I was at Berkeley with Fenenga, both of us as archaeologists for the newly formed CAS directed by Heizer. Up to that time my RBS experience was in the Pacific Northwest with Daugherty under the direction of Phil Drucker and then Douglas Osborne. On August 18, 1948, Drucker wrote me:

> Dear Fritz: I am sending herewith your negatives of some of your Indian friends. I am sorry to have delayed so long, but hope you get them in time to distribute the necessary copies. I might add that I was very much impressed by the one very glamorous Indian you found at Nespelem [on the Columbia River in Washington].
>
> How goes the Pine Flat job? Tell Fenenga I expect to drop by and see you people the latter part of this month. How do I locate your camp?
>
> I have been having a little trouble trying to tie in your negatives with the Photo Records. The difficulty seems to be that some of your descriptions are a trifle laconic. Would you have time to straighten them out for me? I can send you the negatives and a copy of your Photo Records.
>
> Very sincerely,
>
> Phil

While at Nespelem I took a series of photos of Indian families living on the reservation and had promised to send them copies of these photos. The "glamorous" Indian he mentions was a young local woman (non-Indian) who lived near one of the archaeological sites I recorded (I wonder what ever became of that girl?). And regarding my photo records, Drucker, ever the gentleman, did not come right out and say they were rotten, just a "trifle laconic."

Drucker's mention of the Pine Flat job illuminates the fact that the CAS, now that it had been established, was making surveys and excavations for the RBS, and Fenenga and I were now working in California, not in the Pacific Northwest. In a letter dated December 21, 1948, Doug Osborne, as Acting Field Director, RBS, Pacific Coast Area, University of Oregon, Eugene, said: "Thanks for the 'identification' on the last two photographs that I sent. That ends a job roughly comparable, in my estimation, to writing a Ph.D. thesis."

With Fenenga's appointment as Archaeologist, and me as Assistant Archaeologist for the CAS in 1948, a firm association with the RBS program in California became a dominant activity. At this late date, I do not have a clear recollection or control of the data to be specific on the timing and sequence of the projects in California. However, reference to the *Bibliography of Salvage Archeology in the United States* by Petsche (1968) provides a view of what the RBS was doing in California as regards archaeological surveys and excavations. Most prominently cited in the California section of the bibliography are Philip Drucker, Albert Elsasser, Franklin Fenenga, David Fredrickson, C.E. Smith, Adán Treganza, William Wallace, and myself. A number of others, of course, were involved but the foregoing were the most active. It is here, however, that one must comment on Heizer's observation that "the published record is quite minimal if assessed in amount of time and money devoted to it" (Heizer 1978:13). In all fairness to those who made the surveys and wrote the reports, their efforts should be viewed in the light of the budgets appropriated for those tasks. To that point let me cite a letter from Doug Osborne to my brother Harry and me dated June 21, 1949:

> Dear Fritz and Harry,
>
> While we have not the slightest idea whether or not we will be doing survey work this summer, I want to find out just what the old survey hands have planned for the summer so that, in case money does come through for survey work, I can call in a couple of crews immediately.
>
> It looks pretty black to me and my present feeling is that we will get no money for excavation but that we may get some for survey.
> Very best,
>
> Doug

With respect to these RBS projects one must keep in mind that the effort in its totality was a pioneer effort that could only gain sophistication, growth, and acceptance through time. Just as in the development of the automobile, the Model T precedes the Model A, which is followed by the Ford V-8, and so-forth, in that order. As limited as most of these surveys were in terms of present-day standards, one must look at the doughnut and not the hole! The RBS program initiated a tidal wave of concern and action for the nation's patrimony, the effects of which are still being felt with the subsequent passage of various state and federal environmental laws providing a means of protecting, interpreting, and understanding our past.

Part III: Contributions by SI-RBS to the Development of California Archaeology

At the end of WW II, the tempo of archaeological salvage greatly picked up in California as in the rest of the nation. This acceleration was brought about by both federal and state legislation that provided the necessary authority and funding. In every sense of the word, this early archaeological activity in California was a pioneering effort. In retrospect, some aspects of this effort were not up to today's standards. However, the work done provided a foundation for the advancement of knowledge about California's cultural past.

The RBS program ushered in the first archaeological surveys and limited excavations in previously unexplored parts of the state. Earlier research had been focused mainly in the San Francisco Bay area, Sacramento/San Joaquin Delta, and Santa Barbara Channel regions. The RBS fieldwork shed new light on archaeological resources in the Coast Ranges, along the eastern margin of the Central Valley, and Sierra Nevada foothills. The Sierra and North Coast Ranges in particular were virtually unknown archaeologically prior to the RBS surveys. Taken together with what was already known of the Delta, San Francisco Bay, and Santa Barbara regions, the results of the RBS surveys provided a basis for initial comparative studies and statewide syntheses (e.g., Heizer 1964).

The RBS program also gave valuable field experience to young archaeologists who went on to become leaders in the field: Martin A. Baumhoff, David A. Fredrickson, Franklin Fenenga, Francis A. Riddell, Adán Treganza, and Clarence E. Smith, to name just a few. Moreover, the program acquainted these archaeologists with project management and contract administration in the context of government construction projects. The success of this work showed that salvage archaeology and the development of large public works projects could progress together. In many localities, the RBS appraisals were the first of many archaeological projects performed over a span of more than 60 years (e.g., at Dry Creek, Folsom, Isabella, and New Melones reservoirs). The RBS studies were intuitive and the results were biased in favor of large midden sites adjacent to streams; but they were an important start, and they set the stage for the later multi-year, interdisciplinary research projects to follow in the same stream basins.

The early salvage efforts also laid the foundations for what would become cultural resource management (CRM). California's involvement in early salvage archaeology has been chronicled in Nadine Hata's (1992) book, *The Historic Preservation Movement in California, 1940–1976*. It is not possible here to provide details of this development, but suffice to say that the RBS program of the 1940s and early 1950s was the progenitor of what was to follow in CRM in California. The trail blazing by the RBS was a pioneering enterprise that will be forever recognized in California and wherever else it was active in this nation.

Acknowledgments

I am grateful to Kimball Banks and Jon Czaplicki for making available for publication in this chapter of Francis A. Riddell's paper on "The Archaeological Salvage Program in California," prepared initially for and presented in the session on the Interagency Archeological Salvage Program and the River Basin Surveys, co-chaired by Banks and Czaplicki, at the 66th annual meeting of the Society for American Archaeology in 2001. –MJM

CHAPTER 5

The River Basin Surveys in the Columbia Plateau

R. Lee Lyman

Introduction

When Martin et al. (1947:444) published what is a good candidate for the first continent-wide synthesis of North American prehistory, they indicated that "archeological research in the Plateau has not yet developed a prehistoric chronology. . . . On the basis of the available evidence archeological remains in the Plateau do not appear to be very old, nor is there evidence of much cultural change in the prehistoric period." These two observations would guide much of the local archaeological research that began under the auspices of the River Basin Surveys (RBS) the year Martin et al.'s (1947) book was published. That research not only provided the foundations for developing a cultural chronology that spanned the entire Holocene (e.g., Ames et al. 1998), but it ultimately demonstrated that there indeed was some time depth to the human occupation of the intermontane Northwest.

Sometimes referred to as the Columbia-Fraser Plateau, the Plateau cultural area lies west of the Rocky Mountains and east of the Cascade Range. A similar area is occasionally referred to as the "intermontane west." This latter designation also includes the cultural and physiographic Great Basin (and sometimes the Southwest), so it is something of a misnomer for the Plateau. The Columbia-Fraser Plateau is bounded on the north by the upper reach of the Fraser River and by the Ochoco Plateau of central Oregon on the south. The area is drained by the Fraser and Columbia rivers and their tributaries. The RBS work undertaken out of its Northwest office concerned the Columbia Plateau, or basically eastern Washington and adjacent Oregon and Idaho and a bit of western

Dam Projects and the Growth of American Archaeology: The River Basin Surveys and the Interagency Archeological Salvage Program, edited by Kimball M. Banks and Jon Czaplicki, 66–84.

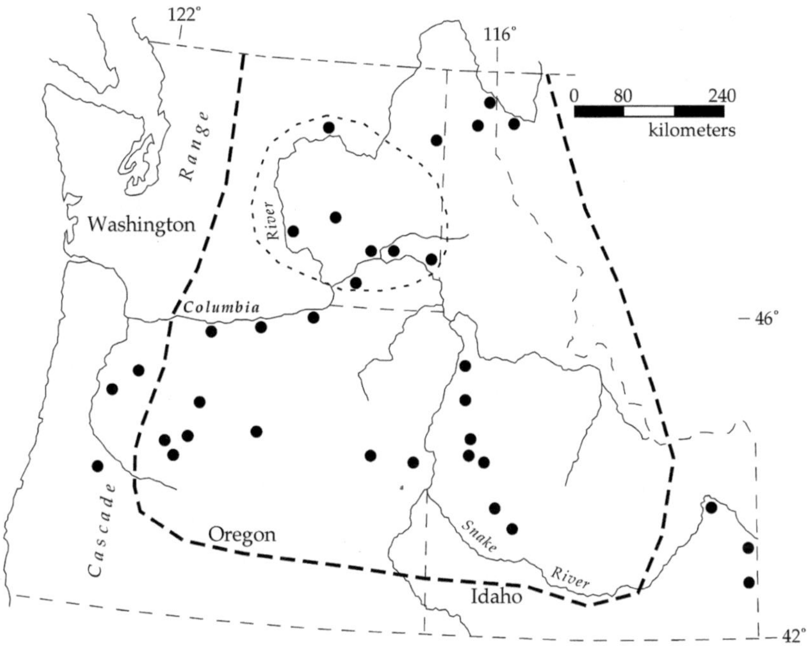

Figure 5.1 Location of RBS projects undertaken through the RBS Columbia Basin Project office in Eugene, Oregon. Heavy dashed line outlines the Columbia Plateau.

Montana (Figure 5.1). The political border between the United States and Canada was the area's northern boundary, a typical acknowledgment that archaeological research is done according to the modern arbitrariness of political boundaries relative to the ancient societies and cultures that are the subject of archaeological scrutiny. Those who work south of the border speak of the southern or Columbia Plateau (or Columbia Basin), and those working to the north refer to the northern or Fraser Plateau.

Much of what we know about the Columbia Plateau's prehistory today has its intellectual roots firmly in the groundbreaking (no pun intended) work of the RBS archaeologists who labored in the area at a time when virtually nothing was known about local prehistory. To be sure, some work had been done previously (e.g., Collier et al. 1942; Krieger 1927, 1928, 1935; Smith 1904, 1906a, 1906b, 1910; Strong and Schenck 1925; Strong et al. 1930), but this work revealed little beyond the general nature of what local archaeological materials looked like, hence Martin et al.'s (1947) terse treatment of the region's prehistory. This situation would begin to change after Luther Cressman joined the faculty at the University of Oregon in 1933; Richard Daugherty went to Washington State University in 1949; and Douglas Osborne joined the faculty at the University of Washington in 1950. These men, as well as several others, had major roles in the RBS and its activities in the Columbia Plateau.

A history of the RBS in the Pacific Northwest has been summarized previously in very brief terms (Sprague 1973, 1984), and I draw heavily on those synopses. But I also draw from the *Smithsonian Institution Annual Reports* published under RBS Director Frank H. H. Roberts, Jr.'s name, archival documents such as letters between various personnel, and published and unpublished RBS reports that provide little known details of the RBS in the Northwest. I begin with a brief outline of some of the historical details of the Northwest office. Then I turn to the analytical work accomplished by RBS researchers to show that not only was local RBS-affiliated archaeological research a product of its time, it provided the basis for establishing knowledge of an early Holocene occupation of the area and for the construction of a detailed cultural chronology. I conclude with a few observations on several of the unique aspects of the RBS program in the Pacific Northwest.

Throughout the discussion, I do two things that should be of historical interest. First, I identify recent (re)analyses of RBS-generated collections. This demonstrates that much of the RBS work, though done more quickly and less thoroughly than we would prefer by today's standards, had and indeed still has a great deal of value with respect to producing knowledge about local prehistory. Second, where pertinent, I list both unpublished RBS reports and published versions. Some of the unpublished reports served as doctoral dissertations for individuals who received extensive field training precisely because of the opportunities provided by the RBS (and the subsequent National Park Service [NPS]) program. This serves to illustrate the value of the RBS program to the discipline in general. Further, the RBS archaeologists trained many of us working today.

Background

The U.S. Army Corps of Engineers (Corps) and the Bureau of Reclamation (Reclamation) developed plans under the *Flood Control Act of 1994* to construct a series of dams and reservoirs in the Columbia River Basin. The Snake River was included in plans for the basin as the Cold War intensified, and demands for hydroelectric power increased. The Columbia River Basin is a physiographic unit made up of the low-lying area of eastern Washington flanked by various highlands (Figure 5.1). The basin is a physiographic area whereas the Columbia Plateau is an indigenous cultural area. A multitude of hydroelectric and irrigation

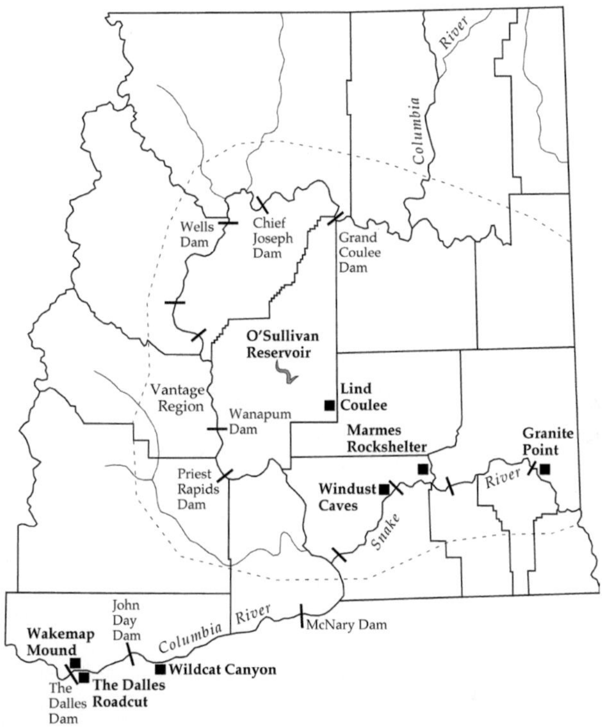

Figure 5.2 Location of hydroelectric dams and selected archaeological sites and locations in the Columbia Basin (outlined by dashed line) of eastern Washington state.

projects was contemplated in the Columbia Plateau, and RBS work took place within the project boundaries of many of them (Figure 5.2). Many of the projects were on the drawing board in 1945 when the RBS was formed; some were subsequently added, others discarded. As it turned out, only a few were ultimately drivers of local archaeological research beyond reconnaissance under the auspices of the RBS.

The RBS in the Northwest operated only from 1947 to June, 1952, when what was referred to as the Columbia Basin Project office closed (Sprague 1973). During the six brief years the office was in operation, however, much of the groundwork was laid for subsequent research in the region under the term of "salvage" archaeology, a term that came to imply work of low quality on a nationwide basis. Some of the most famous sites excavated in the Columbia Basin were initially found by RBS personnel, though the RBS may not have been directly involved in subsequent excavation. And, several of the biggest names in local archaeology cut their teeth on RBS projects.

The RBS in the Northwest

Plans for construction in the river valleys of the Columbia Plateau involved numerous projects. Perhaps the most complex was the "Columbia Basin Project" which had begun in the 1930s with the construction of Grand Coulee Dam (Figure 5.2). Completed in 1941, the dam was constructed to irrigate the arid Columbia Basin where some of the richest agricultural land in the west was found yet was naturally too arid to be productive. The Columbia Basin Project continued through the 1940s and into the early 1950s, but with the onset of World War II, Grand Coulee Dam became a major source of hydroelectric power for the production of aluminum and the construction of Boeing aircraft for the war effort (Simonds 1998). The irrigation portion of the project involved an intricate series of canals, smaller dams, and holding and recycling reservoirs. Some of these smaller construction projects had already been completed by the time the RBS was organized and operating in the area.

At least 35 mimeographed reports were produced by the RBS Columbia Basin Project (Sprague 1984). The actual count may be greater (Lohse and Sprague 1998), but hard copies of some reports cited in the local literature have not been located (Sprague 1984). The reports were printed in limited number to keep sensitive site-location information from getting into the hands of looters. Many, but not all, of these reports concern archaeological reconnaissance of a proposed reservoir. These initial surveys were often rather superficial; reading the reports reveals that often times both banks of a fifty-plus kilometer-long proposed reservoir were (supposedly) examined in four or fewer three- and four-person days. Perhaps not surprisingly, therefore, subsequent surveys have regularly recorded many more sites than the original RBS teams did (Reid and Gallison 1996). For example, Sprague (1973:262) reports "in Lower Granite Reservoir [along the Lower Snake River in southeastern Washington] the initial survey found seven sites (Osborne 1948b), while a later survey reported almost one hundred sites." Similarly, the original RBS survey of the Chief Joseph Dam reservoir (Rufus Woods Lake) recorded 20 sites (Osborne 1949; see also Osborne et al. 1952); there are now more than 300 sites known in the reservoir area.

One reason for the increase in the number of sites recorded was a proposed raise of 10 feet in normal reservoir level that increased the included land area within the Chief Joseph Dam reservoir. More land examined typically means more sites. I discovered another reason for the increase in the site inventory in 1975 when I worked in the area (Lyman 1976). At that time, I interviewed several local landowners who remembered the RBS survey of some 30 years earlier; they showed me some small, unimpressive, and previously unrecorded sites in which they indicated the RBS reconnaissance team was not interested. This reflects what seems to have been the typical mid-twentieth century mind-set of seeking large, deep, artifact-rich sites—a sort of get the most bang for the buck mentality (e.g., Osborne et al. 1952). The RBS team can hardly be faulted for this because they had limited antecedent knowledge and limited resources available for increasing that knowledge. It made sense to go for the biggest sites with the greatest potential for a low cost-high benefit ratio. By today's standards, of course, such is not acceptable; but at the time, it was reasonable.

Authorship of the mimeographed reconnaissance reports was variously listed; sometimes the report was attributed to the field director, sometimes to the surveyor(s). Surveyors included Richard D. Daugherty, Francis A. Riddell, and Franklin Fenenga—all names that local archaeologists are familiar with in light of later work accomplished by these individuals (see below). The first field director of the RBS's Northwest office was Philip Drucker who served from March to June, 1947; he was followed by Douglas Osborne (appointed October, 1948), who departed for the University of Washington in May, 1950. Joel L. Shiner then became acting field director. Again, these are all names with which local prehistorians are familiar: Shiner (1961) for authoring one of the major RBS-derived reports; Drucker (e.g., 1951) for ethnographic work on the Northwest Coast; and Osborne for a significant amount of the published RBS work (e.g., Osborne 1953, 1957, 1967). During the time the RBS operated on the Columbia Plateau, the NPS archaeologist who served as liaison between the NPS and the RBS was Louis R. Caywood, yet another name that is familiar to local archaeologists (e.g., Caywood 1954, 1955).

The base of operations for the Columbia Basin Project was on the University of Oregon campus in Eugene (Sprague 1973). Luther S. Cressman made the arrangements for the "office and laboratory" (Cressman et al. 1960:3) and access to the campus library (Shiner 1961:156). The warrant for such regional field offices was that if a region was to witness the planned extent of construction, then a regional office was mandatory for the efficient support of logistics and personnel (Jennings 1985). It is unclear in archives I have examined why Eugene was chosen. For instance, the Smithsonian Institution-RBS annual reports provide no clarity. Perhaps Eugene was chosen because the University of Oregon (UO) campus had an archaeology program (specifically, a Department of Anthropology), with then nationally known Cressman in residence. Neither Washington State University nor the University of Washington (UW) as yet had such a program; further, neither had on staff an archaeologist working in North America who had a reputation of magnitude equivalent to Cressman's. Whatever the case, archived correspondence between RBS national director Frank H. H. Roberts, Jr. and NPS archaeologist Louis Caywood indicates there was some difficulty setting up the cooperative agreement between the RBS and the UO in 1947–48 (Roberts letter to Caywood, May 16, 1952 [UO-Museum of Natural and Cultural History archives; UO-MNCH, hereafter]). With limited funding, work began in O'Sullivan and McNary reservoirs in Washington. Initial reconnaissance efforts in the McNary Dam reservoir area recorded 120 sites (Drucker 1947a, 1948i), 22 of which were recommended for excavation (Shiner 1961:169). Early reconnaissance efforts also occurred in several other proposed reservoirs (e.g., Daugherty and Riddell 1947a, 1947b).

In September, 1946, the Corps of Engineers transferred funds to the NPS for archaeological and paleontological work on lands outside the Missouri Basin that were to be impacted by Corps projects, including the Columbia Basin (Roberts 1961a, 1961b). Similarly, in March, 1947, Reclamation transferred funds to the NPS for archaeological and paleontological work on their projects outside the Missouri Basin, including projects in the Columbia River Basin (Roberts 1952). In the fall of 1947, the Bureau of Budget ruled that all subsequent requests for appropriations should be included in the Department of the Interior's budget (Roberts 1952). The NPS, which was administered by the Department of the Interior, then transferred funds to the RBS and cooperating museums, historical societies, and universities. Those funds increased from 1947 through 1950, but then decreased the next several years (Roberts 1961a 1961b). Institutions that had previously cooperated directly with the RBS such as the UO had borne various costs (Roberts 1947, 1961a 1961b), so one suspects that for economic reasons some of those cooperating agencies signed cooperative agreements directly with the NPS in the 1950s. It seems equally likely that shrinking budgets prompted the RBS at the end of fiscal 1952 to alter the program. At the end of the fiscal year, a change was made in the plan of operations for the Interagency Archeological Salvage Program (IASP). The work of the RBS was terminated in the Columbia Basin and Pacific Coast areas. With the beginning of the new fiscal year, the direction and supervision of the investigations in those areas were to be under the NPS with its respective regional offices in direct charge. At the close of the year, arrangements were being made to transfer certain RBS personnel to the NPS and for the latter agency to take over the various field headquarters. Shiner's affiliation with the RBS was terminated on June 30, 1952, by transfer to the NPS. The NPS was to keep the RBS office at Eugene open; however, and Shiner was permitted to complete his reports on his work for the RBS (Roberts 1953:62, 66; see also Roberts 1961b:527). Shiner did finish his reports (Shiner 1954, 1961), but the Eugene office closed.

After mid-1952, both archaeological projects originally under the RBS and new projects were under the auspices of the NPS. Paul J. F. Schumacher (Chief, Archological Investigations, Western Service Center, San Francisco) took over as the NPS archaeologist who administered the work (Sprague 1973:262). When news about the upcoming shift in administration became known, Cressman expressed dismay. He had found the presence of the RBS on the UO campus to be "a real benefit to us" (Cressman letter to Frank H. H. Roberts, Jr. Sept. 5, 1951; UO-MNCH). And, simply put, he was not happy about the upcoming shift (Cressman et al. letter to Roberts, June 10, 1952; UO-MNCH).

Cressman was not happy because of personal experiences with the NPS. He told Roberts,

> [o]ne of my objections to the NPS is that it operates very much like a tax commission, telling one what has to be done and when without any idea of discussion of a problem to determine if the proposed program is mutually satisfactory. For instance I never saw a proposed contract for The Dalles dig [Cressman et al 1960] until it was sent to me the other day from San Francisco with the instruction to have the University sign the appropriate number of copies and get it back to them before June 30th. I think I have a kind of long standing emotion of bias against the NPS from earlier contact here when people I came in contact with always reminded me of YMCA boys (Cressman letter to Roberts, June 10, 1952; UO-MNCH).

Although using more tempered words, Roberts tended to agree with Cressman, telling him "I am sure the NPS people in that area do not appreciate how much you people have contributed to the program and what your cooperation has meant to us" (Roberts letter to Cressman, June 17, 1952; UO-MNCH).

Despite the discontinuation of the RBS, what it had started in the late 1940s was critical to the long-term success of local archaeological programs. Thus, although much of what we know about regional prehistory may not be directly attributable to the RBS, it is certainly indirectly attributable to that program. Although my coverage of archaeological work in what follows extends chronologically beyond 1952, most details concern the work done between 1947 and the middle 1950s.

Archaeological Results

In his history of archaeological research in the Pacific Northwest, Roderick Sprague (1973:262) remarked "the contribution of the River Basin Surveys to later archaeological investigations in the Plateau cannot be denied." However, on one hand, some histories of United States archaeology are a bit less than flattering in their discussions of the results of RBS work. Kehoe (1998:126), for example, notes that much of the work accomplished under the auspices of the RBS was "tedious" and "repetitive" (see also Jennings 1985). Fagan (2005:172), on the other hand, notes that much of the RBS work displayed a deep interest in tracking the spatio-temporal distributions of cultural traits manifest as artifact types and very little interest in "how people had lived in the past." As we will see, published results of some of the RBS-related work done in the Columbia Plateau indeed indicate much interest in cultural traits. Given that, it must be remembered that this was the dominant approach to archaeology in the late 1940s and early 1950s, in part, because cultural traits were the fundamental analytical unit of American anthropology at the time (Lyman and O'Brien 2003). Further, had it not been for RBS work, it likely would have been another 10 to 15 years before we learned that the local archaeological record has a temporal depth equivalent to just about anywhere else in the Western Hemisphere. Interestingly, that the local archaeological record may include material older than Clovis, the oft-presumed initial cultural manifestation in North America, is increasingly probable (e.g., Jenkins et al. 2012).

Wakemap Mound/Cultural Traits

On page one of his doctoral dissertation, Caldwell (1956:1) states in the very first sentence "it is axiomatic that in any program of archaeological research the basic problem is oriented toward elucidation of temporal-spatial sequences." He operationalized this axiom by studying the spatio-temporal distribution of a series of cultural traits and using as his "anchor" the stratified site of Wakemap Mound (Caldwell 1956:1). The site (Figure 5.2) was excavated during the summers of 1953 and 1954 with financial assistance from the IASP. The site had long been known to local looters and avocational archaeologists and had been partially excavated by Strong et al. (1930) in 1924 and 1925. Caldwell (1956) excavated more of the site, and his analyses epitomize the culture-history approach to archaeology. Caldwell focused on the distribution of cultural traits manifest as artifact types across four distinct stratigraphic horizons at Wakemap, and the geographic distribution of about 100 traits across 29 sites in the Pacific Northwest, including the multiple components at several sites. According to Caldwell (1956:195), "[t]he trait list provides an objective, non-intuitive mode of comparison that is used with some currency in numerous archaeo-ethnic studies," but he also recognizes that the comparative analyses can be influenced by dissimilar classifications and typologies. His conclusion basically is that Pacific Northwest prehistory is characterized by a "generalized hunting-gathering subsistence base" followed by increasing diversity (decreasing numbers of shared traits across the area) as groups tune in to their local environments without becoming too specialized (Caldwell 1956:265–266). Change basically was "progressive and cumulative" and "largely that of elaboration upon preexisting materials and techniques" (Caldwell 1956:266).

There are a number of things to note about Caldwell's dissertation. First, his advisor was Douglas Osborne, himself a major advocate of studying the spatio-temporal distributions of cultural traits manifest as artifact types. The modified and published version of Osborne's doctoral dissertation (Osborne 1951) on archaeological material collected from the McNary Dam reservoir area under the auspices of the RBS focused on documenting the spatio-temporal distribution of cultural traits (Osborne 1957). Of that study, Sprague (1973:260–261) remarked "while poorly organized and extremely difficult to utilize for comparative purposes, [it] is a landmark in the River Basin Surveys work, since it is a complete inventory of the work accomplished in McNary Reservoir and shows exhaustive historical background research. It has also served as a point of origin for theoretical constructs in Plateau prehistory and the relationship between the Plateau and the Northwest Coast."

The second thing to note about Caldwell's dissertation is that his conclusion about the cultural prehistory of the region was reiterated in somewhat refined form a few years later by Richard Daugherty (1962), who had cut his teeth on RBS fieldwork. In particular, Daugherty (1962:149) concluded that prehistoric culture history of the "intermontane west" (including the Southwest, Great Basin, and Columbia Plateau) was characterized by "cultural conservatism," that culture change occurred "principally by the addition of new elements," and that local cultures began to emerge from a common base during the middle Holocene.

Third, B. Robert Butler did additional work at Wakemap Mound which he used for his Master's thesis (Butler 1960). Butler (1960:96) reported that he and Caldwell were "preparing for publication" a monograph on the site that would include a "complete description" of all materials then recovered from the site and which would be published by the "National Park Service as one of its *Archaeological Research Series*." Unfortunately, that monograph was never published; whether or not it was completed or whether it exists in some form is unknown. Butler (1960) obtained a series of nine radiocarbon dates from the site (Crane and Griffin 1958a, 1958b), and while these were among the first such dates for the Columbia Plateau, they were not *the first*. The radiocarbon dates from Wakemap ranged from 560 to 1090 uncalibrated radiocarbon years before present (RCYBP) and facilitated decipherment of the site's formational history (Butler 1960; Crane and Griffin 1958a, 1958b).

Shallow or Deep Time

The Lind Coulee (45GR97) archaeological site was recorded in 1947 by RBS archaeologists who were informed of its existence by locals (Daugherty and Riddell 1948). When first visited, Richard Daugherty noted that "the nature of the deposits, the depth of the flakes and bone fragments from the surface (ca. 14 ft), and the heavily mineralized state of the bones, suggested . . . the possible existence of an archaeological site containing evidence of Early Man" (Daugherty 1956a:223). Daugherty again visited the site in 1950 while working on the RBS project at O'Sullivan Reservoir (Daugherty 1952). During the later visit, a projectile point fragment and a scraper were found in the exposed coulee wall; these tools demonstrated that the location was in fact an archaeological site, and it was tested in 1950 and excavated in 1951 and 1952 (Daugherty 1956a).

Lind Coulee is significant to local RBS history because it was recorded by RBS personnel; it produced the first radiocarbon dates in the area; and it demonstrated that people had been present on the Columbia Plateau longer than anyone thought at the time (Daugherty 1953, 1956a, 1956b). When Martin et al. (1947:444) commented on the time depth of the archaeological record of the Columbia Plateau, they indicated "archaeological remains in the Plateau do not appear to be very old, nor is there evidence of much cultural change in the prehistoric period." At the time, fluted points, such as Clovis and Folsom were clearly ancient, and were associated with remains of extinct mammals. No such point styles and no such associations between any point styles and extinct fauna were known in the Plateau; so during the 1940s and early 1950s, the basic notion was that the time depth of the Plateau's archaeological record was "no more than two thousand years" (Swanson 1962:1; see also Leonhardy 1968). This concept began to change with the discovery of fluted and other point styles not unlike Paleoindian points elsewhere on the continent (Osborne 1956). However, the points recovered from Lind Coulee were not fluted, and although they were stemmed, they did not look much like classic Paleoindian points from the Plains. Daugherty reported the two radiocarbon dates from Lind Coulee in his dissertation (Daugherty 1953, 1956b) and in the site report (Daugherty 1956a). The dates average 8700 ± 400 uncalibrated RCYBP. In light of new AMS ages on bone collagen, the original assessment appears to be too young by about 1100 to 1600 radiocarbon years (Craven 2003, 2004).

Associated point styles are now referred to as Western Stemmed Points, and how coincident they are with or how much they might pre-date, western fluted points remains unclear (e.g., Beck and Jones 2010; Jenkins et al. 2012). Nevertheless, if it had not been for the work of RBS-affiliated archaeologists (particularly Daugherty), the Lind Coulee Site might never have been studied and the time depth of the local archaeological record might not have been discerned in the 1950s, or even in the early 1960s. In the 1950s, many local prehistorians believed there was no significant time depth to the archaeological record of the Columbia Plateau. Lind Coulee put an end to that belief:

> It had long been my feeling that evidences of culture earlier than that exemplified by the Lind Coulee Site [dated to 8700 14C yr BP] would not be found in this area. I had assumed that in Anathermal [early Holocene] or pre-Anathermal times this region would have been something less than inviting to cultures which seem then to have found their optimum in the southern plains. . . . It came, therefore, as a distinct surprise to me not only to see in a Northwest collection a Clovis fluted point, the first observed from this region, but to be able to locate and handle two others of that category and two more points which I believe to be a Plainview and a Scottsbluff (Osborne 1956:38).

Based on RBS-sponsored work done in the McNary Dam reservoir area, Shiner (1954, 1961) offered conclusions that would be repeated by others in the 1960s and 1970s. For example, he indicated there was no evidence of significant immigration of a foreign culture and the "persistence of old artifact types and the gradual acquisition of new ones tend to confirm a local development of culture" (Shiner 1961:260). Daugherty (1962) would offer a similar conclusion a bit later. Although there was no clear evidence of Paleoindians in the area (that is, there were no known fluted-point sites), Shiner further argued that the earliest peoples tended to have a "relatively simple hunting and gathering economy [that evolved into] one far more complex. During this period, the inventory of material culture grew by addition and diversification, and a superior adjustment to the environment was made. . . . [M]ost characteristic artifacts of the Plateau were probably indigenous" (Shiner 1961:262–263). Perhaps this conclusion should not be too surprising because the perceived absence of a terminal Pleistocene-early Holocene occupation of the Columbia Plateau had only recently begun to be seriously questioned by the discovery of several isolated stone points representing Paleoindian styles (Osborne 1956). The absence of detailed context for any of them, and the lack of an association between them and remains of extinct animals made those points tantalizing at best. But things would quickly change.

Chronology Building

During the 1950s and early 1960s, published versions of RBS archaeological reports from the Columbia Plateau appeared sporadically (e.g., Cressman et al. 1960; Daugherty 1952, 1956a; Mills and Osborne 1952; Osborne 1953, 1957; Osborne and Crabtree 1961; Osborne et al. 1952, 1961; Shiner 1961). Many of these have a rather dry, descriptive tone.

But there are some bright spots, particularly those where chronological issues begin to see some resolution and those demonstrating the presence of an early Holocene occupation of the area.

During the 1950s, suspicions as to the nature of the culture chronology in the Columbia Plateau were beginning to firm up (e.g., Swanson 1956, 1962). Few radiocarbon dates were available, in part because of benign neglect that seems to have been reinforced by doubts about the validity of the technique and a previous dependence on geochronology (Lyman 2000). This situation would change in the 1960s with the excavation of several deeply stratified cave and rockshelter sites. Swanson (1956, 1962) excavated a number of sites originally recorded by RBS personnel, including a deeply stratified deposit at Cedar Cave (45KT8) in 1953–54, but he could not tie the record from that site to the obviously old material from Lind Coulee then being excavated by Daugherty. Similarly, Butler (1961) described an ancient cultural manifestation—the Old Cordilleran Culture—that he suggested was found throughout the Pacific Northwest. Based on the fact that the hallmark leaf-shaped projectile points (known as "Cascade points") of the Old Cordilleran Culture were stylistically similar to laurel-leaf shaped Lerma points of northern Mexico that seemed to date to the early Holocene (Wormington 1957), Butler (1961:63) proposed the Old Cordilleran was the "basal culture" of the Pacific Northwest. Daugherty's (1956a) stemmed points from Lind Coulee could not be placed in time relative to the Cascade points of the Old Cordilleran, so Butler (1961:41) opted to suggest that the two were "coeval." This picture would change shortly, and simultaneously the entire Holocene sequence would be penciled out.

Windust Caves (45FR46) is a series of nine small caves and rockshelters located on the Snake River in south-central Washington (Figure 5.2). Several of these caves and shelters had been looted at the time of excavation in 1959–1961. The RBS did not record the site during the survey of the Ice Harbor Dam Reservoir (Osborne 1948b). Instead, a local farmer informed Daugherty in the late 1950s about the site while the latter was undertaking salvage excavations at sites the RBS had previously recorded. Excavations at Windust Caves were supported by the NPS through the office of NPS archaeologist Paul Schumacher who had by then replaced Louis Caywood. Harvey Rice (1965) directed the excavations, who after testing three of the caves, focused on Cave C (the deposits of Windust Cave H have more recently been sampled for purposes of paleoecological research [Thompson 1985]). The nearly twenty-foot thick sediments within the seventy-foot-long, eight-foot-deep cave revealed a sequence of 10 strata and a tremendous collection of projectile points (Figure 5.3). Rice (1965) assigned ages to the materials based

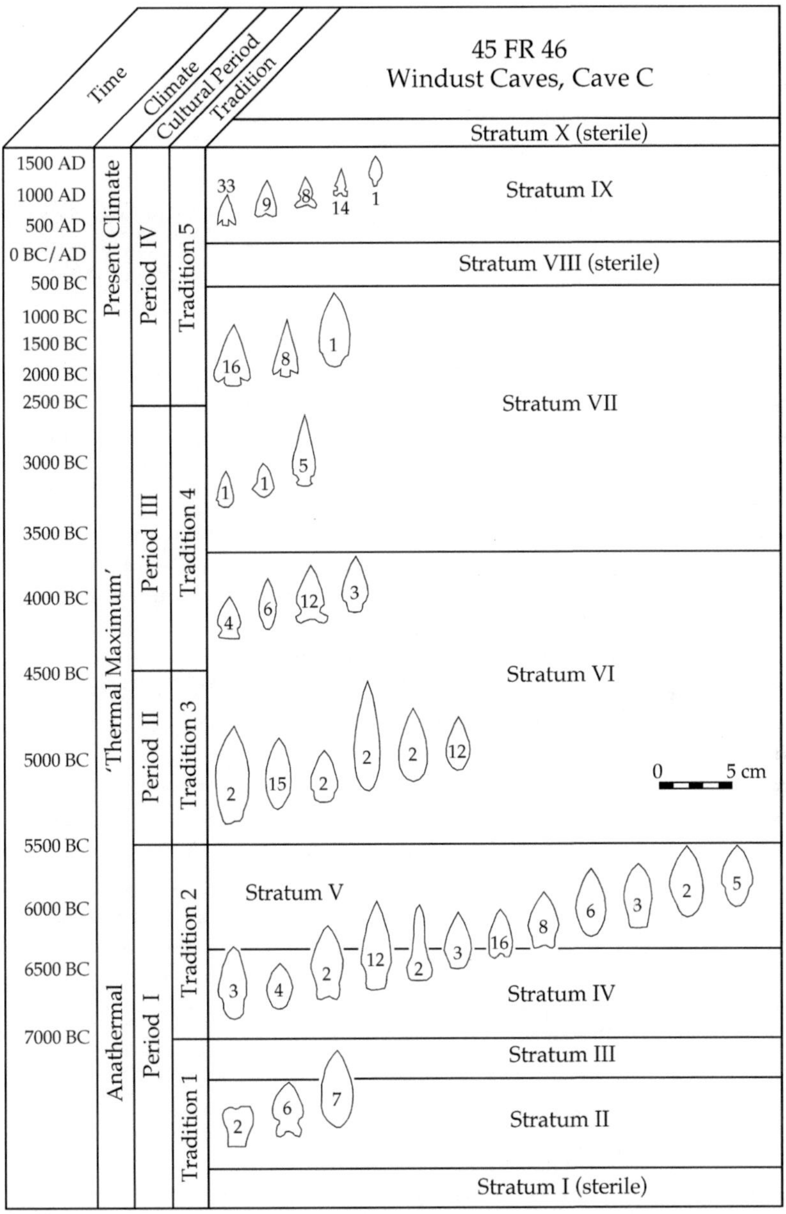

Figure 5.3 Windust Cave C chronology of projectile point types (redrawn from Rice 1965:108). Numbers in point outlines are frequency of the type. Stratigraphic placement represents the center of the vertical distribution of specimens in each type. Size of the outline represents the mean size of specimens within each type.

on typological cross-dating with materials from other sites, especially Lind Coulee (Daugherty 1956a). Windust Cave C revealed that stemmed points chronologically preceded Cascade points; those stemmed points eventually became known as "Windust points" after their clear chronological position at Windust Caves.

Rice (1965) did not procure any radiocarbon dates for his excavations. Recent reexamination of material Rice collected has obtained AMS radiocarbon dates and analyzed previously recovered faunal remains (Jenkins 2011) that substantiates the notion that curation of such collections is worthwhile. The AMS radiocarbon dates were derived from bone collagen, but unfortunately, all are less than 2000 uncalibrated RCYBP and likely are invalid for several reasons. Despite the lack of chronometric indications of age of the materials from Windust Caves, the proposed chronology of projectile point styles from Cave C (Figure 5.3) came to play a significant role in the establishment of a local chronology.

Development of that chronology was facilitated by excavation of two other stratified sites, neither of which had been recorded by RBS workers (Osborne 1948b). One, Marmes Rockshelter (Fryxell and Daugherty 1962; Rice 1969, 1972), not only provided radiocarbon and stratigraphic details but some of the earliest human remains from North America ever found (Fryxell and Keel 1969). The other site is Granite Point (45WT41), an open site. Although excavated at the same time (1960s), Granite Point did not receive the publicity that Marmes Rockshelter did. But it produced radiocarbon dates and confirmed and filled out the stratigraphic sequence revealed at Windust Cave and Marmes (Leonhardy 1970). The result was the oft-cited cultural chronology that Leonhardy and Rice (1970) outlined; not only did local archaeologists subsequently frequently refer to the chronology, so too did textbooks on the prehistory of North America (e.g., Jennings 1989:178–181). Something to keep in mind is that the NPS-sponsored excavation of these sites was a direct result of the initial reconnaissance and excavation work performed under the auspices of the RBS.

Some Odds and Ends

Geographic Coverage

Thus far I have focused on how RBS work and the subsequent related research under the sponsorship of the NPS resulted in a detailed cultural chronology for the Columbia Plateau. A plethora of other work filled in numerous geographic gaps. As it turned out, the major universities in the area—in effect, if not by explicit design—divided up the Columbia Plateau; such that, each worked for numerous consecutive years in one part of the region. The UO worked along the lower Columbia River, downstream of the mouth of the Snake River. Thus, they worked in the reservoirs behind The Dalles Dam (e.g., Cressman et al. 1960) and the John Day Dam. For the latter, many annual reports were never published and are poorly known even among local archaeologists, but a final report was produced some years after the fieldwork was complete. (Dumond and Minor 1983). That report focused on yet another site—Wildcat Canyon—that, when excavated in the 1960s, helped fill out and confirm the cultural chronology then becoming apparent from work at Lind Coulee (Daugherty 1956a), the Vantage area (Swanson 1956), and Windust Caves (Rice 1965).

The UW worked in several reservoir areas of the middle and upper Columbia River that included Priest Rapids Dam, Wanapum Dam (Swanson 1956, 1962), Wells Dam (Browman 1966; Grabert 1968, 1970), Chief Joseph Dam (Osborne et al. 1952), and McNary Dam (Osborne 1957; Osborne et al. 1961). The final report on the Priest Rapids Dam reservoir was completed some years after the fieldwork (Greengo 1986). Washington State University, with Daugherty and his students and colleagues, particularly with Roald Fryxell in the lead, worked for nearly two decades along the lower Snake River, from its mouth east to the Idaho border. The number of reports from that work is difficult to estimate; I have 45 individual titles on my bookshelves. These titles have publication dates younger than 1980 and do not include an unknown number of limited-distribution progress reports. As indicated above, the lasting contributions of all of this work was the development of the first detailed cultural chronology (Leonhardy and Rice 1970) and the demonstration that people had been on the Columbia Plateau since the late Pleistocene (e.g., Rice 1972).

Excavation Unit

Frank Leonhardy once told me that Osborne had popularized the local use of a planning tool known as an "excavation unit" (EU). This tool was also used in the Missouri Basin Project of the RBS (e.g., White and Cooper 1951). I cannot improve upon Osborne's (1949:10) description of an "excavation unit" and thus quote it at length here.

The definition and breakdown of costs for an Excavation Unit are as follows:

> The Excavation Unit is an operational unit designed to provide labor, supervisory personnel, and equipment for the recovery of archeological materials in the field; technical and clerical personnel, laboratory space, and laboratory equipment for processing these materials and preparing them for study; and finally, for the study of the materials and data

recovered, and preparation of a technical report (for which the archeological supervisor will be responsible).

01	Personal services:	
	1 Archeologist–Supervisor for 1 year	$4,480.00
	1 Field Assistant, 4 months at $250 per month	1,000.00
	2 Laboratory assistants, 1 year at $200 per month	4,800.00
	1 Clerk-typist	2,500.00
	10 Laborers, 800 days at $8 per day	6,400.00
02	Travel and per diem	900.00
03	Transportation of things	220.00
04	Communication services	100.00
05	Rents and utility services	600.00
07	Other contractual services	300.00
08	Supplies and materials	600.00
09	Equipment	2,600.00*
	Total cost per Excavation Unit	$24,500.00

*After the first year in the case of extended programs this item would be considerably reduced because most of the major equipment would have been purchased.

Sites were evaluated for their research potential and recommendations made as to how many excavation units should be expended on them. Thus, for the 20 sites the RBS recorded in the Chief Joseph Dam reservoir area, seven were recommended for excavation with EUs ranging from 1/64 to 1/16 per site, and four sites were recommended for testing for a total of 1/64 EU. The total cost was thus 3/16 of an EU, or $4,593.75 (Osborne 1949). Shiner (e.g., 1951) continued to use the EU to provide cost estimates for the work thought necessary within individual project areas. For unclear reasons, a Shiner EU cost $4,500 less than an Osborne EU. Savings were made with respect to equipment and salary for the supervisor; the latter was given a Government rank of GS-7 and a pay rate of $3,825 per year. Laborers were paid the same in both an Osborne EU and a Shiner EU—$8 per day. Ah, when times were simpler. . . .

Reanalyses

One of the perhaps seldom-considered aspects of what the RBS (and subsequent NPS) program accomplished is the preservation (through salvage excavation) of many irreplaceable materials. This aspect is no-better-illustrated than by considering the fact that excavations at The Dalles Roadcut Site (Cressman et al. 1960) produced what at the time was a rather startling discovery: an abundance of salmonid bone in the deep, early Holocene strata. In light of the fact that salmon were a critically important resource ethnographically and that anthropologists thought that perhaps the intensive exploitation of salmon during the late Holocene had facilitated if not prompted the emergence of local ethnographic land use and social patterns, The Dalles Roadcut Site was something of an anomaly. Reanalysis of the original collection (Butler 1993), plus new excavation at the remaining deposits (Butler and O'Connor 2004) indicated that the site was hardly anomalous. Much of this reanalysis and others mentioned above, such as new studies of Lind Coulee, could never have happened without the fieldwork of RBS archaeologists.

Training

Some of the biggest names in the history of local archaeology wrote doctoral dissertations or master's theses on RBS (or NPS) generated collections. Several of these have been mentioned above, including Caldwell (1956), B.R. Butler (1960), Daugherty (1953), Swanson (1956), Rice (1965), Leonhardy (1970), and Grabert (1970). Others include David Cole (1954), who was responsible for much of the work at the John Day Dam reservoir; Roderick Sprague (1959, 1967), who became the doyen of prehistoric burials in the Columbia Plateau; David Browman (1966), who coauthored an early and still-important synthesis of local prehistory (Browman and Munsell 1969); Dave Rice (1972), who provided the first detailed description of the temporally basal local cultural manifestation known as Windust; and Roald Fryxell (e.g., 1963), who was one of the first geoarchaeologists in the nation and was responsible for the excavation of Marmes Rockshelter that revealed a complete Holocene stratigraphic sequence.

Conclusion

In his interim report on the IASP, Frederick Johnson (1951a) presented 23 photographs that reveal something of the nature of the program. Large, open-surface areas have been cleared and are being excavated; heavy equipment such as bulldozers are being used to remove overburden; excavators work with rising reservoir waters lapping at their feet; and dams rise in the background. Of the 23 photos, only two have as their subject archaeological material of the Columbia Plateau. This is in some ways an accurate reflection of the RBS program. A tremendous amount of work was accomplished in other areas, amounts that make the work on the Columbia Plateau appear relatively miniscule.

But as I hope is clear from the preceding too-brief historical sketch, the work that was done under the auspices of the RBS and the groundwork it laid for the subsequent archaeological salvage program of the NPS played a significant role in the development of local archaeology.

Much of what we know today we still owe directly to the RBS program. Further, a lot of other knowledge rests on a foundation of RBS work, even if that knowledge was gained long after the RBS work had faded from memory. It is a fitting tribute to the efforts of Drucker, Osborne, Shiner, Cressman, and all the others who worked out of the Eugene office that they be remembered for their energy and enthusiasm for local archaeology.

Acknowledgments

I'm not sure whether to thank or to curse G. Tom Jones and Ruthann Knudson for the suggestion that I write this chapter, but I do thank Kimball Banks and Jon Czaplicki for the invitation to write it. I struggled with it, but managed to learn a lot putting it together. I hope readers learn something as well. The most important lesson is that those of us who work in the area of the Columbia Basin RBS owe a great deal to our predecessors. Preparation of this chapter was facilitated by the assistance of Mary Collins (Washington State University, Museum of Anthropology), Pam Endzweig (University of Oregon, Museum of Natural and Cultural History, and Museum of Anthropology), and Laura Phillips (Thomas Burke Memorial, Washington State Museum of Natural History and Culture).

CHAPTER 6

Archaeological Salvage at Texas Reservoir Construction Projects: 1945–1969

Edward B. Jelks

The Interagency Archaeology Salvage Program (IASP) in Texas was initiated in 1945, when the Smithsonian Institution (SI) opened a River Basin Surveys (RBS) office in Austin under the direction of Joe Ben Wheat at The University of Texas at Austin (UT). The Texas program was run by the SI (RBS, with SI staff) only from 1945 to 1953, then by the NPS (IASP, with NPS staff) from 1953 to 1958, and by Texas Archaeological Salvage Program (TASP) with UT staff from 1958 to 1971 the program morphed into a cultural resource management format in which a number of different contractors, located at various places in the state, became involved. Wheat carried out the first RBS fieldwork in 1946 when he made a preliminary survey of Addicks Reservoir near Houston and conducted limited excavations at several sites there. The following year, Wheat moved to Colorado, and Robert L. Stephenson replaced him as director. Between 1947 and 1949, Stephenson surveyed about a dozen reservoir projects, but did no digging. I was hired as Stephenson's assistant in 1950. When Stephenson left to work on a Ph.D. at the University of Michigan in 1951, I took over as director and continued in that capacity until joining the faculty of Southern Methodist University in 1965.

Initially, from 1945 to 1953, the program was administered from Washington by Frank H. H. Roberts, Jr. as part of the SI's RBS program. In 1953, the Texas program was transferred to the Region Three office (now the Southwest Regional Office) of the (NPS) where it continued under the auspices of the IASP with Regional Archeologist Erik Reed as administrator.

In 1958, the NPS decided to stop managing the program in-house and began contracting UT for the work in Texas. At that time, the Texas (TASP)

Dam Projects and the Growth of American Archaeology: The River Basin Surveys and the Interagency Archeological Salvage Program, edited by Kimball M. Banks and Jon Czaplicki, 85–94.

was established at UT to handle contracts with the NPS, and I continued directing the program there until 1965.

Dee Ann Story directed that TASP from 1965 to 1969, and Frank Eddy ran it from 1969 to 1970. Later, David Dibble took over from 1970 to 1971. By that time, the salvage program had faded out of existence as a distinct entity. Texas had a state historic preservation officer and a state archaeologist. Other institutions than UT did contract archaeology at reservoir projects that private archaeological contracting firms operated, and cultural resource management (CRM) replaced salvage archaeology as the model for preconstruction archaeology. A number of archaeologists conducted surveys, directed excavations, supervised laboratory workers, and wrote reports for the RBS and the IASP in Texas over the years. Among them were: Bob Alexander, Bob Benfer, Mike Collins, Bill Davis, Mott Davis, Dave Dibble, Lathel Duffield, Frank Eddy, Jerry Epstein, John Greer, Don Hartle, Ken Honea, Bob Humphreys, LeRoy Johnson, Dessamae Lorraine, Bernie McClurkan, Ed Moorman, Parker Nunley, Mark Parsons, Fred Peterson, Elton Prewitt, Dick Ross, Dan Scurlock, Harry Shafer, Bill Sorrow, Dee Ann Story, and Curtis Tunnell.

During the early years, Wheat, and later Stephenson, struggled to get the program underway with severely limited funds. In 1950, the program received the first allocation of enough money for serious excavations. That spring Stephenson, with me as his new assistant, began excavations at Whitney and Lavon reservoirs. This initiated a period of 19 years of intensive salvage archaeology at more that 40 reservoir projects in Texas (Figure 6.1).

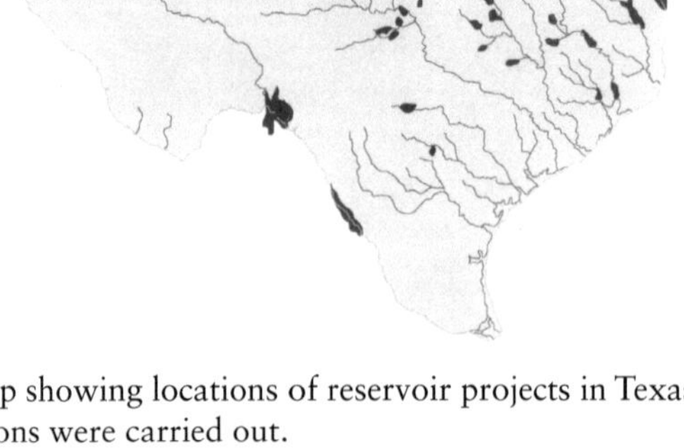

Figure 6.1 Map showing locations of reservoir projects in Texas where surveys and/or excavations were carried out.

The salvage program located and recorded an estimated 1,600 archaeological sites at the reservoir projects. Many were tested by limited excavation, and scores of the more important ones were extensively excavated. These efforts produced a massive accumulation of field specimens (artifacts, biota, soil samples, etc.) together with supporting field records. Field specimens and notes from the early period—1946 to 1954—are curated at the Smithsonian Institution's National Museum of Natural History in Washington DC Those from 1954 to 1969 are curated at the Texas Archeological Research Laboratory at UT's Jake Pickle Research Center.

Field Methods and Strategies

Surveys

Wheat conducted his first surveys of reservoir projects alone. Stephenson did the same, traveling in his own car (as the RBS had no vehicles), camping wherever nightfall happened to catch him, and usually cooking his supper and breakfast on a Coleman stove. We got company transportation in 1950 when a Jeep and a one-ton, stake-body truck were sent down from Washington DC (Figure 6.2). After Stephenson left for Michigan in 1951, I did several small surveys by myself, but before long, it became standard practice to send out two-person survey teams that usually stopped at motels rather than camping out. Because the time available for carrying out a survey was severely restricted by deplorably limited funds, little time was spent walking about looking

Figure 6.2 RBS crew at Lavon Reservoir, 1950. Standing, left to right: E.O. Miller, Cecil Calhoun, Adolph Witte, Calvin Reuter, Bob Stephenson. Kneeling, left to right: Rex Housewright, Ed Moorman, Ed Jelks.

for sites. Instead, we checked out known sites—those recorded in the UT files or reported by avocational archaeologists, artifact collectors, and landowners—and did brief walkovers of high-potential places near access roads. This left up to 50 percent of some reservoir impact areas uninspected. When a site was discovered, we filled out a standard survey form on which we recorded the site's location, property owner, physical description, the types of artifacts found on the surface, its probable cultural affiliations, and its potential for productive excavation. We also performed a random collection of artifacts from the surface.

Excavations

The meager funds allocated each year to the Texas program were spent on the most urgent projects, that is, those where dam construction was underway and inundation of important known sites was imminent. The idea was to collect a sample of data from a "representative" group of known sites in a project area. In the earlier years, there was sometimes enough time and money to sample two to seven sites before water was impounded behind a dam. For example, seven sites were sampled at Whitney, three at Lavon, and three at Texarkana—nowhere near a truly representative sample of the scores of recorded sites inundated in each of those reservoirs. By the late 1950s, the budget improved to the point that we were able to maintain a staff of two or three technicians to process artifacts in the laboratory and three or four archaeologists. We could have two or three crews in the field at the same time.

Even so, there was not enough time, money, or labor to excavate any site completely, so a sampling approach was adopted. Sampling designs were based strictly on the archaeologist's judgment. Randomized sampling and other approaches based on inductive statistics were still a few years in the future. Typically, several test squares would be placed selectively about a site where surface indications suggested that middens or other features might be present. Information gained from digging these squares would determine whether additional excavation was warranted and where it should be located. If a site appeared to have appreciable depth, a test square typically would be dug to a depth below any evidence of cultural materials in order to get a preliminary fix on possible geologic or cultural strata. Other than obvious artifacts, the archaeologists regularly collected animal and plant specimens from excavated sites. Workers were instructed to save all bones, no matter how small, as well as other biota such as fish scales, mollusk shells, seeds, and so on. Later, in the laboratory, these were sorted, and some specimens (e.g., bone splinters with no identifiable diagnostic features) were discarded.

We made a practice of collecting charcoal samples for C^{14} dating, although few of them were ever analyzed as dating laboratories were quite expensive. We routinely saved soil samples from the different strata, feature matrices, and other structural components of a site with the expectation that maybe they contained pollen or other substances that might yield significant data, but few of them were ever examined in the laboratory. In the early 1960s, we began to collect and analyze pollen-profile samples from stratified sites and include the results in the site reports (cf. Johnson 1963). Eventually, collecting pollen samples became standard procedure.

Studies of stone knapping in the United States were just beginning to get under way, and in the beginning, the archaeologists did not collect waste stone flakes or discarded cores unless there was something unusual about them, such as being of exotic material. We saved only those stone objects that were identified in the field as "artifacts," and distinguishable from what we perceived as unimportant manufacturing by-products. About 1958, Jerry Epstein announced that he had found burins in Centipede and Damp caves at Amistad Reservoir (Epstein 1959). No one had ever recognized burins in Texas before, and they looked like randomly fractured waste flakes to us. Jerry had to do some tall talking, but eventually he convinced us that burins were real. Soon, all of our field archaeologists began looking for burins along with other chipped-stone artifacts. As we became familiar with studies of knapping techniques, it became standard procedure to collect by-products of stone knapping, such as reduction flakes, cores, and so on.

Laboratory

From 1945 until the summer of 1956, the program's office and laboratory were housed in an ancient building at the UT's Little Campus complex located a few blocks southeast of the main campus. The UT's extensive archaeological collections, accumulated largely from WPA projects of the 1930s and early 1940s, reposed in dead storage in the same building. Many Texas pottery and projectile-point types had previously been defined, chiefly by Charles Kelley, Alex Krieger, and Tom Campbell; and the thousands of examples of these types in the collections served as a useful reference resource for the salvage archaeologists. Field crews washed, classified, numbered, and cataloged some artifacts on rainy days. Those they did not get to were brought back to the Austin laboratory for processing. The laboratory had plenty of table space to lay out the artifacts from a site where the archaeologist writing the report on the site could examine, classify, and sort them. After survey and excavation reports had been completed, the artifacts were boxed and temporarily stored in a room at the laboratory.

Reports

The archaeologist who conducted the survey prepared a report that included a detailed description of each site located together with descriptions of the artifacts collected from its surface and an evaluation of its potential for further investigation. The survey reports were reproduced by mimeographing or multilithing. A run of 20 to 30 copies of a report was typical. Copies were provided to the SI, the NPS, and UT's Department of Anthropology, and several were retained as file copies. Survey reports were distributed as appropriate to individual archaeologists with a research interest in a particular survey area and sometimes to avocational archaeologists who had cooperated by providing information about site locations during surveys. A detailed report of every excavated site was prepared by the crew chief who directed the excavations. Most of these were published either in Archaeology Series, Department of Anthropology, The University of Texas, in Miscellaneous Papers of the Texas Archeological Salvage Project, or occasionally in the *Bulletin of the Texas Archeological Society.* Often, a site report was provided to the NPS in published form to meet contractual requirements.

Results

Advancements in Knowledge of Texas Archaeology

The following brief description of the state of archaeology in Texas in the 1940s and 1950s will provide a baseline from which to evaluate the extent to which the salvage program advanced knowledge of the native and the early Euro-American inhabitants of Texas. In 1945 when Joe Ben Wheat opened the RBS office in Austin, there were four other archaeologists in Texas who could be termed professionals by virtue of job title and/or advanced degrees in archaeology/anthropology. They were Tom Campbell and Charles Kelley, both on the teaching faculty of the Department of Anthropology at UT; Alex Krieger, a nonteaching UT research scientist; and Bob Stephenson, fresh out of the Marines at the end of WW II who, with no job available as an archaeologist, had opened a neighborhood grocery store in Austin that he managed with his wife, Georgie. Several geologists had been doing scientific field archaeology for years, especially at Paleoindian sites: notably E. H. Sellards and Glen Evans at the Texas Memorial Museum and Grayson Mead at Texas Tech. Historians Curry Holden at Texas Tech and Victor Smith at Sul Ross also were competent archaeologists. To the best of my knowledge, the only university in Texas offering a degree in anthropology at the time was UT, where one could earn a BA and an MA with an archaeology specialty.

The UT archaeologists had arbitrarily divided the state, like ancient Gaul, into three parts, with each archaeologist having proprietary rights to one of the parts. Krieger's primary domain was eastern Texas, but he also held rights in the panhandle area because of his publications on the Antelope Creek Focus; Campbell controlled the coastal region; Kelley claimed a broad area encompassing Central Texas and the Trans-Pecos region. Paleoindian sites, few and far between, were viewed as a special case and were pretty much up for grabs for whoever could find one. However, it was generally recognized that the team of Sellards and Evans had some priority on Paleoindian sites because of their previous work at Plainview, Lubbock Lake, Miami, and other early sites. Krieger also had a long-standing involvement in Paleoindian research, some of it in collaboration with Sellards and Evans.

By the mid-1950s the tripartite archaeological segmentation of Texas had disintegrated, largely because salvage archaeologists were working pretty much all over the state. Also, two of the proprietors pulled up stakes. Kelley joined the faculty at Southern Illinois University in 1950, and Krieger moved to California to become director of the Riverside Museum in 1956. Texas archaeologists had long used the Midwestern Taxonomic System of foci and aspects for classifying archaeological cultural units. Krieger had defined a series of chronologically ordered foci and aspects for eastern Texas. Campbell had done the same for the coastal region as did Kelley for both Central and Trans-Pecos Texas. They also had described a number of pottery and projectile-point types for their respective areas. In the main, the foci and aspects were accurately formulated, but research by the salvage archaeologists led to a number of revisions, especially in Central and West Texas. Sellards and Evans had put together a good basic profile on the Paleoindian period. Eventually, the Midwestern system was replaced by the current system of phases and, in East Texas, by numbered periods, Caddo I–V.

Other universities in the state added archaeologists to their faculties in the 1950s and 1960s, including Jack Hughes at West Texas State, Fred Wendorf at Southern Methodist University, George Agogino at Baylor, Frank Hole at Rice, Rex Gerald at the University of Texas at El Paso, and Bill Mayer-Oakes at Texas Tech. And owing to the rise of CRM archaeology, private archaeological contracting firms were appearing. The number of archaeologists in Texas increased through the years from the five in 1945 to well over 100 today (2012). Some examples of discoveries made by the salvage archaeologists that significantly increased knowledge of Texas archaeology follow. Many other examples could be cited.

Central Texas

Charles Kelley (1947) and Krieger (1946:165–168) had named and briefly defined two late prehistoric foci in Central Texas—Austin and Toyah—that he believed to be contemporaneous. Stratified deposits in two rockshelters at Whitney Reservoir on the Brazos River—the Blum and Kyle sites, excavated respectively in 1952 and 1959/1960—revealed that the Austin Focus predated the Toyah Focus. The sites also provided enough information for a detailed description of the two foci (Jelks 1953, 1962).

Through the joint use of historic documents and archaeological evidence, the Stansbury Site at Whitney, excavated in 1950, was identified as the location of a documented late eighteenth-century Tawkoni village (Jelks 1970). Subsequent excavations at related sites in Texas and Oklahoma produced information that, combined with information from Stansbury, led to the recognition of a complex of native and European trade artifacts affiliated with southern Wichita tribes that was designated the Norteño Focus (Bell et al. 1967; Duffield and Jelks 1961).

North-Central Texas

Stephenson (1952) defined the late prehistoric Wylie Focus, primarily on the basis of excavations at the Hogge Bridge Site at Lavon Reservoir on the East Fork of the Trinity River in 1950. A unique feature of Wylie Focus sites was a large concave depression 20 to 40 meters in diameter in the center of each site. These depressions were interpreted as serving some uncertain kind of ceremonial function.

East Texas

McGee Bend reservoir lies just south of the Caddoan Area proper and was little known archaeologically when 11 sites were excavated there between 1956 and 1960. A chronology for the archaeology of that part of southeastern Texas, from Early Archaic up to the Protohistoric Period, was developed on the basis of those sites (Duffield 1963; Jelks 1965; McClurkan et al. 1980). Key sites that contributed to that chronology were:

- Wolfshead, with an Early Archaic occupation typified by San Patrice points;
- Print Bell, with an occupation typified by pre-Caddoan, sandy-paste pottery;
- Jonas Short, a burial mound containing ceremonial Middle Woodland artifacts related to the Marksville Culture of the lower Mississippi basin;
- Walter Bell, a Late Woodland site of Caddoan affinity; and
- Etoile, a Protohistoric Caddo-related site.

West Texas

Amistad Reservoir flooded scores of sites on the Rio Grande, Pecos, and Devils rivers; most of them were rockshelters. Deposits in many shelters were dry enough that perishables, including basketry, wooden artifacts, and plant parts were preserved. Spectacular pictographs adorned the walls of some. Excavations at several shelters produced large quantities of stone and perishable artifacts along with a wealth of vegetal food remains. Amistad Reservoir contained several stratified sites with

Figure 6.3 Stepped excavations of deep stratified deposits at the Devils Mouth Site, Amistad Reservoir, 1962; dam under construction in the background.

alluvial deposits several meters thick. One of these—the Devils Mouth Site (Figure 6.3)—dated from late Paleoindian to late prehistoric times and produced a series of stone artifacts that were used to create a chronology of artifact types spanning some 10,000 years (Johnson 1961). The reservoir also contains numerous shelters, the best known being Bonfire Shelter (Dibble 1968). This shelter was the locus of two stratigraphically separated bison bone beds. The earlier bed consisted of *B. antiquus* bones associated with Folsom points; the later bed of *B. bison* bones were associated with Late Archaic points.

Many of the shelters at Amistad, including some with important pictographs, were not inundated but stood only a few feet above water level. These sites had been well protected because of their inaccessibility in very rough terrain, and the sheep ranchers of the region vigorously kept trespassers off their property to prevent disturbance of the sheep.

A major concern was that once the lake was filled, a constant stream of boaters would pass by in full view of, and with easy access to, the shelters and the pictographs. Under those circumstances, it was inevitable that over time vandals would deface the pictographs and would dig potholes in the shelters' deposits to look for artifacts. Because of those concerns, the decision was made that salvage excavations would be carried out at those shelters; the pictographs would be fully recorded; and steps would be taken to protect the pictographs and the shelter deposits from vandalism. Two artists, David Gebhard (1960) and Terry Grieder (1965) were hired to record many of the pictographs by photographing and tracing them on large sheets of plastic. Some of the most important pictograph sites are now in Seminole Canyon State Park, opened in 1980, where they are permanently protected.

Collections as a Research Resource

The artifacts, field records, survey reports, and excavation reports constitute an irreplaceable resource of vast proportions. The exact number of times the collections have been used for research purposes is uncertain, but they undoubtedly have been used thousands of times over the past half-century for teaching, research, theses, dissertations, and museum exhibits. Users have included Smithsonian archaeologists; students and faculty from UT and other universities; and researchers from government agencies, private foundations, contracting firms, and others. As I write this in February, 2012, a student at Texas A & M University is using collections and field notes from excavations at McGee Bend Reservoir more than 50 years old as a main source of data for his Ph.D. dissertation.

Student Employment and Training

The salvage program offered field and laboratory job opportunities for students, with the promise of professional employment after graduation, which began to attract more and more students as time went on. Some stayed on to become leaders in Texas archaeology. Dee Ann Story, Curtis Tunnell, Harry Shafer, LeRoy Johnson, Elton Prewitt, and James Corbin were among those who got early field and laboratory training as employees of the salvage program.

Concluding Thoughts

By any measure, the Texas archaeological salvage program was an outstanding success. It added substantially to knowledge of Texas prehistory and history; produced a massive resource of artifacts and field records from archaeological sites that would have been destroyed without any study had the salvage operations not taken place; led to a voluminous library of publications and file reports; and maintained an on-hands training ground for two generations of future archaeologists.

CHAPTER 7

Federal Archaeology in the Southeast

Richard A. Krause

Early Developments: Late 1800s to Early 1900s

For over a century, American laymen and scholars pondered the authorship and use of the monumental earthworks in the Midwestern and Southeastern United States (Silverberg 1967). Yet systematic investigations have been confined to the last nine decades of the twentieth century and the first decade of the twenty-first. These investigations were preceded by the work of curiosity seekers, some with scholarly interests, who made desultory inquiries of archaeological deposits in the region's major river basins. In the early years of American nationhood, when the triumph of free market capitalism was regarded as inevitable (Trigger 1984), interpretations of mound building were compatible with an ethic of manifest destiny and territorial expansion. Native Americans were considered residues of ages past—primitives in the way of progress—and people incapable of the organizational efforts that monumental works entailed even if they were constructed of dirt (Resek 1960:13; Trigger 1986:189–194).

Insofar as Native Americans were responsible for the archaeological record, that record was interpreted as shallow and static. Many scholars saw North America's mounds as clear evidence for the prehistoric presence of an Old World influence (Barton 1787, 1797; Clinton 1820; Gallatin 1836; Kingsborough 1830). Using notions of hyperdiffusionism, they linked North American mound builders to Egypt, Israel, and other classic civilizations. For others, an Old World link was not as clear, although the notion that Native Americans were the descendants of mound-building ancestors was still inconceivable. Instead, they posited a mysterious "mound building race" to explain North America's tumuli (Clinton 1820; Gallatin 1836).

Dam Projects and the Growth of American Archaeology: The River Basin Surveys and the Interagency Archeological Salvage Program, edited by Kimball M. Banks and Jon Czaplicki, 95–114.

Early digging in the Southeast was guided by a tradition of gentlemanly antiquarianism that seems, on occasion, to have lapsed into errant vandalism. C. B. Moore, a wealthy Harvard University graduate, purchased a flat-bottomed steamship to sail the Florida rivers and explore shell middens and burial mounds along their banks. In the summer, he tested sites along the Southeastern waterways. In the winter, he analyzed and reported the results in the *Journal of the Philadelphia Academy of Natural Sciences* (Moore 1894:5–128). In 1899, Moore traveled the Alabama River, and in 1805, he sailed the Black Warrior River. While on the Black Warrior, he tested two mounds and mapped 20 more at the site of Moundville. Impressed by Moundville's size and elaborate artifacts, he returned to conduct additional excavations the following summer (Moore 1907:337–405). Although some adequately described the sites and artifacts they discovered, few were interested in interpreting them.

Nevertheless those who were more "intellectual" did credible work by the standards of the day. For example, Moses Fiske in his 1815 letter to the American Antiquarian Society described the aboriginal remains in the state of Tennessee in great detail (Fiske 1820). Fiske argued that cone-shaped mounds contained burials and that rectangular flat-topped mounds formerly supported buildings. He also defined structure patterns and palisade walls which provided evidence of villages surrounding platform mounds. Fiske did not believe Europeans were responsible for building the mounds (the so-called "Mound Builders"). He called for an attribution substantiated by research. Many of his contemporaries, however, still believed that the burial mounds in the Southeast and Midwest were the result of an extinct race of people. Yet due mostly through the investigative efforts of Cyrus Thomas, the myth of a mound building race was quickly becoming passé. In 1887, Thomas published a short article in the Bureau of American Ethnology on mound building in Tennessee, North Carolina, and Ohio (Thomas 1888). His section on the Cherokee was simply entitled "The Cherokees Probably Mound Builders." His evidence was based on artifacts recovered from mound excavations, on domestic structures, and the analysis of shell gorgets.

In the late 1800s, Joseph Jones of the Smithsonian Institution excavated stone box graves in Tennessee. He accurately described the remains and connected them to Mississippi River groups, such as the Natchez (Jones 1876). Then, too, in the fall of 1877, Frederick Ward Putnam of Harvard's Peabody Museum excavated stone box graves in the vicinity of Nashville. He posited a common ancestry for groups in Tennessee and the central Mississippi River Valley (Putnam 1878). The late nineteenth- and early twentieth-century pioneering efforts of Moses Fisk, C.B. Moore, Joseph Jones, Fredrick Putnam, Cyrus Thomas, and others presaged later twentieth-century archaeological work that identified the authors of Southeastern earthworks as the ancestors of contemporary Native Americans.

By the 1900s, Native American marginality was a well-established fact of reservation life and Amerindians were viewed as better at receiving rather than donating wisdom. Thus, transformations exposed during artifact sequencing were explained by one of several forms of diffusion from a few centers of Native American innovation: usually Mexico, the Southwest, or Gulf Coast of the United States (Adams et al. 1978). The indigenous authorship of North America's mounds was firmly established, and in order to achieve chronological goals, scholars sorted them into classes based on their morphology and evidence of their use. Geometric and effigy earthworks were separated from burial tumuli and frustum-shaped or rectangular platform-shaped mounds. A systematic comparison of mound contents indicated that geometric and effigy mounds were earlier than burial and platform mounds (Silverberg 1967).

The Works Progress Era: The 1920s to the 1940s

The Great Depression of the late 1920s and early 1930s stimulated the emergence of government-sponsored works and building programs that included dam building and archaeological inquiry through work relief. Norris was the first in a series of nine dams to be built on the Tennessee River and its tributaries. To mitigate the damage to the region's archaeological resources, the Works Progress Administration (WPA) and the Tennessee Valley Authority (TVA) combined efforts to create an archaeological salvage program. The 1934 TVA survey and excavation of sites along the Clinch River in eastern Tennessee's Norris Reservoir was proto-typical (Lyon 1996:40–41).

William S. Webb, the chairman of the University of Kentucky's Anthropology department, was appointed TVA's first director of archaeology (Lyon 1996:40). Webb selected Thomas M. N. Lewis as supervisor for the Norris project. George D. Barnes and A. E. Wilkie served as Lewis's field supervisors (Lyon 1996:39–40). The 1934–1936 Norris Basin work identified and explored 23 archaeological sites. In the resulting report (Webb 1938), James B. Griffin analyzed the ceramics (Griffin 1938) and used them in his Ph.D. dissertation (Griffin 1936). William D. Funkhouser (1938) described the skeletal remains, and Florence M. Hawley (1938), one of the Southeast's first female archaeologists, discussed the work done on the area's dendrochronology.

In 1935, Webb returned to Kentucky to direct the state's reservoir projects and Thomas M. N. Lewis replaced him as TVA's director of archaeology. Lewis hired Madeline Kneberg to direct the archaeology lab in Knoxville. She and Lewis conducted the Chickamauga Basin salvage program during which the Dallas, Hixon, Ledford Island, Mouse Creek, and Hiwassee Island sites were excavated. Project personnel explored five platform mounds, 10 village sites, 165 structures, and 2,000 burials; and they made extensive collections of pottery, stone, bone, shell, and copper artifacts (Lewis and Kneberg 1941, 1995). Lewis and Kneberg's seminal 1946 discussion of Hiawassee Island is an archaeological classic (Lewis and Kneberg 1946). The report remains the archaeological "yardstick" for phases in the area. Similarly, the Eva Site in western Tennessee was one of the best-described examples of an Archaic Stage occupation (Lewis and Lewis 1961). Other reservoir projects conducted in the 1940s included Watts Bar, Douglas Lake, Fort Loudon, and Kentucky Lake. Additional TVA reservoir projects were also conducted in the Guntersville, Pickwick, and Wheeler Lake areas of Alabama.

Webb appointed Walter B. Jones to organize and supervise work in the Alabama reservoirs, thereby building on a 1932–1934 University of Alabama Museum of Natural History archaeological survey that had located 237 sites in the Wheeler Basin. Webb chose David DeJarnett as the field supervisor. After some serious delays in acquiring WPA labor and equipment, DeJarnett began fieldwork on January 6, 1934, and continued until July 15th. He supervised the work of 12 field assistants: Robert Adams, Eliot Davis, Kenneth Disher, James Foster, Bennett Gale, D.W. Lockard, Horace Miner, Robert Morrison, J. J. Renger, Alden Stevens, Sidney Thomas, and James White, Jr. (Lyon 1996:47). The crew excavated 19 of 237 sites in the reservoir area, including five burial mounds, five village sites, one cave site, two flint workshops, two shell mounds, two shell mound/village sites, one burial mound and village, and one earth mound. Today these sites would be identified as Mississippian mounds, villages, and cemeteries, Copena burial mounds, and large Archaic shell middens (Lyon 1996:46–50).

As work began in the Pickwick Basin, the Alabama Museum of Natural History—with Walter B. Jones directing and David DeJarnett supervising—built upon the successes of their Wheeler Reservoir Project. Their field supervisors included Harold Andersen from the University of Alabama, H. Summerfield Day from the University of Illinois and Harvard University, Charles Wilder from Oberlin College and Columbia University, Steve Wimberly from the University of Nebraska, James Foster from the University of Kentucky, Theodore Johansen from the University of Alabama, Wayne Kraxberger from the University of Denver, and Carl F. Miller from the University of Arizona. Julie Adcock from the University of Alabama and Harold Dumas from the University of Nebraska were laboratory supervisors. Charles Snow from Harvard and Marshall Newman served the project as physical anthropologists (Lyon 1996:131–135).

In the Pickwick Basin, Webb and DeJarnett were particularly interested in the large shell mounds at the Smithsonia Landing, Perry, Bluff Creek, O'Neal, Mender Scar, Long Branch, Union Hollow, Mulberry Creek, and Georgetown Landing sites. Webb and DeJarnett considered them examples of the earliest inhabitants' attempts to utilize the basin's natural resources. The excavated Mississippian residential mounds and villages in the Pickwick Basin included Seven Mile Island, McKelvey Mound, and Koger's Island. When Webb and DeJarnett wrote the report on the Pickwick Basin, they included an analysis of the pottery by William Haag who considered temper a diagnostic for ordering the area's ceramic wares. He produced a Pickwick Basin ceramic series that included, from early to late, fiber-tempered, sand-tempered, limestone-tempered, grog-tempered, and shell-tempered wares (Lyon 1996:127–137).

Personnel from the Alabama Museum of Natural History began work in the Guntersville Basin on July 1, 1937 (Lyon 1996:137). Carl Miller, Harold Dahms, Charles Wilder, H. Summerfield Day, Theodore Johansensen, and Steve Wimberly were the field supervisors. These archaeologists tested five Copena sites: Rose, Samuels, Columbus City, Roden Mounds, the Hampden Cave; and three large Mississippian sites: McKee Island, Langston, and Gunter's Landing. Other Guntersville sites included Benson tested by Wilder; Snodgrass and Sublet Ferry tested by Wilder and Wimberly; Crow Creek Island and Cox tested by Johansen and Wimberly; and the Rudder Site tested by Johansen (Lyon 1996:38). Prior to World War II, Southeastern archaeologists also tested sites in Georgia, Mississippi, Louisiana, Kentucky, and Arkansas.

There were four separate federal archaeological programs in Georgia: Chatham County, Glynn County, Ocmulgee National Monument, and the University of Georgia state-wide Archaeological Survey. After construction of a runway for the St. Simons Island airport exposed burials, a WPA project was organized in coastal Glynn County. Preston Holder, who had been working in Florida, began work on the site in May, 1936. Twelve exploratory excavations and four trenches revealed 22 burials. Holder also worked on a small mound on the north end of Sea Island where he uncovered eight burials. Ultimately, he exposed more than 200 burials and recovered 21,000 grit-tempered, complicated-stamped pottery sherds. After work at the airport site was completed, Holder surveyed the area for additional sites. As a consequence, he excavated the Charlie King Mound south of the airport. Here Holder found eight group burials and very different pottery, leading him to conclude that the site represented a different culture. He also excavated the Hampton River Mound on Sea Island where he found fiber-marked pottery accompanying 45 skeletons, many of them children and infants. Holder next turned his attention to a large village site at Cannon's Point where he found cord-marked pottery. He also worked at the Sea Island Mound and Sullivan's Fish Camp (Lyon 1996:107–109).

The WPA, in conjunction with the Chatham County commissioners and the Savannah Chamber of Commerce, established a separate archaeological project on the Georgia coast. From September, 1937, to January, 1940, the project focused on the Irene Mound. After he finished the St. Simons Island project, Holder became the supervisor for the Irene Mound excavation where he worked closely with A. R. Kelley of the National Park Service (NPS). When Holder left the project, Seltzer and Kelley had trouble replacing him. They first appointed Vladimir Fewkes as supervisor, a position he assumed in January, 1938. Fewkes was replaced by Claude Schaeffer on June 11, 1938, and Schaeffer was in turn replaced by Joseph Caldwell. Caldwell's workforce consisted almost entirely of African American women. Their work is still recognized as some of the best of the era. (see Classen 1999 for more information on the women at Irene Mound). Caldwell remained at Irene Mound where he worked with prominent Georgia archaeologists, including Antonio J. Waring and Catherine McCann. In the late 1930s, he also visited Stallings Island with Waring where he collected a large number of artifacts. While the project focused on the Irene Site, researchers also tested the Deptford, Bilbo, Oemler, Dotson, Walthour, and Cedar Grove sites. Then too, from 1939 to 1940, Caldwell excavated at Wilmington Island, Georgia (Lyon 1996:109–110).

After vexing difficulties, A. R. Kelley convinced the University of Georgia to sponsor a state-wide archaeological survey. In 1938, the University appointed Robert Wauchope, assistant professor and survey director. While Wauchope was director survey members identified numerous sites in Barrow County and excavated a number of sites in the Coosa-Etowah drainage, including the Run Creek, Free Bridge, Woodstock, Wilbanks Mound, Long, Swamp Creek, and Horseshoe Bend sites. Survey personnel also examined sites in the Georgia Piedmont. The Capt. John Smith Farm and Sandtown sites were excavated. In the Towaliga-Upper Ocmulgee drainage of the Piedmont, survey crews tested the Towaliga, Jackson, Nancy Head, and Eight Mile Creek sites (Lyon 1996:112–115).

A. R. Kelley made Ocmulgee National Monument the focus of NPS archaeology in the Southeast. As early as 1938, Ocmulgee archaeologists organized a laboratory at the site and began processing collections from other sites in Georgia. By 1939, this laboratory had assembled over 35,000 collections containing a million artifacts. In the process, the laboratory became the major repository for NPS collections in the Southeast. Kelley concentrated Ocmulgee's WPA funds and eighty-man crew on excavating the large mound at the site. In his work here, Kelley attempted to provide stratigraphic control for the ceramic sequence he was building. When he was forced to cut the size of his excavation crew, Kelley turned his attention to stratigraphic work at other sites in the area. Gordon Willey directed the work and, in 1937, tested the Lamar, Mound C Village, Southeastern Plateau, Spur No. 2, Tuft Springs 1 and 2, Scott, Napier, Hawking Point, Mossy Oak and Cowart's Landing, Big Sandy No.1, and Lords Ridge sites. Willey continued his survey in 1938 by testing the Swift Creek, Stubbs Mound, Oconee Town, and Cowart's Landing sites. Ocmulgee archaeologists also were involved in test excavations at the Kolomoki Site in southwest Georgia. This site became a state park in 1935 after which Fairbanks excavated its mounds (Lyon 1996:178–187).

Federal archaeology in Florida began in the spring of 1934 when Marshall T. Newman tested the Englewood Mound in west Florida. In November, 1935, Vernon Lamme, the state archaeologist, directed a crew of 90 in the excavation of 11 sites in Hillsborough County. Matthew Stirling provided technical supervision and Preston Holder served as "Straw Boss" for the fieldwork. In January, 1937, a WPA crew of 34 found 48 sites as they surveyed Hillsborough County. After testing the Thomas Mound in 1936, Holder supervised work on the Cockroach Key Site, trenched the Spender and Snavely Mounds, and excavated burials at Cagnini Mound, Branch Mound, Lykes Mound, Jones Mound, and Picnic Mound.

On Florida's east coast, in Broward and Dale counties, WPA personnel excavated the Surfside, Opa Locka1, Opa Locka 2, Opa Locka 3, Golden Glade 1 and 2, and Baily 1 and 2 sites. In 1935, Robert Squires tested the Surfside Site in Miami Beach. Julian Steward replaced Squires, who was, in turn, replaced by D. L. Reichard (Lyon 1996:115–117).

When Webb returned to Kentucky in 1935, he dominated archaeological work in the state. In the summer of 1934, Funkhouser and Webb used Federal Emergency Relief Administration funds to excavate the Adena Mound at the Ricketts Site, but it was not until 1937 that Webb with WPA support organized Kentucky's federal archaeological program. Webb asked William Haag to manage the program and appointed him curator of the museum and instructor in the Anthropology Department at the University of Kentucky. Ultimately Webb added John L. Cotter, Albert C. Spaulding, David B. Stout, Henry A. Carey, Ralph D. Brown, Hans Theodore Edward Hertzberg, and Ivar Skarland to the program. By 1937, four projects employing 135 men were in place. Webb had serious disagreements with WPA administrators over his laboratory size and performance but persevered to focus attention on Archaic sites in the Green River Valley, among them Indian Knoll, Carlson Annis Mound, Read Shell Midden, Chiggerville, Cypress Creek Villages, and Archaic sites in McLean County. Many of the Green River Archaic sites also included Woodland and Mississippian components, the excavation of which was duly reported (Lyon 1996:95–101).

Kentucky's New Deal archaeologists were intensely interested in the state's Adena occupation. The Ricketts Site was the first to be tested. In 1934, Funkhouser and Hagg, with a complement of 60 men, excavated part of the Ricketts Mound. From 1939 to 1941 Elliott tested the large Robbins Mound and a smaller mound nearby. The smaller mound contained a single Adena tomb, the larger contained a burned circular structure containing cremated human remains that had been covered with a mound containing log tombs and extended flesh burials. In the fall of 1937, John L. Buckner and Claude Johnston managed work at the Wright Mounds, and in 1938 James C. Greenacre excavated the C&O Mounds on the Big Sandy River in eastern Kentucky. In 1939, Claude Johnston tested the Morgan Stone Mound, Elliot supervised excavations at the Crigler Mounds, and Johnston began testing the Mount Horeb earthworks, and Drake Mound.

Kentucky personnel also excavated several Fort Ancient sites. Among them was the Hardin Village where nine structures, 250 features, 301 burials, and more than 25,000 artifacts were uncovered. In 1939, C. T. R. Bohannan tested the Fort Ancient-affiliated Bently Site. Other Fort Ancient sites that were tested included Mayo where hearths, pits, middens, rectangular house patterns, and numerous post holes were uncovered. The major Mississippian excavations in the Kentucky Lake area were TVA salvage projects that came later and were terminated by World War II (Lyon 1996:95–101). Webb did not publish a major report on the Kentucky Basin as he had done for the TVA-sponsored Norris, Wheeler, Pickwick, and Guntersville basins. He did, however, publish on Kentucky's Adena (Webb and Snow 1945) and Archaic materials (Webb and Haag 1947).

In 1938, Jesse Jennings began a WPA-sponsored survey of the Natchez Trace in Mississippi. In the southern part of the parkway he established a sequence of Tchefuncte, Marksville, Troyville, Middle Mississippian, Coles Creek, and Natchez sites. In central Mississippi, he found Baytown, Coles Creek, early Mississippian, late Mississippian, and Choctaw sites. In the northeastern part of the state he found pre-Chickasaw and historic Chickasaw sites. Jennings also organized a WPA excavation program and appointed Albert Spaulding to supervise it. Spaulding began by testing the Miller Site. When Spaulding was transferred to Natchez, Jennings took over excavation of the site and then used Spaulding's labor force to excavate sites in the Chickasaw Old Fields near Tupelo. Here three sites were excavated and four were sampled.

Spaulding worked MLe 18, 19, and 990. Jennings dug MLe 53, 54a, and 56, and both directed excavations at MLe 62, a site containing two mounds and a village. As a consequence of these excavations, Jennings was able to describe three pre-Chickasaw occupations, which he termed Miller I through III. Fiber-tempered, plain-surfaced pottery was considered Miller I; sand-tempered, cord-roughened pottery characterized Miller II; and cord-roughened pottery with a different clay body inclusion became Miller III (Lyon 1996:186–187).

James A. Ford's Louisiana program was by most measures the most systematic and productive in the Southeast. Despite working in Mississippi and Georgia, Ford's enduring interest lay in the prehistory of the lower Mississippi Valley. In his early years Ford combined efforts with Henry B. Collins and Moreau Chambers to surface collect and test over 100 sites in Mississippi and Louisiana. Ford's work at the Peck Site, where he found three pottery complexes—Marksville, Coles Creek, and Deasonville—and his continuing contact with Gordon Willey, Preston Holder, Antonio Waring, Jesse Jennings, and Charles H. Fairbanks led him to conclude that pottery classification could be used to sequence archaeological sites and site components. He used pottery decoration to define seven Louisiana archaeological complexes, namely Marksville, the oldest complex that occurs throughout the state; Coles Creek, an intermediate complex in the south; Deasonville, an intermediate complex in the north; and the state's youngest prehistoric complexes: Choctaw, Natchez, Tunica, and Caddo. In 1936, he published a summary of his work and a synthesis of lower Mississippi Valley prehistory (Ford 1936).

Ford organized a three-pronged program consisting of a state supervisor overseeing a major laboratory in New Orleans and two field units at Avoyelles and LaSalle. Gordon Willey was appointed state supervisor and director of the New Orleans laboratory. Robert Stuart Neitzel was hired to direct the Avoyelles unit. He was assisted by Edwin B. Doran, Jr. William T. Mulloy supervised the LaSalle unit, assisted by Arden King. George I. Quimby replaced Willey as state supervisor and lab director in 1939. He supervised the laboratory, and while Ford was away in 1940 and 1941, he managed the field units and the processing of collections. Between Willey's departure and Quimby's arrival, Preston Holder ran the New Orleans laboratory (Lyon 1996:82–85).

The Avoyelles unit started work at the Greenhouse Site in 1938. Greenhouse was selected for excavation because a surface collection of its pottery indicated a significant difference from the nearby Marksville Site. Ford's analysis of the remains at Greenhouse led him to claim that it was a multicomponent site with Troyville and Coles Creek manifestations. He compared Greenhouse with Peck and Troyville and the west Florida sequence described by Willey (1949). In 1941, the Avoyelles unit also tested the Martin Baptiste Place Site, a Marksville period mound, and the mound at the Nick Place Site. The LaSalle unit, directed by William T. Mulloy and Arden King, excavated the Crooks Mound in 1938. Surface collections indicated a Marksville occupation. The large mound at the site was trenched, revealing multiple burials in its center. Mulloy and King also tested the site's small mound where they encountered several additional burials. Similarities with Tchefuncte indicated that the Crooks Site was an early version of Marksville. It also became clear that Troyville, Coles Creek, Caddoan, and the west coast of Florida developments (Weeden Island and Safety Harbor) were derived from the base provided by Marksville (Lyon 1996:90–91).

Before the WPA projects, Louisiana archaeologists considered Marksville the oldest deposits in the state. The discovery of pottery with shapes and decoration that bore similarities to the Stallings Island and Tennessee Shell Heap fiber-tempered wares stimulated archaeologists working in Louisiana to consider a pre-Marksville horizon termed Tchefuncte. Excavation of Tchefuncte sites was the next step. Preston Holder with a WPA crew of 35 began work at Little Woods midden in July, 1939. At the same time, Doran directed work at Big Oak Island and excavated the Tchefuncte Site in Tchefuncte State Park. Louisiana State University (LSU) archaeologists excavated another Tchefuncte site near Lafayette. The final field work consisted of excavations in and around Baton Rouge at the Medora and Bayou Goula sites. From 1939 to 1940, Doran excavated a truncated pyramid and defined a plaza at the Medora Site. The Medora artifacts were identified as components of the Plaquemine culture dated to AD 1300. In 1940–1941, first Doran, then Carlyle S. Smith excavated the Bayou Goula Site south of Baton Rouge. According to Lyon (1996:93–94) "[t]he archaeologists chose the Bayou Goula Site for excavation because the Louisiana WPA project had achieved a fairly good understanding of Marksville, Troyville, and Coles Creek and were developing a picture of the Tchefuncte and Plaquemine cultures, but had no comparable information about the Historic period." With the work at the Bayou Goula Site under their belt, Louisiana archaeologists were developing a picture of the import of the prehistoric Tchefuncte and Plaquemine cultures and the historic Natchez culture. Ford and Quimby, for example, concluded that Tchefuncte was an early culture whose authors were hunters and harvesters who manufactured and used pottery. Plaquemine was a later Mississippian manifestation and Natchez was historic (Lyon 1996:92–93).

Depression-era federal archaeology in the Southeastern United States provided the region's first systematic and evidence-based interpretations of the prehistory of Tennessee, Georgia, South Carolina, Florida, Alabama, Mississippi, and Louisiana. Fieldwork innovations included the creation of photographic records, feature forms and record sheets, large block excavations that exposed post patterns, structure foundations, and the distribution of tombs and interments in grave sites, mound trenching to reveal construction episodes, and the excavation of a mound by removing construction layers. Tennessee's Frank H. McClung Museum, Alabama's Moundville Archaeological Park, and Georgia's Ocmulgee National Monument—all repositories for the collections from WPA and NPS projects—are a lasting legacy of the New Deal.

WPA projects involved large numbers of unskilled laborers supervised by graduate and undergraduate students who became the archaeologists who trained many of our discipline's contemporary practitioners. These individuals include J. Joe Bauxar, Charles Fairbanks, Jesse Jennings, Charles Nash, Robert Neitzel, Georg K. Neumann, Douglas Osborne, Chandler Rowe, Wendell C. Walker, Andrew Whiteford, Gordon Willey, James Ford, Preston Holder, Carlyle S. Smith, David DeJarnette, Madeline Kneeberg, T. M. N. Lewis, William Mulloy, George Quimby, Albert Spaulding, Henry B. Collins, Matthew Stirling, William Haag, Stuart Neitzel, A. R. Kelley, Charles Snow, Marshall Newman, Julian Steward, Florence Hawley, James B. Griffin, and Robert Squires.

WPA-funded archaeology produced impressive results. Yet given the commitment to the idea of Native Americans as receivers rather than creators, the question of the day became whether the people who created states and empires in Mesoamerica colonized portions of the Southeastern U.S., or were the fortified mound centers found here the results of stimulus-diffusion from a Mesoamerican hearth-land? While the routes of trade and travel that would justify claims for immigration or stimulus-diffusion had not yet been identified, few doubted they would be. Most archaeological reports included discussions of similarities between Hopewellian and Mississippian sites in the Southeast and the Mesoamerican social, economic, and political practices they presumably reflected.

The River Basin Surveys Era

From 1940 to 1950

World War II intervened before the issue of Mesoamerican influence could be resolved by the evidence excavation provided. While in some instances laboratory work continued, most of the nation's archaeologists were in the armed forces. Fieldwork was discontinued for the duration of the war, and American archaeology ground to a halt. During the war, the U.S. Army Corps of Engineers was fully engaged in supporting our combat troops. That role changed dramatically at war's end with the passage of the *Flood Control Act* and the establishment of the River Basin Surveys (RBS).

The RBS reintroduced federal archaeology to the Southeast. Although labor and funding were in short supply from 1944 to 1965, RBS personnel worked reservoir areas in Alabama, Georgia, South Carolina, Tennessee, North Carolina, Florida, Louisiana, Arkansas, Virginia, and West Virginia. In 1947, Joseph Caldwell conducted an intensive survey of the Allatoona Reservoir area in the Etowah River basin and located 206 prehistoric sites ranging in date and content from Paleoindian to Mississippian. He recommended excavating 10 sites and testing 33 (Smithsonian Institution 1946–1947:21–22). From 1948 to 1951, Carl Miller and Caldwell worked in the proposed Clarks Hill Reservoir area, which would inundate parts of Georgia and South Carolina. Most of the sites found were surface collected. Twenty were tested, and four were excavated: Rambert Mounds, Fort Charlotte, Lake Springs Shell Heap, and 38MC6 (Caldwell 1953, 1954, 1974; Miller 1949a). The survey produced 7,200 artifacts. Most of them were ceramics and lithics. Pottery types included Stallings Island Plain and Stallings Island Punctate, Dunlap Fabric Impressed, Cartersville Check Stamped, Cartersville Simple Stamped, Deptford Incised, Swift Creek Complicated Stamped, Napier Complicated Stamped, Etowah Complicated Stamped, Savannah Complicated Stamped, Lamar Plain, Lamar Complicated Stamped, and Lamar Bold Incised. The ceramics included types now assigned to the Gulf Formational, Woodland, and Mississippian stages. Projectile points included Lanceolate, Lanceolate Fluted, Dalton, Taylor Side Notched, Kirk Notched, Stanly, Marrow Mountain, Guilford, Yadkin, Hamilton, and Madison. The lithics mirrored the ceramics in age and stage distribution.

In addition to the Clarks Hill Reservoir area, Miller worked for the RBS in North Carolina, Virginia, and Alabama. From 1947 to 1951, he surveyed the Kerr Reservoir area in Mecklenburg, Halifax, and Charlotte counties of Virginia, and Warren, Vance, and Granville counties of North Carolina where he found 92 prehistoric and two historic sites. The prehistoric sites consisted of 34 villages, three Paleoindian campsites, 14 Archaic, Woodland, or Mississippian sites identified as Neo-Indian, and 41 flint workshops. The two historic sites in the area were assigned to the colonial period (Miller 1962). Sampled sites ranged in age from Paleoindian to Late Woodland. Two of the excavations yielded human skeletal remains that were studied by Lucile E. St. Hoyme and William M. Bass. Also in 1947, Ralph Solecki began an RBS survey of North Carolina's Bluestone Reservoir area, and Gordon Willey conducted a survey of the Center Hill Project on Caney Fork River near Baxter, Tennessee. During his survey, Willey identified 39 sites including temple mounds, earth and limestone-covered burial cairns, and occupied caves. The artifacts Willey collected indicated occupations ranging from Archaic to Mississippian. He recommended excavation of the temple mound and one earthen and limestone-covered burial cairn, but the area was inundated before archaeologists could intervene (Smithsonian Institution 1948:61–65).

During the spring and summer of 1948, Ralph Solecki surveyed the West Fork and Bluestone River basins in West Virginia. The West Fork Reservoir is located on the upper Monongahela River in north-central West Virginia. The Bluestone Reservoir is located in southern West Virginia, on the border between West Virginia and Virginia. Both reservoirs are in the heavily dissected Appalachian Plateau with its swift running streams, V-shaped valleys, and narrow river bottoms. The West Fork survey identified 28 small campsites. The Bluestone Reservoir area contained 31 village and campsites, six mound sites, and seven rock shelters. The late prehistoric village sites showed strong affiliations with the Fort Ancient Aspect, and the campsites had some indications of contact with Virginia Piedmont cultures. There were also indications of an Adena occupation (Smithsonian Institution 1948:61–65).

In 1949, University of Arkansas field crews examined and excavated sites in the Bull Shoals Reservoir area. At this time too, NPS personnel surveyed and tested sites in the Jim Woodruff Reservoir area of the Apalachicola River of Northern Florida. In the same year, University of Georgia crews excavated sites in the Alltaoona Reservoir area of the Etowah River, and University of Kentucky personnel continued a pre-existing program of excavation in the Wolf Creek Reservoir area of the Cumberland River and at the Dewey Reservoir area on John Creek (Smithsonian Institution 1949:84–85). In 1950, pursuant to a major excavation program that Caldwell and Miller directed in Georgia's Allatoona Reservoir area, Caldwell established a laboratory at the University of Georgia, and Miller moved to Louisiana for a preliminary reconnaissance of the Bayou Bodcau Reservoir area. In May of 1950, Miller visited archaeological sites at Chester Island and Floyd Island in the Okefenokee Swamp. In June, he excavated a large village and burial site on the Roanoke River in southern Virginia's Kerr Reservoir area. Then too, in 1950, Ralph Solecki surveyed the proposed Deer Creek and Paint Creek reservoir areas in the Scioto River basin near Chillicothe, Ohio (1950:54–59).

From 1950 to 1960

In June of 1951, Carl Miller surveyed the Philpott Reservoir area on the South River in Virginia, and Ralph Solecki investigated the location of a new navigation lock in Morgantown, West Virginia. From November to April of 1951, University of Georgia field crews surveyed the areas to be inundated by the proposed Buford Reservoir on the Chattahoochee River between the Eshowe River and North Oconee drainage. The survey identified two previously unrecorded pre-Lamar Mounds and 29 pre-pottery sites deemed to be Archaic (Smithsonian Institution 1951:61–87). In 1952, work continued in the Allatoona, John H. Kerr, and Clarks Hill reservoir areas. Excavation at the Lake Spring Site in the Clarks Hill Reservoir area produced Archaic materials underlain by Old Quartz Culture lithics similar to those found at lithic stations in the Georgia and South Carolina Piedmont (Smithsonian Institution 1952:62–63).

Miller worked on previously excavated materials from the Kerr Reservoir area in the first half of 1953, and in August he conducted a brief survey of the Demopolis Reservoir basin in Alabama's Black Warrior River valley. He also checked sites in the Grenada Reservoir on the Yalobusha River in Mississippi. From March to June, Miller surveyed and excavated two sites, tested two others, and located 25 more along the portion of the Flint River to be flooded by the Jim Woodruff Dam. One excavated site was a Weeden Island Variant similar to the Weeden Island Complex found at the Lusk Springs site. While Miller worked the dam area, Caldwell began work on several phases of the Weeden Island Complex at Fairbanks Landing (Smithsonian Institution 1953:9–11). During June, 1954, Ripley P. Bullen supervised excavations on the Florida portion of the Jim Woodruff Reservoir area where in several sites he found four superimposed levels. The lowest of them yielded only lithics. The next contained fiber-tempered pottery. The third level had Deptford materials, and the uppermost Fort Walton artifacts (Smithsonian Institution 1954:7–12).

In 1956, Miller began excavating Russell Cave in northern Alabama (Miller 1958:426–437). During the first season, Miller encountered the remains of a small hemispherical coiled fiber basket that contained the charred remains of *Chenopodium* seeds (Miller 1960a:31–32), indicating use of this plant for food as early as 5,000 years ago (Smith 1985). From 1956 to 1958, he excavated the cave floor to a depth of 32 feet and encountered an infant skeleton four feet below the surface and the remains of an adult male eight and one-half feet below the surface. The infant skeleton dated to ca. 5,000 years ago and the adult skeleton to ca. 7,000 years ago (Miller 1957). Artifacts suggest that the occupation spanned at least 7,000 years and that pottery manufacture (Fiber Tempered Stallings Island ware) began as early as 3500 BC (Smithsonian Institution 1956–1957:10).

In the late 1950s, as work on the various dams in the Southeast intensified and engineers anticipated the closing of many if not most of them, the rhythm to RBS archaeological investigations also intensified. Between late October, 1956, and March, 1957, RBS crews excavated a large mound in the Savannah River portion of Hartwell Reservoir. In April,

RBS crews conducted a preliminary survey of Warrior Lock and Dam on the Black Warrior River near Tuscaloosa, Alabama. From March to April, RBS personnel conducted a preliminary survey of the Dardanelle Reservoir area on the Arkansas River in Arkansas where they located 52 sites and recommended testing 23 of them. From October to March, RBS crews also excavated a mound at the site of the Lower Cherokee town of Tugalo on the Savannah River near Toccoa, Georgia. The excavation uncovered four superimposed pottery deposits, establishing a clear continuity between modern and prehistoric Cherokee ceramic production. The excavators also found the remains of earth lodge temples on the upper three mound levels and traces of another type of structure at the fourth and lowest level. The latter rested on deposits identified as late Middle Creek Culture (Smithsonian Institution 1956–1957:76–111).

In the spring of 1957, RBS crews excavated sites in the Dardanelle Reservoir area. From February to June, another survey party excavated two sites along the lower Chattahoochee River in Alabama and Georgia, and from April to June yet another crew excavated two sites in the Hartwell Reservoir area. From February to June, preliminary surveys were conducted in the areas to be flooded by the Columbia Lock and Dam, the Walter F. George Lock and Dam and the Oliver Dam along the portion of the Chattahoochee River that forms the Alabama-Georgia border. One hundred and seventeen sites were found on the Georgia side of the river and 90 on the Alabama side. These sites ranged from multicomponent villages to single mounds to mound groups. The remains of the 1689 to 1697 Spanish fort of Apalachicola and the historic Creek town of Roanoke were noted as historic sites of national significance. Significant manifestations in the Columbia Lock and Dam area included a major mound site, and three Weeden Island village sites. In the Oliver Dam region, 15 sites were recorded, including mounds, early villages, and cave occupations.

From June to December, 1958, seven sites were excavated in the lower portion of the Dardanelle Reservoir area, including a rock shelter that was completely excavated. Two sites yielded early pottery similar to wares found in the lower Mississippi Valley. Two sites were mound/village complexes and one was an historic Cherokee location. Two sites were excavated in the Hartwell Reservoir basin, one in South Carolina, and one in Georgia. The excavated site on the South Carolina side was probably a Cherokee village and mound known as Chauga. The largest site in the basin, a Cherokee mound/village complex tentatively identified as Estatoes lay on the Georgia side of the River (Smithsonian Institution 1957–1958:5–29).

In February, 1959, Harold Huscher began work in the Walter F. George and Columbia Reservoir areas of the Chattahoochee Basin, while Robert

Neuman excavated sites in the vicinity of the Columbia Dam axis in Georgia and tested a mound on the Alabama side of the river. G. Hubert Smith also worked two locations in the Walter F. George Reservoir area; one in Georgia and one in Alabama. From February through June, Neuman worked seven sites on the Georgia side of the river; six of them Archaic occupations and one of them a Historic Creek occupation dating to 1830. On the Alabama side of the river, Neuman excavated a large mound that was being destroyed by wave action. While Neuman was excavating the mound, Harold Huscher worked on Weeden Island sites near Columbia, Alabama. He recovered Weeden Island and Stallings Island pottery, steatite bowl fragments, and large stemmed projectile points resembling those at Macon Plateau.

Huscher also tested three sites; one of them was on the Georgia side of the river that produced European trade goods. This site probably was the remains of an AD 1800 Creek village. Later, Huscher moved upstream to test two sites on the Fort Benning Military Reservation; one on the Georgia side of the river was an early Lamar Site. The second on the Alabama side was a Swift Creek/Weeden Island site with an earlier component that yielded fiber-tempered pottery. Also in 1959, G. Hubert Smith excavated two historic sites; one on the Georgia side of the Chattahoochee was identified as the village of Roanoke. Smith was unable to trace structure foundations here, but did recover a large collection of artifacts from both the European and Native American occupations. On the Alabama side of the river, Smith tested the Spanish fort of Apalachicola. He did not dig in the fort proper, but delimited the extent of the fortifications and tested a nearby Native American occupation. During the same field season, Harold Huscher located and recorded 10 new sites in the Walter F. George and Columbia Reservoir areas and made collections from 46 additional sites (Smithsonian Institution 1958–1959:10–15).

From January to June, 1960, RBS field crews excavated 10 sites in the Walter F. George Dam and Lock area; six in Georgia and four in Alabama. One of the Georgia sites was multicomponent with Creek, early Mississippian, Weeden Island, and Macon Plateau occupations. Two of the Georgia sites consisted of large mounds with associated village areas, but the mounds had been plowed down by the time of the investigations. Several additional sites contained Woodland Stage materials. One of the sites on the Alabama side of the river was a Yuchi village that William Bartram visited in the 1800s. In the summer of 1960, the University of Georgia, the University of Alabama, and Florida State University conducted co-operative projects. A crew from the University of Georgia excavated a platform mound near Stark's Landing, University of Alabama students dug the village site near a large mound in Upper Francis

Landing, and a Florida State University crew investigated Apalachicola and a village near Trinity, Alabama (Smithsonian Institution 1960:8–11).

From September to January of 1961, Harold Huscher tested a series of 15 sites that contained Mississippian and late Creek pottery in the Walter F. George Reservoir area below Eufaula, Alabama. These sites included the Mississippian stage Spann's Landing, the multicomponent Singer Moye that contained fiber-tempered pottery, which today would be identified as Gulf Formational, 1HE51 that contained Deptford and Swift Creek pottery, and several sites near Hard Ridge Creek. To the north, between White Oak and Cheneyhatchee Creek, Huscher tested sites 1BR46, 1BR47, and 1BR10. Site 1BR46 contained late Creek pottery, and 1BR10 contained Gulf Formational pottery (Smithsonian Institution 1961:56–61).

Following a survey of the public use areas proposed for the Walter F. George Reservoir, Huscher examined 24 sites in the reservoir area, collecting artifacts from 21 of them, and excavated two mounds south of Georgetown, Georgia. One of these, the Cool Branch Mound, was an extraordinary example of a large Mississippian burial mound with an accompanying village surrounded by a palisade. In late April, Huscher explored a sub-mound burial pit and gathered data on the palisade walls and general village features at Cool Branch Mound before he turned his attention to nine other sites, six of which he tested. One of the tested sites on the Alabama side of the river near Fort Benning is the presumed location of the last Yuchi-occupied town. A second site on the opposite side of the river in Georgia may also be of Yuchi affiliation (Smithsonian Institution 1961:56–61).

In 1962 and 1963, the various reservoirs in the Southeast were either full or were filling to projected pool levels. Salvage efforts now focused on documenting the damage done by wave action and partial inundation of archaeological sites, as well as emergency work aided by heavy equipment, several instances of which were conducted in the Walter F. George Reservoir area (Smithsonian Institution 1961–1962:8–11). By the mid-1960s, the reservoirs were filled and RBS field work was complete (Smithsonian Institution 1967–1968:8).

Aftermath

As a consequence of WPA and RBS projects in the Southeast, C^{14} dating, stratigraphy, and artifact seriation won the day for determining time depth and sequencing (Ford and Willey 1941; Ritchie 1944). Archaeologists could now frame refutable chronological claims and achieve a moderate consensus on units of analysis, scale of measurement, and scope of inquiry. The prior commitment to the Midwestern

Taxonomic System, a taxonomic hierarchy that excluded time and space from unit definition and operated by the logical principle of set inclusion, was replaced by the "more flexible" Willey and Phillips Phase-Tradition-Horizon system that included time and space in unit definition, operated by the logical principle of multiple set intersections and assigned defined units to one of a series of stages (Krause 1994). The Willey and Phillips (1958) stages included the Lithic, Archaic, Formative, Classic, and Post-Classic stages. The latter two were appropriate for sequencing materials from Meso- and South America but were lacking in the North American sequence. North American archaeologists responded by transforming Willey and Phillips' Stages into Periods without accounting for the fact that stages form a quasi-serial order while periods form a strict serial order. The logical difference between stages and periods created taxonomic problems that persist to the present (Jenkins and Krause 2009).

Federally financed archaeology also created a picture of pervasive change in the details of material culture. Viewed in broad perspective and through large spans of time, the record of change in the Southeast seemed reasonably regular. Big-game hunters and the lifestyle this mode of subsistence engendered preceded a lifestyle that depended on smaller, more scattered game. Increases in hunting and harvesting efficiencies preceded a settled lifestyle and prepared the way for agricultural modes of production. Agriculture, in turn, set the stage for material productions of greater sophistication and complexity. Yet detailed comparisons of artifact inventories, burial practices, domestic and public architecture, subsistence and settlement patterns that RBS work made possible, revealed localized trends and trajectories in social development. While in some areas there was stability in artifact style, in others there was remarkable variability. In some areas there was stability in burial practices, in others there were rapid alterations (Krause 1985:17–39).

In the Paleoindian Period (Willey and Philips' Lithic Stage), surrounds were replaced by ambush hunting in the eastern woodlands while drives were the norm in the grasslands to the west. While geometric earthworks were being constructed in the Lower Mississippi Valley, remains of the dead were in interred in limestone and earthen bluff-top tumuli in and around Saint Louis to the north (Klepinger and Henning 1976:92–170). Linear and curvilinear earthen enclosures were combined with mound burials in Adena manifestations along the Ohio River and its tributaries. Hopewellian sites also combined linear and curvilinear embankments with burial mounds but were marked by at least three different "core" mound-burial features: the charnel house, the burial crypt, and the platform (Brown 1979). While combinations of these three were reported from Hopewell sites of the Miller manifestation in the Tombigbee drainage of Mississippi and Alabama (Jenkins and Krause 1986:114–117), to the east in the Tennessee Valley Hopewellian ideas and practices were expressed as elements of Copena mortuary ritualism (Walthall 1980:119). While in Florida and South Georgia burial mound ritualism and platform mound construction were continued in the Weeden Island Complex (Jenkins and Krause 1986:52), in other portions of the Southeast mortuary programs were simplified rather than elaborated (Krause 2012). It is precisely trends and trajectories such as these, and the contrasts of lifestyle they seem to reveal, that most severely challenge the archaeological imagination. In the mid-1950s, Joseph Caldwell accepted the challenge as he attempted to synthesize the results of WPA and RBS work in the Southeast.

In his doctoral dissertation, Caldwell (1958) presented a compelling and intellectually challenging synthesis of WPA- and RBS-generated archaeological data. He argued that the developmental patterning in the Southeast was qualitatively different from that common to the configuration of culture growth in the "nuclear civilizations" of Mesoamerica and the Old World. While the earlier cultural stages in the Southeast paralleled those in the areas where civilization arose, later Southeastern developments were "of a kind not shared with nuclear regions" (Caldwell 1958:vi). In other words, he argued that Southeastern prehistory was much more than a dim reflection of events that took place in Mesoamerica. There might be, he suggested, developmental regularities peculiar to a given non-nuclear region that inhibited rather than promoted the emergence of a civilization. According to Caldwell (1958:vi), "civilization might, indeed, be something rather special, possibly an abnormality, as it certainly can be an uncomfortable condition of cultural development." There were, he argued, developmental regularities in the forested portions of eastern North America that produced a "non-nuclear culture type" (Caldwell 1958:75). Chief among these were the fine-tuning of a hunting and harvesting lifestyle that Caldwell (1958:6–10) termed Primary Forest Efficiency—a fine-tuning that inhibited rather than promoted the development of states and empires.

In developing his thesis, Caldwell defined three overarching trends in the development of Southeastern cultures: (1) an increasing efficiency in exploiting forest resources (Caldwell 1958:6–18); (2) a Gulf Tradition along and adjacent to the Gulf Coast that served as an agent for the transmission of Mesoamerican ideas and practices (Caldwell 1958:61–64); and (3) a maize economy and attendant radiation of a Mississippian peoples and the spread of Mississippian ideas to regions peripheral to the Mississippian heartland (Caldwell 1958:64–68). His was a challenging synthesis that raised important questions of theory and practice—a challenging synthesis made possible by WPA and RBS work that raised questions not yet adequately answered.

Why states and empires flourished in Latin America but not in North America is an intriguing question. That a satisfactory answer requires reference to the advent of agriculture, technology, and population growth cannot be doubted. To restate the issue: given sufficient time, would North American Indians have become state and empire builders? An affirmative answer requires belief in an orderly march toward civilization and implies that all Amerindian societies responded to the beat of the same drummer, some more slowly than others. A negative answer calls into question the end toward which Amerindian societies were moving. To sharpen the focus a bit: do the Mississippian manifestations in the Southeast, and by extension those of the Mississippi River drainage as a whole, prefigure the emergence of states and empires? If so, then the protohistoric organizational simplification and population decline testified to by the historic record (DePratter et al. 1985:108–121) indicates an interlude, a brief hiatus in the gathering of momentum. States and empires would have been inevitable if Euroamerican conquest had not intervened. If not, then the Mississippian manifestations explored in WPA and RBS research represent a cultural climax, a pinnacle of momentum, an example of the ultimate in the potential of a social, economic, and political structuring, which could, in and of itself, have gone no further. From the latter perspective, the protohistoric simplification of social, political, and economic life was more than an interrupted interlude. The simplification was, instead, a waning of momentum, a dissipation of organizational force, which in the absence of additional—and most probably external—pressures, could not have been reformulated to the end of state and empire building. A definitive answer eludes us, but had there been no WPA and no RBS, the question could not have been asked.

CHAPTER 8

Contributions of the Washington Office to the Archaeology of the Northeast and Mid-Atlantic Regions

Susan C. Prezzano

The River Basin Surveys (RBS) profoundly influenced the development of applied archaeology in North America, especially the techniques often used today in cultural resource management. Equally important are the contributions of the RBS to general archaeological method and theory and the development of culture histories. These contributions, however, are much more recognizable in the western United States than in the Northeast and Mid-Atlantic regions, due at least partially to the greater number of large dam projects in the West, especially in the Missouri River Basin, and to the better funding provided to these projects (see, for example, Roberts 1953:60, 1959:50). Although the impact of the RBS on the development of archaeology of the Northeast and Mid-Atlantic is more subtle, it is no less profound than advancements in the West.

The projects in the Northeast and Mid-Atlantic regions comprise some of the first professional surveys of several river valleys; these surveys included the discovery of important sites that continue to be investigated and continue to have a strong impact on the archaeology of the East. The RBS in the East trained archaeologists who went on to have successful careers, and it spawned the development of long-term relationships between personnel of this federal program and regional and state museums, historical societies, and universities. The RBS encouraged the development of good relations between professional and avocational archaeologists leading to the creation, at least in one instance, of a strong and active archaeological association. Finally, continuing archaeological research at locations, such as the Conemaugh Reservoir, have provided

Dam Projects and the Growth of American Archaeology: The River Basin Surveys and the Interagency Archeological Salvage Program, edited by Kimball M. Banks and Jon Czaplicki, 115–128.

data on the condition of sites within the inundation zone after the U.S. Army Corps of Engineers (2012) (Corps) and private utility companies built the dams. This contribution outlines the impact of the RBS in the East, specifically the Northeast and Mid-Atlantic regions as managed through the Washington DC office, and highlights current archaeological programs based on these surveys.

Federal Projects Prior to the River Basin Surveys

Pre-World War II federal support of archaeology had greater presence in the Southeast than in the Northeast and Mid-Atlantic regions. Several states in these regions, however, had extensive New Deal archaeological projects in the 1930s (Means 2011a, 2011b:29). During this era, Donald Cadzow, State Archaeologist of Pennsylvania, and Mary Butler of the Historical Museum Commission were instrumental in obtaining federal funds and support from the Federal Emergency Relief Administration (FERA) and Works Project Administration (WPA), to excavate village sites and rock-shelters in Bradford, Bucks, Carbon, Luzerne, Montgomery, Venango, and Washington counties with a crew of 400. How many of these projects were implemented is unknown, although the projects in Somerset County in southwest Pennsylvania and projects in Crawford County are best known (*American Antiquity* 1936:234, 1939:159). The majority of the Somerset projects were on Late Woodland Monongahela sites; crews excavated a total of 12 villages from 1935 to 1940 (Butler 1938; Means 1998, 2010).

Another major project, made possible through funds provided by WPA, was Harry Schoff's 1938 excavations of the McFate Site (36CW1) located along French Creek in Crawford County within western Pennsylvania (Burkett and Cunningham 1997). This site consisted of a series of at least three overlapping palisaded villages with 448 pit features that all dated to the Late Woodland (AD 800–1550). Housing development destroyed the site in the 1970s (Burkett and Cunningham 1997:11). One style of pottery discovered at the site has been named McFate Incised, and McFate has become the type-site for the middle Late Woodland Period (about AD 1300) in the Allegheny drainage.

Other New Deal projects in the Northeast and Mid-Atlantic regions include the Dutch Hollow Seneca Site in Livingston County, New York, excavated by William Ritchie (1954) and projects in Delaware, Massachusetts, and New Jersey. In New Jersey, Dorothy Cross performed WPA surveys and excavations in 1936 for the New Jersey State Museum (*American Antiquity* 1937:227, 1938a:280–281, 1938b:168–169). She excavated the famous Middle Woodland Abbott Farm that she (Cross 1956) summarized in a later monograph; the interpretation of this site and its role in the peopling of North American controversy form the basis of many articles (e.g., Byrne and Parris 1987; Cavallo 1983;

Richards 1939; Stewart 1983). Abbott Farm is now a National Historic Landmark. Other New Deal projects included historical archaeological excavations at Jamestown, Virginia, and the Saratoga battlefield in New York (Means 2011b:32).

Beginnings of the RBS in the Northeast and Mid-Atlantic Regions: 1946–1949

For the first several years after the establishment of the RBS, the Washington office administered all RBS projects in the Eastern United States and served as the central office for the RBS program in general. In 1946, Frank H. H. Roberts, Jr. Director of RBS, performed background survey and preliminary research on sites in Virginia within the James River project area with assistance from the avocational Virginia Archaeological Society, thus establishing a pattern used frequently during the surveys of the Eastern United States (Wetmore 1947:66). In 1947, two years after the establishment of the RBS, Carl Miller, a recently hired staff archaeologist at the Bureau of American Ethnology, began the background research for projects in Pennsylvania, West Virginia, Virginia, and North Carolina in the newly organized survey arm of the Washington DC office (Wetmore 1948:65).

Some of these early projects took place along river systems where dam construction had already commenced. One such project was associated with the impoundment of the Roanoke River by the John H. Kerr Dam that created Buggs Island Reservoir, also known as Kerr Lake, along the North Carolina and Virginia border (Harrington 1947; Miller 1947). Thus, the fieldwork comprised true salvage archaeology. The rising reservoir destroyed or inundated several sites recommended for further study before archaeologists could proceed with excavations. Miller performed salvage work from February, 1947 through the early 1950s, during which he identified 94 sites. He recommended the excavation of 14 sites, including pre-contact Native American burials and Paleoindian Folsom sites, before the dam was completed in 1952 (Johnson 1951b:261; Miller 1947, 1949a, 1949b; Wetmore 1948:65–66, 73–74). His research, especially excavations performed in the early 1950s, produced a coauthored final RBS report (Miller 1962) and several articles in the Washington Academy of Sciences Journal, the *Archaeology Society of Virginia Quarterly*, and *American Antiquity* (Miller 1949b, 1949c, 1950, respectively).

By 1948, the number of surveys required Miller and other RBS personnel to establish an office in Athens, Georgia, to direct projects located in the Southeast (Wetmore 1948:61–62; see Krause, this volume). Here, I discuss only projects in the Northeast and Mid-Atlantic regions under the jurisdiction of the Washington office, with the inclusion of several salvage projects directed by universities and museums under contract with the National Park Service (NPS).

Ralph Solecki, while still completing his graduate studies, joined the RBS in 1948 (Stirling 1949:61) and began three months of background research and field work on some of the first projects in the Eastern United States directed from the Washington office. These projects included the Bluestone Reservoir on the Bluestone tributary of the New River within Monroe and Summers counties, West Virginia, and Giles County, Virginia. He also surveyed the West Fork Reservoir Project on the Monongahela River in Lewis County, north central West Virginia (Stirling 1949:62). Solecki identified a large number of sites in the Bluestone Reservoir, including 31 villages, six mounds, seven rockshelters, rock art, and four colonial forts, 28 of which are within the flood zone of the reservoir (Solecki 1948, 1949a, 1949b; Stirling 1949:62–63). Given the short time in the field, Solecki performed very limited subsurface testing with most of the excavations confined to rockshelters and village sites with artifact scatters, especially pottery, visible on the surface (Stirling 1949:62–63). He tested the largest of the rockshelters to a depth of 1.6 m (5 ft.). Solecki recommended that three villages and the large rockshelter be investigated (Solecki 1948; Stirling 1949:63). Sites identified include French Farm (46SU20), Barkers Bottom (46SU3), and Island Creek (46SU3). Artifacts from this 1948 survey are curated at the Smithsonian Institution's Department of Anthropology, National Museum of Natural History in Washington DC (Smithsonian Institution 2012b).

The Bluestone tributary of the New River has remained important in archaeology even after the impoundment of the river. In 1983, Johnson et al. (1983) surveyed the Bluestone Reservoir locale. Johnson later wrote a dissertation on Solecki's excavations there (Johnson 1984). Maslowski (1996) used ceramics that Solecki found to investigate encoding of ethnicity by cordage twist impressions in Fort Ancient pottery (AD 1000–1750). Maslowski (1996) and Johnson and Myers (2004) argue that the twist direction of the cords used to surface treat pottery is associated with specific geographic and ethnic groups during the Late Woodland and Late Prehistoric periods in Pennsylvania. A late phase (AD 1020–1460) of Fort Ancient culture history is named after the Bluestone River.

Solecki (1949a, 1949c) spent less time in the West Fork area but identified 28 "small camp sites." He judged none of these sites to be significant. Artifacts from the West Fork survey, including those from the Jewell Farm Site (46LE1), are curated at the Smithsonian Institution (2012b).

One added benefit of the RBS was the placement of professional archaeologists in locations, especially in rural areas, that had little or no previous archaeological investigations. This increased visibility of archaeologists benefited other sites slated to be destroyed, but not located within the RBS jurisdiction. For example, in 1949 management transferred Solecki from the RBS to a regular staff position at the Bureau of American Ethnology so that he could investigate Natrium Mound in Marshall County, West Virginia. The mound dates to the Adena and was scheduled for destruction by industrial expansion (Solecki 1953; Stirling 1950:64). The West Virginia Archeological Society was formed as a result of these emergency excavations and through the influence of Solecki (West Virginia Archeological Society 2010:1).

In September, 1949, after a Smithsonian Institution expedition to Alaska, Solecki continued with the RBS by visiting the Deer Creek and Paint Creek drainages of the Scioto River System. Baby and Potter (1963a, 1963b) of the Ohio State Museum later investigated these locations further in the early 1960s. In 1950, Solecki took a leave of absence to join an expedition to excavate Shanidar Cave, the famous Neanderthal site in Iraq (Roberts 1952a:61). His investigation of rockshelters during RBS projects most likely provided him with the experience to excavate Shanidar Cave.

The River Basin Surveys of the East and Mid-Atlantic Regions: 1950–1959

Carl Miller continued excavations along the Roanoke River near Clarksville, Virginia (Roberts 1952a:62). He also resumed test excavations in the Buggs Island Reservoir area by focusing on seven sites identified in his earlier survey (Miller 1947) that had good stratigraphic contexts. Archaeological investigation ceased in this drainage with the closing of the dam in July, 1952 (Roberts 1952a:78). In addition, Miller performed a brief survey near Bassett, Virginia at the Philpott Reservoir (Roberts 1952a:62) for which no appraisal report was written.

In late 1950, Solecki, assisted by William Mayer-Oakes, a staff archaeologist of the Carnegie Museum of Natural History in Pittsburgh, investigated three areas slated for dam construction (Roberts 1952a:62). Solecki combined all three of these projects into a single appraisal report (Solecki 1951). One of these projects, the renovation of the Morgantown Locks on the Monongahela River in West Virginia, required very little ground disturbance and Solecki recommended no further archaeological investigations.

The second project involved a rolled-earth flood control dam on the East Branch of the Clarion River in Elk County, Pennsylvania. In late October, Solecki and Mayer-Oakes performed a pedestrian survey of the East Branch of the Clarion River supplemented by interviews with local landowners (Solecki 1951). They did not uncover any archaeological resources and recommended no further investigations. The East Branch of the Clarion River valley flows through a steep-sided, unglaciated valley with very narrow-to-absent floodplain terraces. The lack of farming during historic times limited the amount of ground exposure, and thus the opportunity for landowners to uncover archaeological remains. This led to few archaeological investigations within this drainage until very recently. Data obtained from recent projects (e.g., Prezzano 2013) suggests that considerable archaeological remains exist within this drainage, especially of very early occupations, but that many lie deeply buried and thus outside the limited testing strategy necessitated by minimal funds and personnel of the RBS. The rolled-earth dam was finished in 1952 impounding a considerable proportion of the upper reaches of the Clarion River.

The third project, situated along the Conemaugh River and its major tributary, Black Lick Creek, proved much more productive. The Conemaugh River flows between Indiana and Westmorland counties and is a tributary of the Kiskiminetas River, a part of the Allegheny drainage located in southwest Pennsylvania. Solecki and Mayer-Oakes located eight sites, five (36IN2, 36WM4, 36WM5, 36WM6, and 36WM10) within the planned reservoir. One of the sites, the Johnston Site (36IN2) located near Blairsville, covers more than 4 ha (10 acres). Surface reconnaissance and subsurface testing by Solecki and Mayer-Oakes retrieved a considerable number of artifacts—including pottery and lithics—from intact components extending below the plow zone. In addition, the landowner amassed an extensive collection of artifacts from years of farming. Most of the artifacts derive from a Late Woodland (Late Prehistoric) Monongahela occupation (AD 1050–1635). Solecki suggested that the Carnegie Museum of Natural History perform further excavations as a part of their multi-year survey of the upper Ohio drainage. The upper Ohio archaeology project resulted in a major monograph that included descriptions of the RBS sites that Solecki and Mayer-Oakes uncovered (Mayer-Oakes 1955).

The Johnston Site has remained a focus of archaeological investigations since its discovery. In late 1952, Don Dragoo of the Carnegie Museum of Natural History, under contact with the NPS, directed two months of extensive excavations at the site. Dragoo's excavations uncovered post molds from multiple stockades, house patterns, storage pits, burials, and considerable faunal material (Dragoo 1955; Guilday 1955). Recent interpretations by several archaeologists have identified elements of both northern Late Woodland McFate and southern Monongahela cultures at this site (Chiarulli et al. 2001; George 1997; Johnson 1999). George (1978, 1997) selected Johnston as the type-site for the Johnston Phase (AD 1250–1580) of the late Middle Monongahela Period.

Chiarulli et al. (2001) used the extensive discoveries at this site as evidence of the rich archaeological resources in upland areas within western Pennsylvania. They also used the discoveries to justify the continuation of archaeological surveys associated with cultural resource management projects within upland areas. Over the past 10 years, archaeologists and students from the anthropology program at Indiana University of Pennsylvania (IUP) further confirmed this assessment through multiple seasons of excavations at Johnston. The excavations have led to many conference papers and laster's theses (e.g., Mitchell 2011).

One fruitful outcome of these recent excavations is a flood damage assessment of the Johnston Site since the impoundment of the Conemaugh River in 1952. Early impact assessment by archaeologists prior to completion of the dam assumed that this site would be under water once the Conemaugh was dammed (Chiarulli and Neusius 2009:5). The site, situated on a high terrace, however, is not within the reservoir proper, but in the broader river catchment area about 14 miles from the dam. Johnston has remained above water since the 1950s, except during high water floods. The initial filling of the reservoir and subsequent floods in the 1970s led to the deposition of 1 to 2 meters of alluvium over Johnston, unintentionally sealing and protecting the site (Chiarulli and Neusius 2009). Excavations by IUP from 2005 through 2012 revealed that artifacts and features, including the stockade, remained intact. The rapid sealing of Johnston resulted in the site escaping scouring or erosion by water action. However, recent retrieval of a higher percentage of smaller-sized faunal material than that collected by Dragoo may indicate erosion of bone by water action; although this difference may be a product of retrieval bias by Dragoo (Chiarulli, personal communication 2012). These results provide data for the assessment of other sites flooded within reservoir systems. Artifacts and data from the Johnston Site are located in the Department of Anthropology, National Museum of Natural History, Smithsonian Institution, Washington DC; the Carnegie Museum of Natural History, Pittsburg; the Department of Anthropology at Indiana University, Indiana, Pennsylvania; and the Pennsylvania State Museum, Harrisburg.

By the mid-1950s, the federal government had reduced RBS funding, and operations in many regions were transferred to the NPS (Roberts 1953:60, 1954:66–88, 1955:32–41). Solecki spent more time on the excavations of Shanidar Cave in Iraq or working for the Missouri River Basin Surveys (Roberts 1954:67). He resigned in 1954 to resume graduate studies and direct the Columbia University Shanidar excavations. Apart from monitoring the flooding of the Kerr Reservoir and writing up reports by Miller, few archaeological investigations were performed in the Mid-Atlantic and Northeast regions until 1956–1957, when money was finally available for archaeological surveys outside of the Missouri River Basin (Roberts 1955:34, 1956:7–8, 1957:48–49, 1958:44, 1959:50, 1960:61). Except for a survey of Sutton Reservoir, where no sites were uncovered (Roberts 1961a:76), at the end of the decade no surveys or excavations occurred in the Mid-Atlantic and Northeast regions that were directly administered through the Washington office.

The limited funds available necessitated that personnel at museums, universities, and historical societies conduct survey and excavation projects for many proposed dam projects. These included projects in Massachusetts, Vermont, New York, New Jersey, Pennsylvania, Ohio, and Indiana. For example, the Ohio State Archaeological and Historical Society assumed responsibility for all reservoir projects in that state, starting as early as 1953 (Roberts 1954:87, 1955:41, 1956:17). The Indiana Historical Society also performed surveys in potential reservoir project areas and produced reports on the results (Roberts 1954:87, 1955:41, 1956:17). In several cases, the archaeologists worked on a volunteer basis and performed emergency excavations at sites slated to be destroyed by projects other than dams (Roberts 1956:17). Thus, the organization and placement of archaeologists in the field for large dam projects provided the personnel to mobilize quickly and address impending threats to other archaeological resources.

RBS in the East and the Rise of the Environmental Movement: 1960–1970s

Federal and private projects in the 1960s and 1970s included efforts to build some of the largest reservoir systems. These projects were planned against the backdrop of the rise of the environmental movement associated with a change in the public sentiment for large dams and the establishment of state and federal laws protecting the environment and cultural heritage. Many of the archaeological projects associated with the dams fell to the efforts of universities, museums, and state agencies rather than to staff members of the Washington office, a trend begun as early as the 1950s. The Tocks Island Reservoir is a prime example of this trend in the federal dam projects of the 1960s.

Devastating hurricanes in the mid-1950s led to the revival of the Corps's 1930-era plan to build a dam near Tocks Island on the Delaware River. This multi-purpose dam would have flooded a forty-mile stretch of the Delaware, making it the largest dam project in the East at that time (National Park Service 2012). In 1965 under congressional authorization, the federal government began condemning and acquiring properties for the purposes of creating the reservoir and developing the surrounding area into the Delaware Water Gap National Recreation Area (Kinsey 1972:xx; Wright and Rohrer 2002:18). The dam and reservoir system would affect the states of New Jersey, New York, and Pennsylvania.

Professional excavations began in 1959 and continued until 1975 through at least 10 contracts between the NPS and the New Jersey State Museum, Seton Hall University, and Franklin and Marshall College (Kinsey 1972:xx). Prominent Eastern Woodlands archaeologists, such as Herbert Kraft, Fred Kinsey, Patricia Marchiando, and David Werner participated in these excavations by running field schools and volunteer excavations (Kinsey 1972). Kraft, Kinsey, and others surveyed and excavated at least 108 important sites, most located within the planned reservoir. These sites date from Paleoindian through historic times. Important sites include the predominantly Late Woodland Miller Field Site (28Wa1) (Kraft 1972), the Transitional Archaic Zimmerman Site (36PI14) (Werner 1972), and the protohistoric Bell-Browning Site (29SX19) (Marchiando 1972). The New Jersey State Museum under the direction of Dorothy Cross and William Sloshberg investigated deeply buried Archaic components at the Pahaquarra Site (28Wa6) by experimenting with mechanical equipment (Cotter 1964:238). Kraft (1976) prepared a report of the site.

Sentiment against the reservoir led to a grassroots effort to halt the project despite the fact that the federal government had already acquired considerable property along the shores of the Delaware. The ballooning cost of the dam, the public opposition, and the lack of foundational bedrock led to the eventual cancellation of the project (Albert 2005). The acquired land became the Delaware Water Gap National Recreation Area, and it had at least 458 archaeological sites documented within the area. Several sites here became part of the Minisink Historic District, which received National Historic Landmark status in April, 1993 (Wright and Rohrer 2002:18). In the end, the extensive excavations of the Tocks Island Project provide the framework for the culture history of the Delaware drainage.

In the western part of Pennsylvania and north into New York State, the Corps planned the Kinzua Dam on the Allegheny River as a direct consequence of severe flooding episodes in the city of Pittsburgh in the early twentieth century. While the Corps developed reservoir construction plans in the late 1930s, a significant flood in 1956 revived this large project. Congress appropriated funds in 1958; the Corps began construction in 1960, and the dam was completed in 1965. Don Dragoo and Stan Lantz of the Carnegie Museum of Natural History performed archaeological surveys and excavations through the mid-1970s even after impoundment of the Allegheny (Dragoo 1977; Dragoo and Lantz 1967, 1971, 1973, 1975). They worked in areas exposed during release of water.

Unlike the large field crews available in the East for projects such as Tocks Island, the more rural nature of northwest Pennsylvania and southern New York and the comparatively few universities available for field schools and academic directors led to smaller crews and minimal projects. Dragoo and Lantz identified many archaeological sites, particularly of the Late Woodland/Late Prehistoric Period, and performed extensive excavations at the Onoville (30CA5) (Dragoo and Lantz 1967) and Kinzua sites (36WA53). Both revealed house patterns and numerous pits. The Carnegie Museum of Natural History worked on sites along the upper Allegheny even after the Corps closed the floodgates and the reservoir filled. Dragoo and Lantz obtained some of the first radiocarbon dates from sites in the upper Allegheny (Crane and Griffin 1972:169). Dragoo and Lantz later grouped the Late Woodland sites into a culture named the "Allegheny Iroquois." While some of the data from these projects are curated in the Carnegie Museum of Natural History and the Pennsylvania State Museum, much of the results of excavations associated with the Kinzua Dam is still in private hands and unpublished. This area was comparatively unknown archaeologically prior to the surveys by the Carnegie Museum of Natural History, and with much of the upper Allegheny flooded, it remains under-researched except for archaeological projects during the New Deal, the work of avocational archaeologists, and recent surveys by the Allegheny National Forest.

One unfortunate aspect of the Kinzua Dam Project was the displacement and removal of the Cornplanter Iroquois communities and cemeteries along the Allegheny near the New York border. These communities derive from primarily Seneca populations who moved south after the Revolutionary War to live along the Allegheny on land deeded by the United States government to Cornplanter, a Seneca leader (Hoover 2006). Today, the Allegheny National Forest manages the land surrounding the reservoir.

In the 1960s, especially after the passage of the *National Historic Preservation Act* in 1966, the pace quickened in regard to archaeological surveys and excavations run by universities, museums, and historical societies in potential reservoir areas and supported with NPS contracts. Additionally, volunteer salvage projects occurred in Pennsylvania, New York, and Ohio (Roberts 1961a:77). Some examples of these projects included those in Pennsylvania conducted by Jacob Gruber of Temple University who surveyed the Aylesworth and Fall Brook reservoirs and salvaged sites in Maiden Creek, Cowanesque, Blue March, and Blanchard reservoirs. In 1966, Gruber investigated the Antonio Site (36-Ti-30) in the Cowanesque Reservoir (Cotter 1966:902; Crozier 1973). In 1964, Dragoo of the Carnegie Museum of Natural History surveyed a segment of the Cumberland River for the Corps in addition to surveying three small reservoirs on the Monongahela River in western Pennsylvania (Cotter 1964:237).

During the 1960s, historical societies and museums continued salvage investigations of reservoir systems in Ohio. In 1961 and 1962, Raymond Baby and Martha Potter (Baby and Potter 1963a) of the Ohio State Museum, under contract with the NPS, surveyed the Paint Creek and Deer Creek drainages near Chillicothe, Ohio; these were the proposed reservoir systems that Solecki originally investigated in the late 1940s. They identified three sites (33HI6 and 33HI7 1 and 2), mostly Hopewellian mounds, within the proposed Paint Creek Reservoir. Deer Creek was impounded in 1968 (Ohio Department of Natural Resources 2012a) while the Paint Creek Reservoir system was completed as late as 1974 (Ohio Department of Natural Resources 2012b). During the survey of these drainages, Baby and Potter (1963a) also investigated Mill Creek, Salt Creek, and the Big Darby drainages.

The Big Darby drainage (Baby and Potter 1963b) contained numerous mounds. Preliminary investigation of the Edward Sidner Mound of the Adena culture revealed intact burials and floors. The Ohio Historical Society investigated the O.C. Voss Mound in the mid-1960s (Baby et al. 1966), a Late Woodland burial mound. After a contentious battle with the City of Columbus that planned to impound the creek, the Big Darby Creek was designated a National Wild and Scenic River due to its diverse biota.

In Maine, the Robert Abbe Museum received an NPS contract to survey the location of the massive Corps Dickey Dam Reservoir Project along the St. John's River, a project that continued until 1967 (Cotter 1966:900). Sanger (1977) submitted a report on the archaeological resources. Congress de-authorized the Dickey Dam Project in 1985 (Day 1981:13).

In 1962, the Washington office sent Miller (Roberts 1963:64) to survey two sections of the Roanoke River near Smith Mountain and Leesville in Bedford, Franklin, and Pittsylvania counties, Virginia. A private utility company, the Appalachian Power Company (APC), a subsidiary of the American Electric Power Company, had already begun construction on two dams here. Miller identified 52 sites, with 51 being within the reservoir system. Of these, Miller recommended four for further investigation, three within the Smith Mountain Basin and the fourth in the Leesville Basin. These sites primarily date to the Early, Middle, and Late Woodland periods (Roberts 1963:64). Miller excavated several sites, including Hales Ford (44FR15), 44FR2, and 44PI1, with more extensive excavations performed in September, 1962, at the Smith Mountain location. He completed the archaeological project in July, 1964 (Stephenson 1964a:86). The APC completed dam construction at Smith Mountain in 1967 and at Leesville in 1963. Artifacts from these excavations are curated at the Department of Anthropology, National Museum of Natural History, Smithsonian Institution (Smithsonian Institution 2012b).

One of the last archaeological projects in the East directed from the RBS Washington office was the planned damming of the New River in Grayson County, Virginia, and Alleghany and Ashe counties, North Carolina (Ayers 1965). In 1962, APC proposed two dams that would create a reservoir system at least 50 miles-long. Harvard Ayers surveyed the proposed reservoir system for the RBS in May, 1965, and discovered 20 sites including open-air sites and a rockshelter; he summarized the findings in a brief appraisal report (Ayers 1965). Ayers recommended additional testing at seven locations and excavations of three sites. During his survey, a test excavation at one of these sites, 44GR105, yielded extensive Woodland artifacts and features below the surface. Site 44GR118 contained heavy artifact concentrations on the surface.

In 1969, as a result of doubling the size of the Blue Ridge Project by APC, the federal government required additional archaeological surveys under the new *National Historic Preservation Act*. Charlton Holland of the University of Virginia, in a brief, unpublished survey for the APC (Holland 1969, cited in Woodard 2006:107), discovered an additional 42 archaeological sites, some dating back 10,000 years B.P. In a situation similar to the proposed Tocks Island Reservoir Project, local landowners and environmentalists organized to halt the construction of the dams. Discovery of the numerous archaeological sites provided further justification for the protection of the New River, especially with the passage of the *National Environmental Protection Act* in 1969, and played a significant role in the final decision to defeat the dam (Woodard 2006:3, 107–109). Thus, archaeological sites discovered by the RBS led to the protection of the cultural heritage of the New River.

Contributions of the River Basin Surveys to the Archaeology of the Northeast and Mid-Atlantic Regions

The River Basin Surveys and the earlier excavations of the New Deal Era in the Northeast and Mid-Atlantic regions advanced the archaeology of the Eastern United States by providing an early framework of culture history, especially in areas that had undergone little previous archaeological research. In some locations, such as the upper Allegheny in northwest Pennsylvania, these federal programs and more recent projects run by the Allegheny National Forest comprise some of the few professional archaeological studies. Many of these projects retrieved data on settlement and community patterns, the first chronologies based on radiocarbon dating, stylistic seriation of pottery, and some of the first professionally excavated sites for key periods. Early and Middle Woodland mounds in Ohio, rockshelters in Pennsylvania and Virginia, deeply buried Paleoindian and Archaic sites in New Jersey, Pennsylvania, and Virginia, and large Late Woodland/Late Prehistoric villages in Pennsylvania, New Jersey, and Virginia comprise the huge array of cultural resources discovered and investigated. Prior to the enactment of modern cultural resource laws starting in 1966, these projects provided some of the only funding available in this region outside of private grants and limited state and federal grants.

Because the RBS did not widely circulate some of the early reports, these projects remain underutilized by archaeologists. Recent projects, such as this volume and Means' (2013) survey of New Deal projects, help to raise the visibility of this important period in North American archaeology. Also helpful is that the Smithsonian Institution (2012b) is systematically placing RBS data online, making some of the information from early excavations of famous sites easily available (Table 8.1). These data include catalog cards and digitized images of some of the artifacts from the RBS projects that were directed from the Washington office.

Table 8.1 Selected projects of the River Basin Surveys in the Northeast and Mid-Atlantic states directed from the Washington office

Project Name	Location	Year	Number of Sites Discovered	Project Director	National Museum of Natural History Accession Numbers[1]
John H. Kerr (Buggs Island) Reservoir	Roanoke River, Mecklenburg, Halifax, and Charlotte Virginia and Warren, Vance, and Granville, North Carolina	1947	94	Carl Miller	00344878:476896-479202[2]
West Fork Reservoir	Upper Monongahela, Lewis County, West Virginia	1948	28	Ralph Solecki	00187540:398058-398075
Bluestone Reservoir	New and Bluestone Rivers, Mercer, Monroe, and Summers Counties, West Virginia	1948	42	Ralph Solecki	00187541: 389076-398288
East Branch Reservoir	Clarion River, Elk and Mckean Counties, Pennsylvania	1950	0	Ralph Solecki	None
Conemaugh Reservoir	Conemaugh River, Indiana and Westmorland Counties, Pennsylvania	1950	8	Ralph Solecki	00202537:404572-404577
Morgantown Locks	Mononghela River, Monongalia County, West Virginia	1950	0	Ralph Solecki	None
Deer Creek	Ross County, Ohio	1958	At least 3	Ralph Solecki	00224549:420598-420604
Smith Mountain/ Leesville	Roanoke River, Franklin, Bedford, Pittsylvania, Campbell and Roanoke Counties, Virginia	1963		Carl Miller	00358976: 483753-489059
Blue Ridge Project	New River, Grayson, County, Virginia and Alleghany and Ashe Counties, North Carolina	1965	20	Harvard Ayers	00325555: 4774196-474332

[1]The records are accessible at the Department of Anthropology, National Museum of Natural History, Smithsonian Institution online database at http://collections.nmnh.si.edu/search/anth/

[2]The first number is the Smithsonian accession number and the second is the catalogue number; the latter can be used to search the Smithsonian Institution data base.

Data from the New Deal archaeological projects are curated in several locations, many that are easily accessible. For example, the projects in Somerset and Crawford counties are at the Pennsylvania State Museum, as are some of the data from the Tocks Island Project. These endeavors lay the groundwork for future comprehensive research projects to uncover more of the early history of archaeology in Eastern North America.

Acknowledgments

The author would like to thank the editors, Kimball Banks and Jon Czaplicki for the invitation to participate in this important volume. Brady Clemens, a graduate student in the Library Science program at Clarion University performed much of the searching for obscure sources.

CHAPTER 9

Upper Colorado River Archaeological Salvage Projects

Don D. Fowler

The 1,450-mile-long Colorado River and its tributaries (Figure 9.1) drain 246,000 square miles in parts of seven western states: Wyoming, Utah, Colorado, Arizona, New Mexico, California, and Nevada. Under the Colorado River Compact, signed by the seven states in 1922, the drainage is divided into Upper Basin and Lower Basin segments. The dividing point is Lee's Ferry at the foot of Glen Canyon and the head of Marble Canyon. The river is the most controlled stream in the United States. It provides significant portions of the electricity and municipal and irrigation water used by over 40 million people residing inside and outside its basin. Major Upper Basin tributaries include the 730-mile-long Green River, the 165-mile Gunnison River, the 250-mile-long Dolores River, and the 383-mile-long San Juan River. Major Lower Basin tributaries are the 315-mile-long Little Colorado River, the 162-mile-long Virgin River, and the 649-mile-long Gila River with its 200-mile-long tributary the Salt River.

The Bureau of Reclamation

The arid West, especially west of the Rocky Mountains, has been the scene of "water wars" since the 1870s. Two adages sum them up. The first, attributed to Mark Twain, is, "Whiskey's for drinkin', water's for fightin' over;" the second is, "Water flows to money." The competition among the states for municipal and irrigation water, plus the need to control periodic disastrous floods along the Lower Colorado River, was a major factor in the formation in 1902 of the Federal Reclamation Service, as an arm of the U.S. Geological Survey. In 1907, the Service became an independent agency within the Department of the Interior. In 1923,

---

Dam Projects and the Growth of American Archaeology: The River Basin Surveys and the Interagency Archeological Salvage Program, edited by Kimball M. Banks and Jon Czaplicki, 129–138.

Figure 9.1 Colorado River, Major Dams, and Drainage Basin. Drawing by Patricia de Bunch.

the Federal Reclamation Service became the Bureau of Reclamation (Reclamation). By law, Reclamation functions only in the 17 contiguous western states.

The Reclamation Service's enabling legislation authorized five projects. Thereafter, projects had to be proposed and authorized individually. By 1917, 22 projects had been authorized and were underway (Davis 1917). Currently (2012), Reclamation manages 180 projects on the Colorado and other western rivers.

Planning for large-scale dams on the Colorado River began in 1902 under the leadership of Arthur Powell Davis, chief engineer and later director of the Federal Reclamation Service. Davis and his compatriots recognized the value of using high dams for hydropower generation to provide municipal electricity and power to pump large amounts of water for irrigation and municipal uses. The technology to generate large quantities of AC power and transmit it over long distances had been developed in the previous decade. Between 1902 and 1924, as many as 12 dams were proposed for various locations along the Colorado River, including four or more in the Grand Canyon. The first major authorized project was Boulder Canyon Dam, later renamed Hoover Dam.

Salvage Archaeology in the Colorado River Basin

Reclamation projects triggered a dozen salvage archaeology programs in the Colorado Basin between 1931 and the late 1960s. Herein, we focus on major Upper Basin projects, specifically the Glen Canyon Project (GCP), the Navajo Project, the Flaming Gorge Project, and the Dolores Archaeological Project (DAP). The earliest Lower Basin salvage project was conducted between 1931 and 1946. Sites were excavated along the lower Virgin River and its Moapa Valley tributary in southern Nevada. These sites were to be drowned by an arm of Lake Mead behind Hoover Dam. The work, funded by the National Park Service (NPS), was conducted by the Southwest Museum using Civilian Conservation Corps labor but was not reported for three decades (Shutler 1961). The sites were a complex of pithouses and above-ground storage structures called "Lost City." The complex became the type site for Western Kayenta Anasazi, or Virgin Branch Anasazi–Puebloan sites dating from ca. AD 600 to 1200 (Lyneis 1992). The NPS built the Lost City Museum in Overton, Nevada, as a repository for the collections and records. Since the 1970s the museum has been a branch of the Nevada State Museum system. Since the early 1980s, all major projects in the Lower Basin were done in the early decades of "Cultural Resource Management" (CRM) that evolved out of the salvage archaeology of the River Basin Surveys ([RBS]; see McManamon, this volume).

The Upper Colorado River Basin Archeological Salvage Program

In April, 1956, Congress authorized the Upper Colorado River Storage Project. The project involved the construction of dams in Glen Canyon on the Colorado; in Flaming Gorge on the Green River; and Navajo Reservoir Dam on the San Juan River. The siting of the Glen Canyon Dam was a compromise between Reclamation and various environmental groups, especially the Sierra Club. The original plan called for a dam on the Green River that would flood portions of Dinosaur National Monument. The Sierra Club and many other environmental groups objected strenuously. Plans for proposed dams within Grand Canyon National Park were also staunchly opposed and finally abandoned. The environmental position was that national parks and monuments were inviolate. The battles (Harvey 1994) marked a major turning point in ongoing relationships between environmental groups and federal agencies in the West that continue to be negotiated in the twenty-first century.

Federal funding for the archaeological salvage projects was channeled through the Santa Fe office of the NPS under the Interagency Archeological Salvage Program (IASP). The parties agreed that the Museum of New Mexico/Laboratory of Anthropology would handle the Navajo Reservoir Project, the University of Utah (UUtah) would oversee the Flaming Gorge Project, and UUtah and the Museum of Northern Arizona (MNA) would jointly handle the Glen Canyon Project. Contracts were signed with each institution in early 1957, although some Navajo Reservoir work had been done in 1956.

The contracts specified that historical and ecological, as well as archaeological, studies were to be undertaken in all three reservoir areas. The UUtah Biology Department undertook the ecological studies. The UUtah, Department of History, conducted the historical archaeological studies for the Glen Canyon Project. Historical studies for the Navajo Reservoir Project were integrated into the overall archaeological program. The cost for the three projects was just over one million dollars.

The Flaming Gorge Project

The Flaming Gorge Dam created a narrow, 91-mile-long reservoir on the Green River in northeastern Utah and southwestern Wyoming. For most of its length, the reservoir is less than one mile wide. The UUtah conducted archaeological surveys during portions of the summers of 1959, 1961, and 1962 (Day and Dibble 1963). One hundred twenty-one sites were recorded. Most were camp sites on dunes or juniper-covered knolls. No sites showed any significant depth and no test excavations were conducted. The sites were interpreted as hunting, fishing, and gathering camps visited intermittently by small groups of Middle Archaic, Late Archaic peoples, and Fremont, and dated from ca. 2000 BC to historic times.

The Navajo Reservoir Project

The Navajo Reservoir Project on the San Juan River was carried out from 1956 to 1963 by archaeologists from the Museum of New Mexico under overall supervision of Fred Wendorf, the museum's associate director. Alfred Edward (Ed) Dittert, Jr. was director of field operations. The major field supervisors were Frank W. Eddy and James J. Hester. Joel Shiner, who was acting director of the Columbia River Basin Survey Program from 1955 to 1960 (Lyman, this volume) before moving to the Museum of New Mexico, contributed to the study of historic Navajo sites in the general project area. James Schoenwetter, who was then just starting a major career-long palynological research program, contributed centrally to paleo-environmental studies.

The Navajo Dam is sited just below the confluence of the San Juan and Los Piños rivers. The reservoir extends up Los Piños Canyon and up the San Juan River to and into the lower reaches of the Piedras River. Surveys were conducted along river terraces and on some adjacent mesa tops. Several hundred sites were recorded, and some two dozen were excavated that ranged in age from Late Archaic (ca. 3000 to 500 BC), through Basketmaker II to late Pueblo II (1 to ca. AD 1050), in the Pecos Classification. Numerous Navajo sites were recorded, dating from AD 1500 to 1700, as well as scattered Hispanic sites dating from AD 1600 to the present. The project and results are summarized in Eddy (1966, 1972). The collections and records are housed at the Laboratory of Anthropology in Santa Fe.

The Glen Canyon Project

The Glen Canyon Project (GCP) ran from June 1957 to July 1963. By agreement, UUtah was responsible for both banks of the Colorado River in Glen Canyon and its tributaries above the San Juan River confluence up to full pool level (Figure 9.2). The MNA was responsible for the left banks (downstream orientation) of Glen Canyon and the San Juan River canyon up to full pool level. The parties agreed that sites in the Navajo Mountain-Rainbow Plateau uplands south of the San Juan River and southeast of Glen Canyon below the confluence (Figure 9.2) would potentially be impacted by boaters and recreation activities once the lake filled, hence salvage operations were required there.

Despite a glitch in Congressional funding, both institutions had crews in the field from July until mid-November, 1957. Full funding

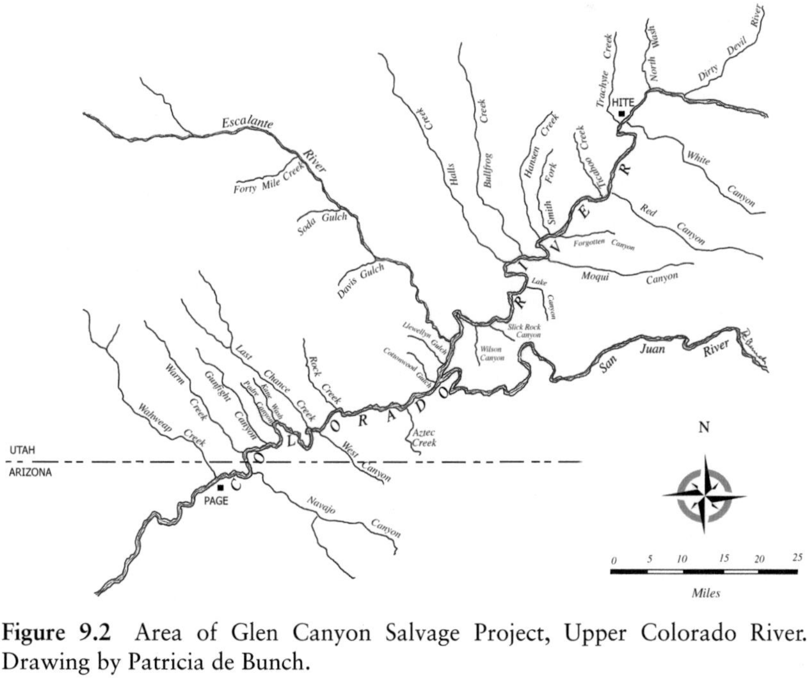

Figure 9.2 Area of Glen Canyon Salvage Project, Upper Colorado River. Drawing by Patricia de Bunch.

came through for the following years. UUtah also had grants from the National Science Foundation and the Wenner-Gren Foundation for Anthropological Research to support work on the Kaiparowits Plateau, the upper Virgin River, and at the Coombs Site (now Anasazi State Park) near Boulder, Utah (see below). The UUtah portion of the Project was directed by Jesse D. Jennings, chair of the Anthropology Department. At the outset, he decided that students would be employed wherever possible; hence, the three-month field seasons were tied to the academic year. Work at the Coombs Site at Boulder, Utah, 1957–1960, was carried out through archaeological field schools supervised by Robert H. Lister, of the University of Colorado (UC). The MNA ran longer field seasons since the museum was not tied to the academic year.

The Glen Canyon country was very difficult to access. The Colorado River could only be crossed at two locations. A small two-vehicle-per-trip ferry crossed the river at Hite at the upper end of Glen Canyon and Navajo Bridge, 173 river miles downstream, just below Lee's Ferry. Adjacent to the dam site, the Glen Canyon Bridge was completed in 1959, greatly facilitating work on the dam and cross-canyon logistics for all concerned. Access to many tributaries of the Glen and San Juan main canyons was possible only from the rivers. Access to the upper reaches of some of the longer canyons was by pack trains or four-wheel drive vehicles on dirt roads and often across slick rock. UUtah used pack trains for two seasons on the Kaiparowits Plateau (1958, 1961) and two seasons (1960, 1961) in the upper reaches of Lake and Moki canyons. The MNA used pack trains extensively in the uplands around Navajo Mountain and on Cummings Mesa. The MNA supplied an excavation party on No Name Mesa by parachute drops, and on Piute Mesa partially by helicopter.

The MNA and UUtah river crews were out for two to three weeks at a time, with no contact with the outside world. The MNA contracted with Frank Wright, an experienced professional river runner, to run boats specially designed for Colorado River rapids downstream from Bluff, Utah (on the San Juan) or Hite in Glen Canyon. UUtah used flat-bottomed "Arkansas Traveler" boats with outboard motors. The relatively slow water flow and small rapids in Glen Canyon proper made it possible to travel both up and down stream. In 1959, 1961, and 1962, military surplus rubber rafts were used as freight boats, running downstream. They were periodically resupplied by the motor boats and launched from just below Hite.

The MNA maintained a field lab for three seasons. UUtah had a processing lab at the university in Salt Lake City. Field crews shipped in artifacts and records whenever they got to a supply-point town (Blanding, Kanab, or Escalante). These artifacts were processed and ready for analysis by the time the crews came in from the field in September. Analysis and writing was done during the school year, ideally by field crew chiefs, and reports were in press by the following May. The MNA issued major reports every other year. The principal UUtah field supervisors were David S. Dibble, Don D. Fowler, William D. Lipe, and Floyd W. Sharrock. The principal MNA field supervisors were David A. Breternitz, Christy Turner, II, Alexander J. Lindsay, Jr., and the team of William Y. and Nettie Adams. C. Gregory Crampton, a history professor at the UUtah, conducted major archival and field studies of historical archaeological sites in the Glen and San Juan canyons, most of which were associated with two gold rushes in the 1890s and the 1930s (see Crampton 2009).

The UUtah and MNA published over 200 papers and monographs between 1958 and 1968. The UUtah published a project summary in 1966 (Jennings 1966), while MNA published a final summary of its work in 1969 (Lindsay et al. 1969). Some 200 individuals, including 28 Navajo Indians, worked on the project. Twenty graduate and undergraduate students went on to receive masters or doctorates in anthropology—and one doctorate in astrophysics. Flaming Gorge and the Utah portion of Glen Canyon collections and records are housed in the Natural History Museum of Utah at the UUtah. The MNA collections and records are housed in the Easton Collections Center, Museum of Northern Arizona, Flagstaff. A detailed history of the Glen Canyon Project is in Fowler (2011:242–348).

The Delores Project

The Dolores Archaeology Project (DAP) was triggered by the construction of the McFee Dam and Reservoir on the Dolores River in southwestern Colorado. Fieldwork ran from June, 1978 to December, 1985 (Breternitz 1993; Lipe 1999; Wilshusen et al. 1999). Unlike earlier salvage projects in the Upper Basin, the DAP is arguably the first post-RBS/IASP project reflecting the new CRM approach to federal archaeology. However, it borrowed much of its organization from the earlier projects as many of the principals had participated in these projects. Reclamation funded the project directly, in part, because by 1978 the agency had its own staff archaeologist, Ward Weakly. Further, both the *National Historic Preservation Act of 1966* and the *Archeological and Historic Preservation Act of 1974*, required federal agencies to conduct and fund cultural resource investigations to mitigate the effects of their construction projects.

The UC was the prime contractor and Washington State University (WSU) was the major subcontractor. David Breternitz of UC was the principal investigator, and William Lipe and Timothy Kohler, both of WSU, were associate investigators. Lipe and R. G. Matson had been running a long-term research program on Cedar Mesa in southern Utah since 1971, and Lipe would later become associate research director of the Crow Canyon Center, outside Cortez, Colorado. Breternitz, who had worked on the Glen Canyon Project in 1957–1958, brought major organizational skills he had developed as a participant in the 1965–1980 Mesa Verde National Park-University of Colorado cooperative archaeological research and salvage program. The DAP cost $9,990,000. Over the life of the project, some 450 individuals were employed in various capacities. Over 1,600 sites were recorded and 120 subjected to some level of testing or excavation.

The DAP was very innovative. By 1978, the "new archaeology" had influenced various ways of approaching, manipulating, and thinking about archaeological data and their relationships. From the outset, DAP created a standard set of data forms. The forms, modeled on those used by UUtah, were coded to facilitate computer data entry and manipulation. The DAP field office had a ten megabyte hard drive, two remote terminals with tape backup, and direct access to a mainframe computer at CU—astounding resources for the time. The DAP supported an extensive program of environmental and paleo-environmental research that was closely coordinated with the archaeological excavation program. The program maintained a year-round field office and processing laboratory north of Cortez, Colorado. Like the Glen Canyon Project, artifact and other data processing went forward simultaneously with the fieldwork, so that analyses and report writing could be done during non-field seasons.

The project produced a 130-megabyte computer data set, several hundred unpublished reports, and 13 published volumes. All the field notes, data, reports, project correspondence, and approximately a million artifacts are curated at the Anasazi Heritage Center (Center), Dolores, Colorado. Reclamation built the Center and turned it over to the Bureau of Land Management. The Center contains a museum and major public outreach program facilities and serves as a repository for artifacts (by 2012, some 3,000,000), records, and data generated by DAP and post-DAP CRM projects in the Four Corners area.

The majority of the sites recorded and excavated by DAP were Ancestral Puebloan, dating between AD 600 to 900. Building on those data, subsequent work in surrounding Mesa Verdean Puebloan sites has demonstrated how complex life was from ca. 400 BC to AD 1300, as populations moved from place to place adapting to changing micro- and macro-environments and finally abandoning the area in the late AD 1200s. (See Ortman [2012] for a major summary and synthesis.)

The Legacy

This volume is devoted to the impact of reservoir salvage archaeology on the development of American archaeology after World War II. In that context, several individuals need to be recognized for the continuing impacts they had. Jesse D. Jennings began working for the NPS in the late 1930s. After World War II, he returned to the NPS and was assigned to Lincoln, Nebraska, as principal liaison between the NPS, the RBS office in Washington, DC, and the various local universities and museums that participated in the salvage program. He became intimately familiar with how the reservoir projects were funded and operated and was able to help create some administrative and fieldwork changes that made the projects work better.

In 1947 and early 1948, the NPS asked Jennings to move to Washington, ultimately to succeed Jesse Nusbaum as Departmental Consulting Archeologist (DCA). Instead, Jennings became chair of the fledgling Anthropology Department at the UUtah and recommended John C. Corbett for the job. Corbett was hired and in time succeeded Nusbaum as DCA (Jennings 1994). Thus, when the Upper Colorado River Basin Project salvage program came into existence, Jennings was able to bring his experience and archaeo-political networks to bear in formulating and running the project. Before, during, and after the Upper Colorado projects, he was centrally involved in building the UUtah anthropology department into a major mid-level university department, and he founded a campus natural history museum, which is now the Natural History Museum of Utah.

To help jump-start the Glen Canyon Project, Jennings hired UC's Robert H. Lister, a well known and respected Southwestern archaeologist. Lister and his wife, Florence, participated in the project from 1957 to 1959. In 1965, Lister and Corbett negotiated a fifteen-year cooperative agreement between UC and Mesa Verde National Park to conduct research and salvage within the Park. The program was, in time, directed by David Breternitz. In 1971, again with Corbett, Lister became director of the University of New Mexico (UNM)-NPS Chaco Canyon Project (www.chacoarchive.org/cra), which lasted 15 years. The project is now housed in the Frank Hibben Center for Archaeological Research at the University of New Mexico, which is named after archaeologist and long-time faculty member Frank C. Hibben. Dozens of students who became leaders in southwestern archaeology from the late 1970s to the present worked on one or both of these projects, as well as the DAP.

Fred Wendorf, after two years of rehabilitation from severe wounds suffered in World War II, began working in the Southwest in 1947–1948. In 1950, he was hired by Jesse Nusbaum, then NPS DCA, to develop the first large-scale pipeline survey and salvage project in the U.S. In 1953, with the New Mexico Highway Department, he initiated the first highway salvage project in the U.S. Both pipeline and highway programs became models for similar programs in many states (Wendorf 2008:57–73). In 1961, he began doing salvage archaeology in Egypt (see below); in 1964 he moved to Southern Methodist University where, with Joel Shiner and others, he helped build a major mid-level anthropology department (Wendorf 2008:107–112, 147–156).

Alexander Lindsay, Jr., the principal MNA field director for the Glen Canyon Project, continued on as director of a highly successful cultural resource management program at MNA. Lipe and Lindsay (1974) edited the summary volume from the 1974 Denver conference of about 100 American archaeologists (including the present author) at which salvage archaeology morphed into cultural resource management (see McManamon, this volume).

After the Navajo Dam Project, Ed Dittert (1922–2006) became Curator of Archaeology at the Laboratory for Anthropology and later its director. In 1967, he was hired to help build an anthropology department at Arizona State University. He brought in James Schoenwetter and Christy Turner as key faculty members. The department is now a major program within the School for Evolutionary and Social Change. Frank W. Eddy and James Hester went on to distinguished careers at the UC.

Colorado River Salvage Programs and the Aswan Dam

It is generally accepted that reservoir salvage archaeology began with the building of the first Aswan Dam on the Nile in Egypt between 1898 and 1992 (Brew 1961). In the late 1950s, Egypt announced plans to build a new Aswan High Dam, which was constructed between 1960 and 1970. The dam triggered a massive internationally based archaeological salvage program. William Y. and Nettie Adams, who worked on the Glen Canyon Project from 1957 to 1959, became UNESCO archaeological advisors to Nubia, and later major figures for 30 years in salvage work related to the dam, the reservoir, and other developments in Nubia. Fred Wendorf recruited Frank Eddy, James Hester, Philip Hoebler, and Joel Schiner, all veterans of Colorado River salvage projects, for an Aswan-Nile salvage program he began in Egypt in the 1960s and then expanded into adjacent areas. Both Adams and Wendorf went on to work with and influence numerous European, Egyptian, and Nubian colleagues, and many American students. Their African work is best told in their own words (Adams 2009:131–228; Wendorf 2008).

In short, the Colorado River salvage programs helped trained a generation of archaeologists and set the methodological stage for what became known as cultural resource management. The programs also ultimately had an international impact. Many of those who managed the programs went on to lend their expertise and experience to salvage projects in other countries.

PART III

THE RBS AND IASP IMPACTS ON AMERICAN ARCHAEOLOGY

CHAPTER 10

Contributions to Physical Anthropology by the Smithsonian Institution's River Basin Surveys

Vanessa A. Mirro

The projects of the Smithsonian Institution's River Basin Surveys (RBS) unearthed thousands of human remains, particularly in the Upper Missouri River Basin. Often the physical anthropologists and archaeologists recovering skeletal remains from RBS sites worked at a breakneck pace under emergency conditions, in part because of limited timeframes and funding (Jennings 1985; Lehmer 1971; Owsley and Jantz 1994). Despite the salvage nature of the recovery efforts, excavations of human remains and subsequent laboratory studies were carried out with precision. This attention to detail, meticulous methodology, and curation of RBS collections have enabled physical anthropologists to continue to utilize the data recovered from RBS sites to address bioarchaeological questions. Further, innovative techniques for the recovery of human remains that RBS physical anthropologist William M. Bass introduced were eagerly adopted by a new generation of physical anthropologists in the 1960s and 1970s. This new generation subsequently influenced the rapidly growing field of forensic anthropology. The RBS provided some of the most (now) well-known physical anthropologists access to an unprecedented sample of skeletal remains and this exposure contributed to their success and influence in the discipline today.

The Physical Anthropologists Involved with the RBS

In *Skull Wars*, David Hurst Thomas's account of the Kennewick Man controversy, Thomas lists eight prominent scientists that sued to obtain access to the Kennewick remains (Thomas 2000). Three of these scientists, Douglas Owsley, Richard Jantz, and George Gill, either as graduate

---

Dam Projects and the Growth of American Archaeology: The River Basin Surveys and the Interagency Archeological Salvage Program, edited by Kimball M. Banks and Jon Czaplicki, 141–150.

students or during the early years of their careers, worked on RBS sites or collections under the direction of Bass. In 1956, the Smithsonian Institution retained Bass to assist in analyzing RBS burials curated at the National Museum of Natural History, as well as the recovery of human remains for the Missouri Basin Project in North and South Dakota. As Bass recalls, they needed him "for help analyzing the flood of Native American skeletal material that was pouring in from the Great Plains" (Bass and Jefferson 2003:19). Although human remains were recovered from many sites, Bass investigated four large cemetery sites: Larson (39WW2), Leavenworth (39CO9), Mobridge (39WW1), and Sully (39SL4) (Owsley and Jantz 1994:3).

Bass began work on his dissertation at the University of Pennsylvania in 1956, which resulted in the first major synthesis of human skeletal remains from the Great Plains (Bass 1961). His dissertation was based on RBS skeletal data and was published as the first Memoir of the *Plains Anthropologist* in 1964 (Bass 1964). Between 1956 and 1970, Bass recovered and analyzed hundreds of human remains from along the Missouri River. During these investigations, Bass introduced his crews to new recovery techniques, such as the use of mechanized equipment to remove topsoil over cemetery sites to expedite the recovery of large samples of human remains. Miraculously, Bass did so without impacting the graves or the skeletal material (Bass 1963). At the Sully Site, for example, Bass and his crew were able to excavate 300 graves in the summer of 1960 using this approach, which was 10 times the number they excavated manually during the previous summer (Bass and Jefferson 2003).

In the late 1960s, Bass accepted a position at the University of Kansas, where he continued his research on skeletal remains from sites in the Great Plains while gaining recognition in the field of forensic anthropology. In the early 1970s, he moved on to become head of the Department of Anthropology at the University of Tennessee (Ubelaker and Hunt 1995:730). In the late 1970s, Bass's focus shifted to forensic anthropology, the analysis of skeletal remains in a legal setting, partially as a result of the increasing concern in the Native American community, particularly among Plains Indians, regarding the excavation and analysis of human remains from archaeological sites. Bass became a key figure in the development of American forensic anthropology, and trained some of today's top forensic anthropologists (Ubelaker and Hunt 1995:729, 732).

Prior to becoming department head at the University of Tennessee, Bass had recovered more skeletal remains than any other physical anthropologist in the United States (Benedict 2003:11–12). In his memoir, Bass estimated that over the 14 summers he worked in the Great Plains for the RBS, and he excavated 4,000 to 5,000 burials (Bass and Jefferson 2003:31).

This experience led him to publish one of the most widely used references on human skeletal remains, *Human Osteology: A Laboratory and Field Manual*, which is now in its fourth edition and is favored by physical anthropologists and students worldwide (Bass 1995).

During his tenures at the University of Kansas and the University of Tennessee, Bass provided his graduate students the opportunity to work on large collections of skeletal remains from RBS sites. Due to his experience, the collections available for study, and the talent and caliber of his graduate students, many became prominent and well-known physical and forensic anthropologists. His students included Judy Suchey, Richard Jantz, Douglas Owsley, Terrell Phenice, George Gill, Douglas Ubelaker, David Hunt, Hugh Berryman, and Patrick Willey (Ubelaker and Hunt 1995:731). Some of these students went on to become division heads and curators of physical anthropology and collections at the Smithsonian Institution's National Museum of Natural History.

Many of Bass's students participated in excavations at various RBS sites, or based their thesis and dissertation research on human remains recovered from the larger cemetery sites, including Larson, Leavenworth, and Mobridge. Bass directed Douglas Owsley, a graduate student at the University of Tennessee, to construct a demographic profile of the Arikara based on the remains of more than 700 individuals from the Larson Site. During his post-doctoral work, Owsley continued to research skeletal material recovered from RBS sites in the Great Plains while at the same time working under Bass on various forensic cases. In 1987, Owsley was hired as curator of physical anthropology in the Department of Anthropology, National Museum of Natural History based on his extensive experience with human skeletal remains from the Great Plains. Since 1990, he has served as the Division Head for Physical Anthropology.

Owsley went on to work on many high-profile cases, including Jeffrey Dahmer's first victim, the recovery and identification of Waco Branch Davidian compound members, historical skeletal remains from Jamestown, and the Pentagon victims of the 9/11 attacks. Owsley is perhaps best known for his involvement in the landmark lawsuit over Kennewick Man, a 9,600-year-old skeleton that was found eroding out of the bank of the Columbia River in Kennewick, Washington, in 1996. He, along with seven prominent scientists (including former Bass students, Jantz and Gill) sued the federal government to allow for the scientific study of the ancient human remains. Owsley and others had observed that the physical features of the Kennewick Man bore little resemblance to those of Native Americans. Owsley and Jantz drew conclusions about tribal affiliation based on their experience in developing one of the world's most comprehensive craniometric databases which included RBS skeletal data (Jantz 1977).

The Influence of the RBS Program on the Discipline of Physical Anthropology

The recovery of skeletal remains from RBS sites coincided with major theoretical and methodological changes occurring in physical anthropology. During this time, there was a general shift away from racial determinism, espoused by physical anthropologists Aleš Hrdlička and Earnest Hooton, to a biocultural approach that paralleled the trends occurring in the field of archaeology (Armelagos and Van Gerven 2003).

The development of modern American physical anthropology was grounded in the typological approaches to the study of human populations prevalent in the nineteenth and early twentieth centuries. This approach was largely in response to theories developing in the European and American scientific communities regarding human origins, race, and evolution (Little and Kennedy 2010). Physical anthropological studies relied on human crania, which could be studied and measured with great precision, to define the races scientifically and make inferences on intelligence based on cranial capacity. In this era of "skull science," Native American skeletal remains were in great demand. This demand led to the looting and desecration of graves by nineteenth-century scientists to support theories on racial typologies using craniometry. This concept dominated the field of anthropology until the 1950s. During this time, the discipline was greatly influenced by Hrdlička, Hooton, and Franz Boas. Both Hooten and Hrdlička believed in the fixity of human races; Boas, contrarily argued against the typological racial classifications based on cranial morphology and measurements and "led physical anthropology away from the taxonomy of race toward the study of human biology and culture" (Thomas 2000:105).

Boas' departure from his contemporaries, Hooten and Hrdlička, was not embraced by the physical anthropological community until the 1950s, when Sherwood Washburn ushered in "the new physical anthropology." This change was in response to a paradigm shift resulting from developments in evolutionary science, particularly in genetics and functional anatomy. Washburn argued against the emphasis on typologies and classification systems, and for a more evolutionary approach (Relethford 2010; Washburn 1951). His groundbreaking paper was, in part, influenced by the Cold Spring Harbor Symposium in June, 1950, entitled "The Origin and Evolution of Man," which focused on the themes of population rather than typology (Stini 2010:176–177). The symposium was attended by top anthropologists, geneticists, and evolutionary biologists, including Wilton Krogman, Bass's advisor at the University of Pennsylvania.

The physical anthropological studies conducted under the RBS reflect this theoretical shift away from the typological and classificatory approach to a more interdisciplinary approach to the understanding of human behavior, biology, and history. In the appendix of *River Basin Surveys Paper No. 25: Archaeology of the John H. Kerr Reservoir Basin, Roanoke River, Virginia—North Carolina*, Hoyme and Bass (1962) report on the human skeletal remains from the Tollifero (Ha6) and Clarksville (Mc14) sites. Lucile E. St. Hoyme was associated with the Division of Physical Anthropology of the National Museum of Natural History where she was promoted to the Curatorial Series as assistant curator. Although some vestiges of the Hrdlička and Hooton's influence are evident in their report, such as detailed information on differences in populations based on cranial indices (citing Hrdlička's methods), the report demonstrates a more rigorous interpretation of human remains within a regional context. Through a comparative study of the skeletal remains from these two sites, Hoyme and Bass (1962) investigated topics such as disease, demography, subsistence, warfare (e.g., scalping), and cultural practices that included cranial modification, possible cannibalism, and burial practices. As such, their study reflected a more theoretically-motivated approach to exploring the two populations than evident in preceding studies.

One of the biggest criticisms of the RBS was the immense backlog of unpublished manuscripts and incomplete reports; many collections lacked any form of report or manuscript (Thiessen 1999). This reporting problem was largely a result of scheduling issues and inadequate funds that hindered the ability of researchers to conduct post-field analyses in a timely manner and produce detailed reports and manuscripts. Results from a number of RBS investigations were presented in the *River Basin Surveys Papers of the Bureau of American Ethnology Bulletins*. However, the majority of the reports on skeletal remains was often brief and purely descriptive, and, with few exceptions, such as the Hoyme and Bass study, failed to provide interpretation or synthesis.

Although the RBS program was terminated in 1969, its collections continued to be studied by researchers in various institutions, including the Smithsonian Institution, universities, archaeological societies, state historical societies, and archaeological research centers of the National Park Service, such as the Midwest Archeological Center (Thiessen 1999). The most comprehensive reports on the sites were completed in the years following the termination of the program. For instance, former RBS archaeologist Richard Johnston documented the results from the McLure Site (39HU7) more than 15 years after the fieldwork was completed. The McLure Site, an Arikara village along the bank of the Missouri River in South Dakota, was excavated in 1965; however, no report was prepared at that time (Johnston 1982). Jantz and Owsley (1982) undertook the osteological analysis of the human remains for Johnston's report.

Similarly, Bass examined the skeletal material from the Fay Tolton Site (39ST11) at the Smithsonian Institution in 1959, but for various reasons, the data were not reported until W. Raymond Wood and anthropology students from the University of Missouri published the report in 1976 (Wood 1976). Bass and his former students Jantz and Hugh Berryman interpreted the human skeletal remains based on morphological and anthropometric data obtained at that time, and they were able to determine population affinities based on crania that fell within the morphological range of historic Mandan Indians (Wood 1976:43). The craniometric studies conducted on the Fay Tolton crania reflect an attempt by Jantz and others to recognize the value of craniometry, which fell out of favor following the rise of the new physical anthropology (Jantz 1977). Craniometric studies enabled researchers to define the temporal and spatial patterning of cranial variation in a particular area to better understand population movements, ethnic origins, and microevolution using statistical methods derived from measurements of the cranium. Jantz tracked down Boas's data sheets of anthropometric measurements of thousands of Native American remains for comparative studies on skeletal remains (Little 2010:68). However, it was the data collected during the RBS program that were invaluable, largely because of the large number of skeletal remains from the Great Plains available for study, particularly from cemetery sites (Table 10.1).

Under the direction of Bass, Owsley worked on a large collection of human remains from the Larson Site in South Dakota as part of his Master's thesis at the University of Tennessee (Owsley 1975). The Larson Site is a Protohistoric village dating to the mid to late eighteenth century that RBS crews excavated in 1963, 1964, and 1966. Bass and his students from the University of Kansas excavated the village cemetery from 1966 to 1968 (Owsley et al. 1977).

For his thesis, Owsley conducted a demographic study of the Larson remains, comparing the mortality rates between males and females. However, in subsequent research, Owsley discovered that human remains from the village area were scattered across the house floors rather than being formally interred as the remains from the village cemetery. The demographic profile of the remains from the village differed from those

Table 10.1 RBS cemetery sites yielding large skeletal samples*

Site Name	Number	Approximate Date	Total Burials
Mobridge	39WW1	1600–1650	650
Sully	39SL4	1700–1750	566
Larson	39WW2	1750–1785	706
Leavenworth	39CO9	1803–1830	285

* Adapted from Jantz (1977:163).

from the cemetery. The cemetery reflected a normal mortality profile for the Larson population, whereas the village yielded a disproportionately higher number of adolescents and young adults and fewer adult females compared to males. Evidence of scalping, burning, decapitation, and other forms of trauma were observed on the remains recovered from the house floors. Drawing from ethnographic and ethnohistoric studies, and comparing the age distributions of the remains in the village to settlement patterns of the Arikara, Owsley determined that a massacre occurred in the spring or the fall. Based on osteological, archaeological, and ethnographic data, Owsley hypothesized that warfare was responsible for the spatial and demographic patterning of human remains in the village area (Benedict 2003; Owsley et al. 1977).

Owsley, Berryman, and Bass adopted a forensic-like approach in their investigation of the human skeletal remains from the Larson Site and reconstructed a single event—a massacre—with unprecedented precision because of their interdisciplinary approach to their research. Their approach stands in stark contrast to the descriptive approaches of their physical anthropologist predecessors. Moreover, the skeletal remains from the Larson Site are a unique sample for demographic study because they represent one of the largest collections from a single site in the Northern Plains that was occupied for a relatively short period of time (Owsley and Bass 1979:145).

Owsley, Jantz, and others who studied RBS collections had the advantage of looking at entire skeletal populations because of the availability of skeletal remains recovered from large cemetery sites. Access to large skeletal populations is rare these days largely because of state and federal legislation regarding the disposition and repatriation of Native American remains, especially the *Native American Graves Protection and Repatriation Act* (NAGPRA). In his memoir, Bass notes:

> As the number of graves we'd excavated mounted into the hundreds, a remarkable picture began to emerge from the prairie earth. For the first time in the history of Great Plains archaeology, we had large, documented samples of an entire tribe's skeletal remains, from birth to old age. For the Arikara, we realized, life was harsh, violent, and often very brief (Bass and Jefferson 2001:29).

The studies conducted for the RBS made a significant contribution to physical anthropology, particularly in the Great Plains. A major publication by Owsley and Jantz (1994) on studies of human skeletal remains from the Plains relies heavily on the physical anthropological work conducted during the RBS program, particularly the large cemetery sites such as Larson, Sully, Leavenworth, and Mobridge. *Skeletal Biology of the Great Plains: Migration, Warfare, Health and Subsistence* presents the results of bioarchaeological research of Plains Indians from the Over Collection (500 skeletons from 37 sites) that was to be reinterred (Owsley and Jantz 1994). However, because of small sample sizes of remains from the various sites in the collection, the researchers relied on RBS collections for comparative purposes, thereby enhancing the utility of both data sets (Owsley and Jantz 1994). The various authors in the volume had access to data collected during the RBS program by Bass and others or were able to gain access to skeletal remains from RBS sites for their studies (e.g., Cole 1994; Cole and Cole 1994; Ruff 1994; Tuross and Fogel 1994). A second publication, *Skeletal Biology and Bioarchaeology of the Northwestern Plains*, edited by Gill and Weathermon (2008), also references RBS collections but to a much lesser extent than the Owsley and Jantz publication. This difference is in part because of geography but also because of the amount of skeletal data from the University of Wyoming Human Remains Repository.

These two publications along with others that utilize the skeletal data collected during the RBS program demonstrate how these collections continue to contribute to bioarchaeological studies today, particularly in the Great Plains. The availability of the collections to physical anthropologists in the years following the RBS enabled the use of modern technology in the analyses of the remains, such as trace element and stable isotope studies for reconstruction of diet. For instance, Tuross and Fogel (1994) prepared collagen from the ribs of 36 individuals recovered from the Sully Site by an RBS field crew to obtain stable isotope values. They concluded that there was a substantial reliance on bison meat relative to other food sources, which is consistent with the archaeological data, and they determined weaning patterns and reliance on breast milk (until one year of age) from isotopic values of infant skeletal remains.

Studies such as these demonstrate that the availability of large, well-documented skeletal samples enable researchers to apply modern techniques, such as chemical analysis of bone, that were not available at the time of the RBS. They also highlight current research avenues being pursued in bioarchaeological investigations, such as infant weaning, sexual variation in skeletal morphology, subsistence, and disease history (Ubelaker 1994). In sum, the skeletal material and data collected during the RBS program continue to increase our understanding of human adaptation, subsistence strategies, demographics, warfare, and health of past populations.

Concluding Remarks

NAGPRA, signed into law in 1990, required human remains and funerary objects, sacred objects, and objects of cultural patrimony housed by museums, federal agencies, and federally funded institutions to be repatriated to affiliated tribes. Many of the RBS skeletal collections housed at institutions, such as the Smithsonian Institution, Midwest Archeological Center, and the North Dakota Heritage Center have been, or are in the process of being repatriated. To some, this seems an enormous loss to the scientific community. In the introduction to Gill and Weathermon's 2008 volume, Gill states:

> The fabulous era for bioarchaeological exploration of the late 1960s and 1970s happened through a developing sophistication in recovery methods coupled with a minimal degree of legal and bureaucratic constraint on scientific endeavors. This combination of events was enhanced by an enduring attitude of public excitement and scientific optimism, lingering from an earlier era in Western civilization, that placed an unfettered value on the kind of scientific information acquired from grave contexts and old skeletons (Gill 2008:2).

Implied in this statement is the contention that the ability to conduct bioarchaeological research on Native American populations has been severely restricted due to increased legal and bureaucratic constraints, namely NAGPRA.

However, NAGPRA does not preclude scientific analysis. The scientific value of some remains may outweigh the need or desire to repatriate. Section 7(b) of the Act states:

> (b) SCIENTIFIC STUDY—If the lineal descendant, Indian tribe, or Native Hawaiian organization requests the return of culturally affiliated Native American cultural items, the Federal agency or museum shall expeditiously return such items unless such items are indispensable for completion of a specific scientific study, the outcome of which would be of major benefit to the United States. Such items shall be returned by no later than 90 days after the date on which the scientific study is completed.

Further, §10.10(c) of the implementing regulations 43 CFR Part 10 states:

> §10.10 Repatriation
>
> (c) Exceptions. These requirements for repatriation do not apply to:
>
> (1) Circumstances where human remains, funerary objects, sacred objects, or objects of cultural patrimony are indispensable to the completion of a specific scientific study, the outcome of which is of major benefit to the United States. Human remains, funerary objects, sacred objects, or objects of cultural patrimony in such circumstances must be returned no later than ninety (90) days after completion of the study.

These clauses leave the door open for scientific analysis but they put the onus on physical anthropologists (and archaeologists) to demonstrate and justify the need for analysis. Analysis must be negotiated. The law and its regulations also allow for the possibility that tribes may request analysis of their ancestral remains. Again, this leaves open the opportunity for physical anthropologists, but they must negotiate with the tribes as to the nature and extent of such investigations.

In this regard, NAGPRA encourages a collaborative effort between Native Americans and physical anthropologists (and, again, archaeologists). Such collaborative efforts require open communication and mutual respect and are necessary if the Native American community, physical anthropologists, and archaeologists are to get past the animosity and distrust that has built up over the years. For their part, physical anthropologists and archaeologists must recognize that they need to give back to the Native American community. After all, modern Native Americans are the descendants of those skeletal remains and the makers of those artifacts that physical anthropologists and archaeologists study.

The key may be to develop research programs through consultations that draw on native perspectives and actually have value for native peoples. There are recent examples of federally funded excavations of Native American burial populations that have undergone thorough scientific analyses prior to repatriation (e.g., Perry et al. 2010). These studies have directly benefited from recovery and analytical methods developed under the RBS program and the bioarchaeological expertise that was cultivated under that program. But these studies also benefited from native perspectives on origins, identity, and cultural affiliation. For indigenous peoples, the issue of having the right to maintain and protect ceremonial objects and human skeletal remains is at the forefront of their concerns. Nevertheless, through effective consultation, communication, respect, and trust, it may be possible to conduct analyses and build on the tremendous data that were collected as part of the RBS.

CHAPTER 11

Missouri Basin Projects and the Emergence of Historical Archaeology on the Great Plains

William B. Lees

Developments in the early twentieth century were instrumental in the emergence of the discipline of historical archaeology in North America and other regions of the world. Earlier archaeological encounters with historic sites dating as far back as the British Boundary Commission in 1796 were certainly precursors (Cotter 1993:5–6). However, the large projects and programs of the 1920s and 1930s brought a scale and frequency to historical archaeology that was foundational for a modern historical archaeology that is increasingly defined in global terms (Lees 2008).

In 1928, archaeological excavation began at Colonial Williamsburg as part of the massive heritage tourism project that resulted in the reconstruction of much of the town. Focused on architectural reconstruction that used cross-trenching as a fundamental discovery method, extensive areas of Williamsburg were explored in the ensuing years. Work remained narrowly focused on discerning details about the construction and appearance of individual buildings (Poole 2012).

Between 1949 and 1957, Ivor Noël Hume worked for London's Guildhall Museum. His job was to recover archaeological remains during the rebuilding of London after World War II. He quickly recognized the importance of Post-Medieval archaeology and informing the public about his findings (Noël Hume and Miller 2011). It was not until 1958, when Noël Hume was hired to direct archaeology at Williamsburg, that the lives of people became a topic of interest for Williamsburg archaeology (Poole 2012). Nonetheless, the work conducted at Williamsburg prior to Hume's arrival helped illustrate and validate the utility of archaeology for sites of the historic period.

Dam Projects and the Growth of American Archaeology: The River Basin Surveys and the Interagency Archeological Salvage Program, edited by Kimball M. Banks and Jon Czaplicki, 151–166.

Between 1933 and 1942, a series of programs collectively known as New Deal archaeology brought archaeology to much of the United States, including most of the Southeast and the Great Plains (with the exception of Kansas) (Hawley 2006; Means 2011b:31). The Works Progress Administration (WPA), Civilian Conservation Corps (CCC), and other work relief agencies used survey and excavation as a means to put unemployed and often unskilled men to work. The focus of this work was often prehistoric sites, but historic sites also benefited.

In 1934, investigation of the original site of Jamestown began as a CCC project. The major goal was the recovery of information about the site's building remains (Pykles 2011). In 1936, J. C. Harrington took over and developed a holistic archaeological approach to the site and invited the public to see first-hand the open excavations and what was being learned through a properly constructed program of archaeological research. Harrington was born in 1901 in Michigan and received a Bachelor's degree in Architectural Engineering from the University of Michigan in 1924. He was exposed to heritage studies during a required summer practicum that he took in New Mexico. Following graduation, he returned to New Mexico and worked on a number of archaeological projects with colleagues he had met during his practicum. When the Great Depression cut short his budding architectural career in 1932, Harrington returned to school at the University of Chicago and in 1935 received a Master's degree focused on archaeology. He began doctoral studies at Chicago but never completed the dissertation due to the demands of his career (Miller 1998).

In 1936, the National Park Service (NPS) asked Harrington to take over archaeological excavations at Jamestown. Work on the site had come to a halt due to disagreements between a crew excavating buildings led by an architect, and a crew excavating "everything else" led by an archaeologist who was not allowed within three yards of a foundation. The NPS saw Harrington's background in both architecture and archaeology as a means to reconcile the various needs of this site under a single director. Results proved the wisdom of this decision. Harrington remained at Jamestown until 1946 and then became the Eastern Regional Archeologist for the NPS until his retirement in 1965 (Miller 1998).

Harrington's pioneering work at Jamestown was influential and is the basis for some calling him "the father of historical archaeology." His work was part and parcel of why the Society for Historical Archaeology named their lifetime achievement award in his honor (Miller 1998:1). Jamestown was not the only archaeology conducted on historic sites as a result of the New Deal, but it was certainly the most important in defining the emerging discipline of historical archaeology.

New Deal work at Fort Laramie, Wyoming, is illustrative of a project in the Great Plains conducted concurrently with Harrington's work at Jamestown (Walker 2011). This work, begun in 1937 by G. Hubert Smith of the NPS, was initially intent upon defining the archaeological resources at this site (Minnesota Historical Society 2012a; Smith 1939; Walker 2011). Smith was born in South Dakota in 1908, but grew up in Minneapolis, Minnesota. He attended the University of Minnesota and graduate school at the University of Chicago between 1939 and 1941. During college he worked for the NPS at Fort Laramie until 1941. Additional work on this site, which had become the Fort Laramie National Historic Site, continued after World War II under the auspices of the River Basin Surveys (RBS) (Minnesota Historical Society 2012a; Walker 2011).

Following World War II, Smith worked for the Minnesota Historical Society and the NPS. In 1951, he began a seventeen-year career with the Smithsonian Institution's RBS Missouri Basin Project, stationed in Lincoln, Nebraska. He died suddenly in 1972 in Lincoln and is remembered as a leader in the emerging field of historical archaeology because of his work on historic sites during the Great Depression, his work with the RBS, and for his mentoring of others (Minnesota Historical Society 2012a; Walker 2011).

The archaeological work on historic sites prior to World War II, and in particular the work conducted as part of the New Deal, established a role for historical archaeology in an emergent federal archaeology program that the *Historic Sites Act of 1935* legitimized. Although the New Deal ended in 1942, the *Flood Control Act of 1944* laid the groundwork for further growth of a federal archaeology program of which a significant component would be the RBS. There is little question that the *Serviceman's Readjustment Act of 1944* (GI Bill of Rights) was also critical in providing returning veterans with college educations, helping to prepare the professionals that would be needed in the Missouri Basin and elsewhere in the 1950s and beyond.

Flirting with Historical Archaeology: The Direct-Historical Approach

Archaeologists became interested in sites dating to the historic period on the Central and Northern Great Plains shortly after the turn of the twentieth century. Inspired by rich ethnographic knowledge of plains-dwelling Native Americans such as the Pawnee, archaeologists sought to use sites occupied by these peoples during the early historic period as a bridge to understanding sites of the prehistoric past.

Ethnographic analogy had been commonly used to explain archaeological remains, but in the Central and Northern Plains, archaeologists believed there was direct-historical continuity between the people of history and prehistory. Using the method that became known as the direct-historical approach, archaeologists sought to address questions of the origins of the prehistoric Plains populations by working from the historic known to the unknown of the protohistoric and prehistoric periods. Earle has referred to this as "analogy by historical continuity" (Earle 2008:195). This approach was quickly adopted by archaeologists working in other regions (Lyon 1996).

The direct-historical approach was developed to address a specific archaeological problem, and was firmly seated in the anthropological thinking of the day. The direct-historical was one of the fundamental approaches used in the archaeology of the Great Plains—primarily the Central and Northern Great Plains—prior to World War II, and it continues to be important to this day. Significant discussions of this approach are found in now classic works such as Strong's *Introduction to Nebraska Archeology* and Wedel's *Introduction to Pawnee Archeology*, published in 1935 and 1936 respectively (Strong 1935; Wedel 1936). Wedel, in 1938, and Strong, in 1940, published articles that substantially served to formalize this approach following work on the Central Plains Pawnee and in the northern Great Plains (Strong 1940; Wedel 1938). It was in Wedel's 1938 article that the term direct-historical approach was first formally presented (Willey and Sabloff 1974:114), although he attributed its first practical application to his mentor, William Duncan Strong (Wedel 1977:3).

Wedel was born in Newton, Kansas, in 1908, and spent a good part of his youth searching for and collecting prehistoric artifacts (Wedel 1977). He received a BA in 1930 from the University of Arizona where he had joined his "relic-hunting partner" and neighbor, Emil Haury, as a student (Wedel 1977:3). He then moved to Lincoln for graduate school and received an MA from the University of Nebraska in 1931. At Nebraska he studied under William Duncan Strong, who had joined the faculty in 1929 (Wedel 2006). Strong had been trained in ethnography while at the University of California, Berkeley, and brought that to his archaeological study of the Central Plains.

While at Nebraska, Wedel gained substantial experience working with Strong at sites in the Central and Northern Plains, but in 1932 he moved to the University of California, Berkeley, where he received a Ph.D. in 1936 (and where he too became well schooled in ethnography). In that same year, Strong moved to a new job at the Smithsonian Institution. After a short stint at the Nebraska Historical Society following graduation from Berkeley, Wedel also joined the Smithsonian Institution where he remained until his retirement in 1976 (Gradwohl 1996). Between 1946 and 1950, Wedel directed the Missouri Basin Project of the RBS (Smithsonian Institution 2012a).

As defined by Wedel and Strong, the direct-historical approach focused on establishing a baseline from which cultural and chronological sequences could be developed for the protohistoric and prehistoric past. According to Wedel:

> It was believed that by isolating and clearly defining the archeological characteristics of the historic peoples a whole series of sites could soon be removed from the category of unknowns; and furthermore, that a comparison of materials so identified with earlier remains in the region might open lines of attack which would permit the establishing of a time sequence extending "from the known historic into the unknown prehistoric" (Wedel 1938:1).

At the end of his career, Wedel summed it up this way:

> Methodologically, it has involved in the Plains area, (1) the isolation of historic sites, preferably those of identifiable tribes; (2) a definition of their cultural complexes; and (3) establishing the basic criterion of continuity by carrying the sequences back in time to progressively earlier periods and cultures. To these aims I would add another, namely, the use of ethnographic analogy in searching for the possible uses and meanings of archaeological artifacts and features (Wedel 1977:7).

In the context of pre-World War II archaeology, which to a large degree focused on problems of culture chronology and cultural-historical reconstruction (Jennings 1986), the direct-historical approach was intended to address one specific question: the cultural identity of inhabitants of sites occupied during prehistory. Under this methodology, investigated sites were selected because of the information they could provide for the resolution of this question. The focus was clearly on the problem rather than the site. The direct historical approach can be seen, therefore, to have developed as a vital part of the general anthropological approach to the prehistory of the Great Plains and to have operated via the scientific method.

Certain characteristics of the direct-historical approach are important in understanding its relation to the subsequent development of historical archaeology on the Great Plains. Most important among these are, first, that the direct-historical approach relied fundamentally on the historical archaeological record but was itself concerned primarily with problems in prehistory. Second, the direct-historical approach focused on Native American sites to the exclusion of all others.

The use of historical archaeology in addressing problems in prehistory represents, significantly I believe, an unbiased use of the historical archaeological record for general anthropological problem solving. Archaeologists, such as Wedel and Strong, who were working on the Great Plains before World War II did not dichotomize prehistoric and historical archaeology, but rather were led by the pursuit of a problem, enabling them to embrace any pertinent data or approach. This approach was not universal, of course, and to be clear, Wedel and Strong had little interest in sites of the historic period other than the few that directly informed their studies using the direct-historical approach.

In the direct-historical approach, prehistory and history were correctly and necessarily viewed as a continuum with both history and prehistory contributing differently to the resolution of a common problem within anthropology: cultural identity. Of interest, this is a position that was called for by Kent Lightfoot in an article published in *American Antiquity* in 1995 almost 50 years after Wedel and Strong had so successfully introduced the direct-historical approach into Plains archaeology (Lightfoot 1995). As Lightfoot suggests, neither the direct-historical approach nor any other anthropological theory has resulted in archaeological practice where the line between history and prehistory is anything different than a dividing line; a proverbial "Berlin wall" of North American archaeology.

The focus on Native American sites was likewise an artifact of the problem on which the direct-historical approach was centered, but it is nevertheless important in understanding subsequent development on the Great Plains. Prior to World War II archaeologists had little interest in questions pertaining primarily to the historical period. Interest in historic Native American sites was itself fueled by the critical importance of these sites for studies of the prehistoric past using the direct-historical approach.

While today the direct-historical approach might be seen as an historical archaeological methodology that developed out of anthropology, the approach was hardly perceived this way in the 1930s. As such, it provided little help in dealing with the non-Native American historic sites threatened by the archaeological emergency that developed on the Great Plains following World War II. Although with a much weaker link to anthropology, the New Deal work at Fort Laramie that G. Hubert Smith conducted was in fact a better model for what was to follow. If nothing else, this work helped prepare Smith for his emergence as a leader and mentor in historical archaeology in the Missouri Basin and beyond.

The River Basin Surveys

The importance of the *Flood Control Act of 1944* for the development of the federal archaeology program is discussed in detail by Thiessen, et al., and McManamon in this volume. For the Central and Northern Great Plains, the act resulted in an ambitious program of congressionally mandated water resources control projects known as the Pick-Sloan Plan. Pick-Sloan resulted in the planning for and construction of reservoirs within the Missouri River Basin. In 1946, the newly formed Interagency Archeological Salvage Program began an unprecedented program of survey and excavation on these projects, which were categorized by major reservoirs and sub-basins (Jennings 1985:281).

Work on reservoir projects was done both by staff of the Smithsonian Institution's RBS Missouri Basin Project and by archaeologists affiliated with partnering institutions. Initially, archaeologists with partner institutions such as the University of Nebraska State Museum, University of Kansas, and the State Historical Society of North Dakota voluntarily conducted salvage excavations within reservoirs that were by the late 1940s under construction. In 1950, through a program of cooperative agreements with the NPS, partner institutions received funds to support field expeditions and report preparation (Jennings 1985:286–288).

Construction of Fort Peck Dam on the Missouri River (Montana, Fort Peck Lake) by the Public Works Administration (PWA) between 1933 and 1943 was the precursor to the main stem projects that would follow, although no archaeological work accompanied construction of this project. Fort Peck Dam was followed by five major dams on the mainstem: Garrison Dam (Lake Sakakawea, construction begun 1947 and completed 1953), Oahe Dam (Lake Oahe, 1948, 1962), Big Bend Dam (Lake Sharpe, 1959, 1964), Fort Randal Dam (Lake Francis Case, 1946, 1956), and Gavins Point Dam (Lewis and Clark Lake, 1952, 1957) (United States Bureau of Reclamation 2012). Archaeology conducted at the numerous smaller reservoirs on Missouri River tributaries, such as Wilson and Kanopolis lakes on the Smoky Hill River in Kansas, probably resulted in collectively more work than that undertaken on the larger main stem projects.

The scope of archaeological work that emerged following World War II was unprecedented and confronted headlong the capacity of the existing archaeological infrastructure. Formidable challenges that faced Plains archaeology at this time included: limited knowledge of archaeological resources—both prehistoric and historic—on the Great Plains; the imminent, wholesale loss through inundation of previously untouched resources along massive stretches of major river valleys; the limited timeframe available for archaeological studies; a shortage of trained professionals; and, at times, severe budgetary constraints (Jennings 1985; Lehmer 1971).

While these problems had important effects on all resources endangered by the Pick-Sloan reservoirs, the impact on historical resources was profound in part because there were few advocates among the archaeological community being assembled to deal with the crisis. Another factor that is difficult to assess but of certain importance is the relative lateness of large-scale Euro-American settlements of the upper Missouri River Basin, especially as seen through the lens of the late 1940s and early 1950s.

Illustrative of this is a statement by historian Merrill J. Mattes about historic sites along the upper Missouri: "steamboat landings, villages, missions, and other communities of fairly recent origin might bear such similarity to still-existent communities that archeological findings might not be rewarding" (Mattes 1960:12).

The challenge facing the archaeological community was the need to consider a resource base that was soon to be destroyed. In many cases, this situation provided an important opportunity for large-scale intensive investigations of an existing problem or topic of study. Exemplary of this opportunity was research on the late prehistoric and protohistoric village agriculturalists of the Middle Missouri. Sites of this period had been a focus of research that had been conducted prior to the interagency work, and which received a somewhat disproportionate but not unjustified amount of the effort connected with the post-World War II boom. Although based on what would soon be seen as a small number of projects conducted by a handful of archaeologists, there existed a legacy of research, and archaeologists with standing research interests were ready to seize the opportunity offered by the funding that soon washed over the Plains.

Regardless, prior to World War II and the RBS, relatively few professional archaeologists had been working in the Missouri River Basin; notable of course are Strong and Wedel, and the likes of businessman-turned-archaeologist, A.T. Hill of the Nebraska Historical Society. Wedel opened up the Lincoln research office of the Smithsonian Institution's RBS Missouri Basin Project in 1946 and continued to have a role in the Missouri Basin. Other than Wedel and a few others, however, most of the players following the war were new to the Plains (Thiessen 2012).

The onset of the massive projects of the Pick-Sloan Plan and the resulting archaeological work caused a great influx of archaeologists into the Missouri Basin who worked directly for the Smithsonian Institution, as well as for partner institutions. Answering the call were the likes of W. Raymond Wood, Carlyle S. Smith, Marvin Kivett, Ralph Solecki, Thomas A. Witty, Jr., Roger Grange, Wesley R. Hurt, Carl Miller, W.W. Caldwell, James F. Deetz, John L. Champe, Carl Chapman, Paul L. Cooper, E. Mott Davis, James Gunnerson, Thomas Garth, Preston Holder, Jack T. Hughes, Donald J. Lehmer, George Metcalf, Alan R. Woolworth, and Robert L. Stephenson, among others. Many of these individuals only spent a few years in the Missouri Basin before going on to distinguished careers elsewhere. Many of them, including G. Hubert Smith, Deetz, Chapman, Garth, Carlyle Smith, Wood, Grange, Mills, Miller, and Woolworth made their mark at least partly in historical archaeology although not necessarily on the Plains (Hawley 2006).

Lacking in the Missouri Basin at the onset of the RBS was precedence for historical archaeological research on Euro-American sites and other sites falling outside the scope of the direct-historical approach as it had been used historically. Euro-American sites represented the majority of the historic period archaeological resource base in any region but had received only limited attention, such as G. Hubert Smith's work at Ft. Laramie. The general lack of interest or concern for Euro-American sites on the part of archaeologists working in the Basin during the early twentieth century naturally carried over into the post-World War II era. For an area that had few practicing professionals prior to the RBS, and given the incipient nature of historical archaeology elsewhere in the country, this finding is not unexpected.

When faced with the loss of the cultural resource base of much of the Missouri River Basin, however, the need for historical archaeology survey and salvage was not ignored. This may not have been the outcome without the emergence of historical archaeology during the 1920s and 1930s through highly visible projects at Jamestown and Williamsburg, and through work on the Plains at places like Fort Laramie. Historical archaeology pioneer J. C. Harrington, who in 1946 had left Jamestown for a new position with the NPS as Eastern Regional Archeologist, may also have influenced the inclusion of historical archaeology as part of the new program of the RBS. The influence of historian Merrill J. Mattes on historic sites in the archaeological work within the Missouri Basin was certainly of importance as well.

Mattes was without question one of the preeminent western historians working during the mid-twentieth century. He was born in 1911 in Congress Park, Illinois, and graduated from the University of Missouri and the University of Kansas where he earned a Master's degree in 1933 (Saxon 1996). In 1935, he began a life-long career with the NPS with his first post as park superintendent at Scotts Bluff National Monument, Nebraska (Saxon 1996). He served as acting superintendent and then as acting historian for Fort Laramie from 1938 to 1946. Significantly, between 1946 and 1966 Mattes was the NPS Region Two Historian and was stationed in Omaha, Nebraska (Wyoming Authors Wiki 2012). His position as historian corresponded with the period when the RBS was most active in the Missouri Basin. In 1967, after a career working closely with archaeologists and on archaeological projects, he became a member of the first board of directors of the Society for Historical Archaeology; he was one of two historians on this board (Jelks 1993:11).

Partly because historians had been involved in research on the reservoirs prior to the beginning of most large-scale excavations that accompanied the onset of NPS cooperative agreements in 1950, and partly because of a lack of interest and an already overloaded work schedule among archaeologists, the impetus for this historical archaeological research did not come from anthropology but rather the field of history.

As a result, while anthropologists were involved in all phases of site survey and salvage for prehistoric and protohistoric sites, historians provided the baseline from which the survey for and salvage of Euro-American sites was to be conducted.

Given the importance of the NPS cooperative agreements with partner institutions for emergency excavations conducted in the 1950s and 1960s, it is also important that in 1950 Mattes was placed in charge of NPS's role in the Missouri Basin salvage program. His responsibilities involved entering into critical cooperative agreements with partner institutions, thereby solidifying the relationship between early historical survey and excavation funded through the NPS (Lehmer 1971:2–3).

A typical approach to the historical archaeological resources of a reservoir project started with a detailed documents search conducted by an historian to identify the possible historical archaeological resources within the project area. An early and representative example of this approach is the historical overview of the Fort Randall Reservoir (now Lake Francis Case), South Dakota. Based on detailed reviews of documents and literature and a few brief field expeditions, Mattes prepared a comprehensive report on the historic sites of this reservoir project in 1948 (Edwards and Mattes 1948). Mattes identified a total of 120 historic sites within the project area (Table 11.1); it is important to note that Mattes recommended that 75 of the sites—well over half—held little potential for archaeological research (e.g., 35 historic streams and 25 historic islands) (Mattes 1949). Needless to say, "comprehensive" is a relative term, but it also reflects the conventional wisdom of the late 1940s and the field of history as to what constituted an historic site.

This and similar reviews of historic sites imposed severe biases on the recognized historical resource base of a project area. A review of the 120

Table 11.1 Historic sites identified through historical research within the Fort Randall Reservoir, South Dakota (Mattes 1949:477)

Site Type	Number
Military post sites	3
Trading post sites	6
Existing towns and Indian agencies	4
Abandoned post offices, towns, settlements, and steamboat landings	8
Lewis and Clark camp sites, 1804 and 1806	15
Churches and cemeteries	4
Steamboat wrecks	6
Historic islands	25
Historic streams	35
Miscellaneous historic landmarks	6
Railroad and highway bridges	3
Total	120

historic sites identified for the Fort Randall Reservoir, for example, shows that Mattes focused on sites with a high historical visibility. A number of significant historical resource types with lesser historical visibility, such as early Euro-American homestead settlements and early reservation period Native American settlements, are not included in the list. But in evaluating the list, it must be remembered that concepts of history and historical importance have changed in the past half-century and to complain that those working then did not share current predispositions is unfair. What is important, however, was the transference of design level research decisions concerning historical sites from archaeologists to historians.

The resource assessments prepared by historians during this period served as the baseline for determining which sites were in need of excavation. Since funds were not available for the archaeological investigation of all historic sites identified by these comprehensive studies, decisions on which sites to evaluate were required and were based on three criteria: (1) the degree of historical significance; (2) the extent of the available documentation on the site; and (3) whether or not a documented site could actually be located in the field.

To illustrate the operation of this selection process in the Missouri Basin, Mattes states:

> The Lewis and Clark expedition was unquestionably of epic importance, hence the explorers' camp sites satisfy the first requirement. However, it would be difficult to make a case for archeological work at such sites since they would be virtually impossible to pinpoint. Steamboat landings, villages, missions, and other communities of fairly recent origin might bear such similarity to still-existent communities that archeological findings might not be rewarding. Three classes of sites, representative of significant frontier eras, offered the greatest promise. These were the trading posts, the military posts, and the early Indian agencies [Mattes 1960:12].

This approach to site selection, as well as the approach to the identification of sites in the reservoir areas, illustrates the significant role of the historian in the post-World War II archaeology of the historic period. Essentially, the historian would identify potential sites, and from this universe, no matter how limited, sites would be further prioritized for excavation.

Another significant and related problem revolved around the involvement of the archaeologist in this process. In most cases, the archaeologist became involved only after sites had been prioritized, and served to find sites selected for excavation, to determine if excavation was feasible, and to design and execute the excavations. The result was that the archaeologists involved in the excavation of a site were not charged with the investigation of any specific problem. Rather, they were usually charged with excavating a site only because it was important in history. In retrospect, we can conclude that the vast majority of historic sites investigated during this period—especially Euro-American sites—were excavated first and foremost because they were material illustrations of history. To be certain, there was a disconnect between the site as representation of historical event or process and the site as a source of new information on these events or processes or of the lives of people involved (some of whom were historically disenfranchised).

Given the context of the Missouri Basin prior to the onset of the RBS and the general context of this work within North American archaeology of the late 1940s through the early 1960s (before the onset of the "new archaeology"), this approach is not unexpected. Although these projects would have been approached differently even a few decades later, it is commendable that there was some level of reasoned, systematic archaeological consideration of historic sites as early as the 1950s. This cannot always be said of some projects undertaken later—after the onset of new archaeology in the late 1960s—where historic sites were not considered at all. In fact, even to this day historic sites within project corridors are often selected for excavation using criteria little different than those used in the Missouri Basin over a half-century ago.

Regardless of the process whereby sites were selected for excavation, or the loss of sites now seen as important that may have been excluded from consideration, the historic sites that were excavated during 20 years of intensive salvage in the Missouri Basin represent a significant body of historical archaeological research. One need only look at the work on historic sites by the likes of G. Hubert Smith, Alan R. Woolworth, W. Raymond Wood, and Carlyle S. Smith to understand—by example—the importance of the RBS Missouri Basin Project as the beginning of professional contribution to historical archaeology. That contribution continued well beyond the end of the Missouri Basin Project.

The career of G. Hubert Smith was introduced earlier in this chapter. His list of historical archaeology projects conducted with the RBS represents an extremely impressive contribution as was his leadership in historical archaeology within the RBS and beyond. His scholarship includes work at the former La Verendrye National Monument, Lake Sakakawea (Garrison Dam), North Dakota (his 1951 report was edited by W. Raymond Wood and published by the University of Nebraska Press in 1980 [Smith 1980]); the Fort Pierre II trading post and Fort Stevenson in Lake Sakakawea (Garrison Dam), North Dakota (Smith 1960a, 1960b); Like-A-Fishhook Village and Fort Berthold in Lake Sakakawea (Garrison Dam), North Dakota (Smith 1972); and Fort Manuel in Lake Oahe (Oahe Dam), South Dakota (published with John Ludwickson in 1981 by the South Dakota Archaeological Society [Smith and Ludwickson 1981]; Minnesota Historical Society 2012a). In addition, he worked in the Big Bend Reservoir (Lake Sharpe), South Dakota. Smith was also detailed from the Missouri Basin Project to the National Capitol Parks to investigate Old Stone House in Georgetown, outside of Washington DC, in 1957; to excavate Fort McHenry National Monument and Historic Shrine in 1958; and to work at the Colonial National Historical Park in 1960 (Smith 1957, 1968; Thiessen 1999).

By his own admission, Alan Woolworth's interests in historical archaeology originated from his mentor G. Hubert Smith (Woolworth 2006). Woolworth was born in 1924 in South Dakota and grew up in a home where history was greatly valued. After service in World War II, he entered South Dakota State College in 1946. Upon deciding to study anthropology, he transferred to the University of Nebraska where he studied under John Champe. While a student, the RBS Missouri Basin Project office moved into the same building as the department, and Woolworth was in close company with the archaeologists working out of that office. Woolworth left the Plains in 1950 to pursue his Master's degree at the University of Minnesota but returned and became staff archaeologist with the State Historical Society of North Dakota from 1952 to 1957 (Minnesota Historical Society 2012b; Woolworth 2006).

From his arrival at Nebraska in the late 1940s until he left the Missouri Basin in 1957, Woolworth was heavily involved in RBS field projects. Because of his penchant for history and historical archaeology, he was involved in significant historical archaeology projects at Like-A-Fishhook Village and Kipp's Post in Lake Sakakawea (Woolworth 1956; Woolworth and Wood 1960). After leaving the Missouri Basin, he conducted decades of historical archaeological research at Grand Portage National Monument in Minnesota.

W. Raymond Wood was a native of Nebraska and arrived at the University of Nebraska in 1949 with the intention of studying anthropology (though not archaeology). Upon receiving his Bachelor's degree in 1954, Wood was hired by Alan Woolworth to work at Like-A-Fishhook Village and Kipp's Post. Wood pursued his Master's degree at the University of Nebraska starting in 1955 and graduated in 1956. After continuing to work on RBS Projects with Woolworth and others, he went on to pursue his Ph.D. at the University of Oregon-Eugene in 1958. His dissertation was on the Huff Site in Lake Oahe, and he received the Ph.D. in 1961. After a short stop at the University of Arkansas Museum, he joined the University of Missouri-Columbia to direct the River Basin Archaeology program within the department of anthropology (Wood 2006).

Wood and Woolworth became life-long friends and colleagues after their work at Like-A-Fishhook Village and collaborated on the report for Kipp's Post (Woolworth and Wood 1960). Wood stayed at the University of Missouri until his retirement and has continued to be a significant force in the archaeological and ethnohistorical study of the historic period within the Missouri River Basin—a passion that began in 1954 in North Dakota. Examples of his ethnohistorical and historical archaeological scholarship rooted in the RBS abound (e.g., Wood 1993, 1996, 2006, 2008, 2011; Wood et al. 2011).

A native of New York, Carlyle Smith graduated from Columbia University in 1938, where he was exposed to historic sites archaeology. Upon graduation, he promptly set off with William Duncan Strong to do field work in North Dakota. He began graduate studies at Columbia University the following fall, and returned to work on a WPA project with A. T. Hill in Nebraska the following summer. He continued graduate studies after service in World War II and was awarded the Ph.D. from Columbia University in 1949. In 1947, he joined the faculty of the University of Kansas where he remained until his retirement in 1980 (Grange 1990).

Smith's work in South Dakota from 1950 to 1959 under the auspices of the University of Kansas provides a good example of the contributions of a partner institution working under an NPS cooperative agreement. Although the sites he excavated between 1950 and 1952, such as Talking Crow in Lake Francis Case (Fort Randall Dam) were late prehistoric in nature, Smith developed an expertise in historical archaeological analysis that allowed him to contribute to historic components of sites he excavated, as well as sites excavated by others, such as Like-A-Fishhook Village. This and other historical archaeology that Smith conducted while at the University of Kansas, such as that at a house that Confederate guerilla Quantrill burned during his 1863 sacking of Lawrence, led to the presentation of the J. C. Harrington Medal for lifetime scholarly achievement by the Society for Historical Archaeology in 1989 (Grange 1990).

An important footnote to Smith's work in South Dakota was his excavation at the Deerfly Site in Lake Francis Case in 1953. Based on earlier survey data, this site was thought to be a late prehistoric earthlodge village. Smith immediately found it to be a nineteenth-century Dakota settlement. Although he moved his field party to another site because of this mis-identification, it was not until he had conducted and brought to conclusion credible excavations at the site. Although he never wrote a report on this site, the collections and associated records were properly organized and curated at the University of Kansas and were eventually analyzed and published by the author (Lees 1985).

The careers of G. Hubert Smith, Alan Woolworth, W. Raymond Wood, and Carlyle S. Smith are examples of the heritage of historical archaeology that came from the RBS Missouri Basin Project. The contribution by these individuals and others named earlier came from work on the main stem reservoir projects along the Missouri River. Other contributions to historical archaeology were made as well on reservoir projects on tributary basins; many of these projects were constructed in the latter half of the Missouri Basin Project and, like some of the projects on the main stem, were handled by partner institutions working under cooperative agreements.

Prelude to the Present

The end of the RBS work in 1969 corresponds roughly with the onset of the new archaeology that changed the landscape of North American archaeology forever. Due in large part to the attention given to historic sites by anthropologically trained archaeologists during Great Depression, New Deal programs and during the RBS, the emerging discipline of historical archaeology was poised to take part in the discussion about what archaeology should become.

Led by key contributions such as that by Fontana (1965), Deetz and Dethlefsen (1967), Cleland and Fitting (1968), Schulyer (1970), South (1971), and Leone (1973), as well as by the debate surrounding Dollar's position in a 1968 paper (Dollar 1968, see also Cleland and Fitting 1968; Fontana 1968), historical archaeology emerged as a dynamic subdiscipline within anthropology. This emergence was also signaled by the formation of the Society for Historical Archaeology (SHA) in Dallas in 1967 (Jelks 1993). It is noteworthy that the 22 SHA Fellows named on the creation of the SHA (defined as "those with positions of power in historical archaeology") included Carl H. Chapman, Merrill J. Mattes (also elected to the first SHA Board of Directors), Carlyle S. Smith, and G. Hubert Smith. Their inclusion is testimony of the importance of the Smithsonian Institution's RBS Missouri Basin Project in the discipline (Society for Historical Archaeology 1967a, 1967b).

In the original 2001 conference paper, in which this chapter has its origins (Lees 2001), I argued that the direct-historical approach in the Great Plains started a trajectory toward an anthropological historical archaeology that was somehow interrupted by the archaeological emergency of the post-World War II era. While the epistemological relationship of the direct-historical approach to historical archaeology may have been overlooked, I here depart from my earlier positions. In fact, the historical archaeology conducted as part of the RBS was to some degree informed by the work conducted through the direct-historical approach in the 1930s and, more importantly, by the anthropologically trained archaeologists who sought answers through the use of archaeology on historic sites during the 1930s, regardless of theoretical approach.

In fact, the archaeological emergency caused by the *Flood Control Act of 1944* did not interrupt the trajectory of historical archaeology. Rather, it allowed the trajectory to continue because of a dramatic expansion in the practice of historical archaeology across North America and because of the equally dramatic growth in the number of archaeologists who were interested in and trained in excavation of historic sites. This happened because historic sites archaeology was a part of the RBS through the insistence and leadership of historians and archaeologists alike (and especially Merrill J. Mattes and G. Hubert Smith). Had this not been the case, the discussion regarding the place of historical archaeology that happened in the 1960s would have had far fewer voices, and its outcome may have been drastically different or delayed. For historical archaeology, it is therefore the growth of human intellectual capital that may have been the greatest result of the RBS.

While the approaches to the selection of historic sites for excavation can be seen to have biased and limited the data that survives from massive areas of the Missouri Basin, the fact that historic sites were considered at all, and that they were considered somewhat systematically, is rather miraculous against the backdrop of the times. For the most part, the sites that were investigated were reported, and their collections and field records remain a viable research resource today. The sites that were investigated were significant, some profoundly so, and represent an important comparative data set that has become more and more useful as the discipline of historical archaeology has matured.

What did not end in 1969 when the RBS came to a close was the federal archaeology program (Jennings 1985). The RBS ceased because the federal archaeology program had grown to such an extent that the RBS was no longer needed as individual agencies were taking over archaeological responsibilities that they had come to accept as a result of the archaeological response to the *Flood Control Act of 1944*. New legislation passed within the decade of the end of the RBS program would continue the transformation of the federal archaeology program into the program that we know today simply as cultural resource management (see McManamon, this volume).

In 1969, the Smithsonian Institution's RBS Missouri Basin Project office became the NPS Midwest Archeological Center (MWAC). With a redefined mission focused on providing archaeological services to National Park units, MWAC has continued to provide leadership in research in historical archaeology, some of it game-changing (e.g., Scott et al. 1989). As was the case with RBS historical archaeologists, MWAC archaeologists have also been leaders within the discipline of historical archaeology as evidenced, in part, by leadership within the SHA.

In sum, it is unlikely that the discipline of historical archaeology would be where it is today without the transformative effects of the RBS. Perhaps due to the significance of the level of effort expended in this area, it is unlikely that historical archaeology would be the same without the contribution and leadership of the historical archaeologists working within the Missouri Basin between the close of World War II and 1969.

CHAPTER 12

Paleontological Significance of the Missouri River Basin Surveys

John W. Hoganson

Introduction

The scientific importance of paleontological resources along the Missouri River and elsewhere in the American western frontier was envisioned by Thomas Jefferson, third president of the United States and pioneering paleontologist. This vision is evident in Jefferson's 1803 letter of instruction to Meriwether Lewis in preparation for what is now known as the 1804–1806 Lewis and Clark Expedition. Jefferson directed Lewis to record: "the animals of the country generally, and especially those not known in the U.S.; the remains or accounts of any which may be deemed rare or extinct" (Jackson 1962:63).

Lewis and Clark did observe and collect fossils on their journey up the Missouri River. They were the first to collect fossils in the Missouri River Basin for scientific purposes, although Native Americans had been collecting fossils for hundreds perhaps thousands of years for ornamental and spiritual purposes (Hoganson and Murphy 2003). In his September 10, 1804, journal entry, William Clark recorded the discovery of a forty-five-foot-long skeleton of what he thought was a fossilized fish. This fossil has since been determined to be the remains of a plesiosaur, a marine reptile, found in the Cretaceous Pierre Formation in present day Gregory County, South Dakota (Moulton 1987). In North Dakota, the Expedition collected petrified and carbonized wood from the Paleocene Sentinel Butte Formation near their 1804–1805 Fort Mandan winter camp in McLean County (Hoganson and Murphy 2003). On July 25, 1806, Clark noted large fossil rib fragments along the Yellowstone River a few

Dam Projects and the Growth of American Archaeology: The River Basin Surveys and the Interagency Archeological Salvage Program, edited by Kimball M. Banks and Jon Czaplicki, 167–179.

miles from Pompey's Pillar in eastern Montana that he also thought to be the remains of a large fish. These fossils were likely dinosaur remains from the Cretaceous Hell Creek Formation (Moulton 1993).

The Missouri River and many of its tributaries became major corridors for scientific expeditions to explore the American West following the Lewis and Clark Expedition, and significant fossils were collected during many of these. For example, the first named marine reptile from the American West, *Mosasaurus missouriensis*, was collected from the Late Cretaceous Pierre Formation in the Big Bend area of Gregory County in central South Dakota during the 1832–1834 expedition led by the German Prince, Maximilian of Weld (Patrick et al. 2007).

The Lewis and Clark Expedition began a tradition of federal government sponsored scientific expeditions to the west particularly those by the United States Geological Survey and eventually including the River Basin Surveys (RBS). Ferdinand V. Hayden, later geologist in charge of the United States Geological and Geographical Survey of the Territories, participated in or led geological expeditions to the western frontier during the mid-1800s at which time large numbers of significant fossils were discovered. Legend has it that Hayden was given the Indian moniker "the man who picks up stones while running" when he was fossil collecting in dangerous areas of Indian County during the time of unrest and upheaval (Lehman 1973). These expeditions resulted in monographs describing and illustrating hundreds of fossils from Cretaceous and Tertiary formations (Leidy 1869; Meek 1876). The famous competitive dinosaur bone hunters, Edward Drinker Cope and Othniel Charles Marsh, and others actively collected fossils in many areas of the western frontier in the 1800s including the Missouri River Basin. Cope even extended his research into remote areas of what is now North Dakota, where he collected dinosaur fossils from the Cretaceous Hell Creek Formation near Fort Yates, Sioux County in 1833 (Fiorillo and Daeschler 1990).

The RBS program was inaugurated in 1945 to undertake surveys to identify and salvage archaeological and paleontological resources in areas to be flooded by the construction of dams planned by the Bureau of Reclamation and U.S. Army Corps of Engineers (Thiessen 1999). Architects of the RBS were well aware, based on earlier paleontological investigations, that many areas in the Missouri River Basin were paleontologically sensitive, such as those mentioned above, and that important fossil sites would be found along the soon to be impacted water ways. It should be no surprise that concern for loss of scientifically important paleontological resources was included in the initial development of the RBS and that surveys to identify these resources were part of the program from its inception. It is, however, surprising that paleontological investigations soon became secondary to the mission of the RBS and were terminated early in the program.

Missouri River Basin Paleontology Surveys, 1947–1953

Theodore E. White, an experienced vertebrate paleontologist from Harvard University's Museum of Comparative Zoology was hired in April, 1947, to conduct paleontological surveys for the Smithsonian Institution's RBS, particularly in the Missouri Basin Project area. This was an admirable task because the Missouri River Project area included all or parts of Montana, North Dakota, South Dakota, Colorado, Wyoming, Kansas, and Nebraska (Figure 12.1). In 1947 and 1948, White visited 78 proposed reservoir areas in the Missouri River Project area to determine if paleontological resources would be impacted by reservoir development (Table 12.1) (White 1948). He suggested that 19 of these reservoir areas required additional paleontological inventories and recommended the amount of time that should be spent doing the investigations. White (1952a, 1953) later added additional reservoirs to this list (Table 12.1). Lack of fossil-bearing rocks, lack of or poor rock exposures, and better exposure of fossil-bearing rock units outside the reservoir area were criteria that White used to determine if reservoir development would cause a serious loss to paleontology. White noted that at several of the potential reservoir sites excavation for dam construction could potentially expose paleontological resources and that these construction areas should be periodically inspected (Table 12.1).

Results of White's annual paleontological fieldwork from 1947 through 1953 summarized here was reported in the "River Basin Surveys Papers" in *Bureau of American Ethnology Annual Reports, Bureau of Ethnology* (in the Smithsonian Institution) *Interagency Archeological Salvage Program*, unpublished interagency reports, and in a few professional publications. Reconnaissance of potential reservoirs for paleontological resources was the focus of White's activities during 1947 and 1948 (White 1948; Figure 12.1, Table 12.1). He was assisted by John C. Donohoe, a student from Montana State University in 1947 and by Donohue and Ernest L. Lundelius, a student from the University of Texas in 1948. White also excavated dinosaur fossils from the Jurassic Morrison Formation in the Middle Fork Reservoir and mammalian and turtle fossils from the Eocene Wind River Formation in the Boysen Reservoir areas in Wyoming during that time. In 1949, White again collected fossils at the Boysen Reservoir site. He also spent considerable time at the Canyon Ferry Reservoir in Montana collecting a diverse assemblage of mammalian fossils from Oligocene and Miocene deposits

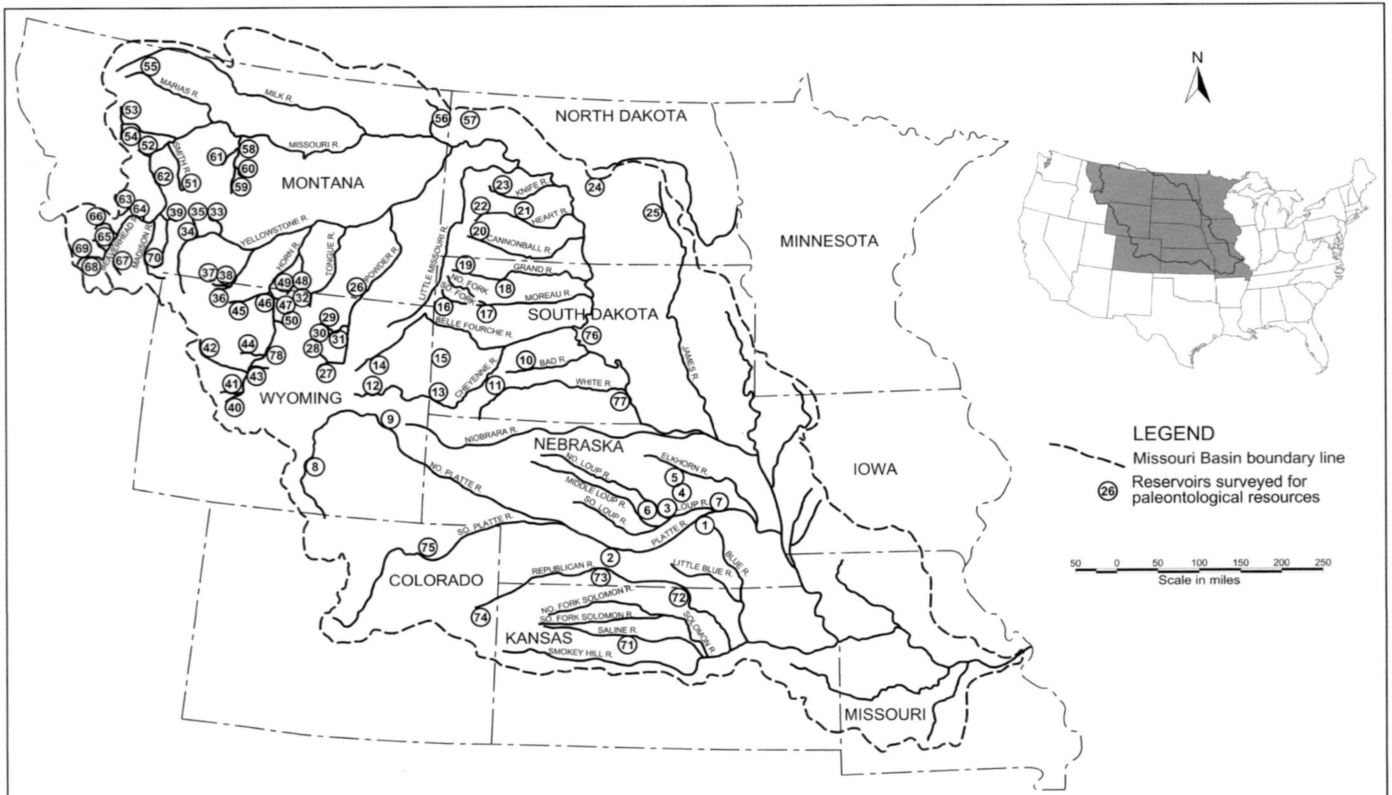

Figure 12.1 Location of Missouri River Basin reservoirs where paleontological surveys, led by Theodore White, were done from 1943 through 1953. Modified from White (1948) with additions from White (1952a, 1953).

Table 12.1 List of Missouri River Basin reservoirs where paleontological surveys were done from 1947 through 1953

Missouri River Basin Paleo Salvage Plan—White 1948				
Basin/ Reservoir	**Location**	**Age**	**Formation/ Group**	**Man Days Needed for Salvage**
Lower Platte Basin				
Bellwood	Polk Co., NB	Pleistocene	Peorian loess	None
Plum Creek	Gosper, Dawson, and Frontier Cos., NB	Pleistocene	Peorian loess	None
Cushing	Howard Co., NB	Pleistocene	Peorian loess	None
Loretto	Boone Co., NB	Pleistocene	Peorian loess	None
Clearwater	Antelope and Hold Cos., NB	Pleistocene	Peorian loess	None
Davis Creek	Valley, Sherman, Greely, and Howard Cos., NB	Pliocene	Oglalla	50–60*
Shell Creek	Platte and Colfax Cos., NB	Pleistocene	Peorian loess	None
North Platte Basin				
Kortes	Carbon Co., WY	–	Granite	None
Glendo	Platte Co., WY	Paleozoic-Pliocene	Several	None
Bad River Basin				
Philip	Haakon Co., SD	Cretaceous	Pierre	40–60
White River Basin				
Rocky Ford	Washington Co., SD	Oligocene Pliocene	White River Undiff.	250–300
		Pleistocene	Undiff.	
Cheyenne River Basin				
Edgemont	Weston Co., WY	Cretaceous	Pierre	50–60*
Angostura	Fall River Co., SD	Jurassic	Unkpapa Morrison	None
Keyhole	Crook Co., WY	Cretaceous	Several	None
Deefield	Pennington Co., SD	–	Metamorphic	None
Moreau River Basin				
Bixby	Perkins Co., SD	Cretaceous	Lance	None*
Green Grass	Dewey Co., SD	Cretaceous	Pierre	60–80*
Grand River Basin				
Blue Horse	Corson Co., SD	Paleocene	Fort Union	60–80*
Shadehill	Perkins Co., SD	Paleocene	Fort Union	None*
Cannonball River Basin				
Cannonball	Grant Co., ND	Paleocene	Fort Union	None*
Heart River Basin				
Dickinson	Stark Co., ND	Paleocene	Fort Union	None*
Heart Butte	Grant Co., ND	Paleocene	Fort Union	None*

(*Continued*)

Table 12.1 (*Continued*)

Basin/ Reservoir	Location	Age	Formation/ Group	Man Days Needed for Salvage
Knife River Basin				
Broncho	Mercer Co., ND	Paleocene	Fort Union	None*
Sheyenne and James River Basins				
Sheyenne	Benson, Eddy, Pierce, and Wells Cos., ND	Pleistocene	Undiff.	None*
Jamestown	Foster and Stutsman Cos., ND	Cretaceous	Graneros	None*
Powder and Tongue River Basins				
Moorhead	Campbell Co., WY Powder R. Co., MT	Paleocene	Fort Union	100–120
Middle Fork	Johnson Co., WY	Cretaceous Jurassic Triassic	Cloverly Sundance Morrison Chugwater	80–120
Smith	Johnson Co., WY	–	Granite	None
Triangle Park	Johnson Co., WY	–	Granite	None
Willow Park	Johnson Co., WY	–	Granite	None
Bull Creek	Johnson Co., WY	Paleocene	Fort Union	None
South Fork	Sheridan Co., WY	–	Granite	None
Upper Yellowstone Basin				
Sweetgrass	Sweetgrass Co., MT	Paleocene	Livingston	None*
Mission	Park Co., MT	Paleocene	Livingston	50–60
Antelope	Park Co., MT	Paleocene	Livingston	None
Sunlight	Park Co., MT	Paleozoic	Undiff.	50–60
Hunter Mountain	Park Co., MT	–	Granite	None
Thief Creek	Park Co., MT	–	Granite	None
Bridger	Gallatin Co., MT	Paleocene Mesozoic Paleozoic	Livingstone Undiff. Undiff.	None
Big Horn Basin				
Onion Flats	Fremont Co., WY	Cretaceous	Mancos	None*
Soral Creek	Fremont Co., WY	Triassic	Chugwater	None
Du Noir	Fremont Co., WY	Eocene	Wasatch	None*
Bad Water	Fremont and Natrona Cos., WY	Eocene	Wind River	80–100*
Anchor	Hot Springs Co., WY	Jurassic Triassic	Chugwater Sundance	60–80
Oregon Basin	Park Co., WY	Cretaceous	Cody Mesa Verde	None*
Kane	Big Horn Co., WY	Cretaceous Jurassic Triassic	Colorado Sundance Morrison Chugwater	None

(*Continued*)

Table 12.1 (*Continued*)

Basin/ Reservoir	Location	Age	Formation/ Group	Man Days Needed for Salvage
Red Gulch	Big Horn Co., WY	Jurassic	Sundance Morrison	None*
		Triassic	Chugwater	
Little Horn	Big Horn Co., WY	Triassic Pennslyvanian	Chugwater Tensleep	None
Yellowtail	Carbon and Big Horn Cos., MT; Big Horn Co., WY	Triassic Pennslyvanian Mississippian	Chugwater Tensleep Madison	None
Lake Solitude	Big Horn Co., WY	–	Granite	None
Upper Missouri Basin (NE Section)				
Newland	Meagher Co., MT	Precambrian	–	None
Wells	Lewis and Clark Co., MT	–	Granite	None
Wilson	Lewis and Clark Co., MT	Cretaceaous Jurassic	Undiff.	None
Nilan	Lewis and Clark Co., Mt	Cretaceous	Undiff.	None*
Tiber	Toole and Liberty Cos., MT	Cretaceous	Colorado	80–100*
Medicine Lake	Roosevelt and Sheridan Cos., MT Williams Co., ND	Paleocene	Ft. Union	60–80
Crosby	Divide Co., ND	Pleistocene	Undiff.	None
Hobson	Judith Basin Co., MT	Jurassic Permian Pennsylvanian	Sundance Phosphoria Quadrant	50–60
Ross	Judith Basin Co., MT	Cretaceous	Colorado	60–80
Snowy	Fergus Co., MT	Cretaceous Jurassic	Undiff. Sundance Undiff.	60–80
Stanford	Stanford Co., MT	Cretaceous	Colorado	None*
Upper Missouri Basin (SW Section)				
Canyon Ferry	Broadwater and Lewis and Clark Cos., MT	Pleistocene Pliocene Miocene Oligocene	Undiff.	160–200
Terry	Deerlodge Co., MT	–	Granite	None
Whitetail	Deerlodge Co., MT	–	Granite	None
Kelley	Beaverhead Co., MT	Carboniferous	Undiff.	None
Apex	Beaverhead Co., MT	–	Granite	None
Landon	Beaverhead Co., MT	Tertiary	Undiff.	None
Clark Canyon	Beaverhead Co., MT	Paleozoic	Undiff.	None
Brenner	Beaverhead Co., MT	Oligocene?	Undiff.	50–60
Taylor Fork	Madison Co., MT	Paleozoic	Undiff.	None

(*Continued*)

Table 12.1 (*Continued*)

White 1953				
Basin/ Reservoir	**Location**	**Age**	**Formation/ Group**	**Man Days Needed for Salvage**
Smoky Hill River Sub-basin				
Cedar Bluff	Trego Co., KS	Cretaceous	Niobrara Carlisle	None
Republican River Sub-basin				
Lovewell	Jewell Co., KS	Cretaceous	Niobrara Carlisle	None*
Trenton	Hitchcock Co., NB	Pleistocene	Undiff.	None
Bonny	Yuma Co., CO	Pliocene	Yuma	None*
South Platte River Sub-basin				
Narrows	Morgan Co., CO	Pliocene	Yuma	None*
Missouri River Sub-basin				
Canyon Ferry	Broadwater and Lewis and Clark, Cos., MT	Miocene Oligocene	Undiff.	More Collecting Needed
Oahe	North Central SD South Central ND	Cretaceous	Pierre	More Collecting Needed*
Fort Randall	South Central SD	Cretaceous	Pierre	More Collecting Needed
Bighorn River Sub-basin				
Anchor	Hot Springs Co., WY	Triassic Permian	Undiff.	None

* White suggested that fossils may be encountered during dam excavations and that construction operations should be inspected periodically.

(Figure 12.2). Later that year he searched for fossils, with little success, in the Cretaceous Carlile Formation at the Cedar Bluff Reservoir in Kansas.

Delay in allocation of funding in 1950 limited White's fieldwork. He was assisted in the field by Prentiss Shepherd, Jr., a student from Harvard University, and William C. Harrup, Jr., a student from Columbia University. White's party explored for fossils and were rewarded with few finds in Permian and Triassic formations in the Anchor Reservoir area, Wyoming, and Jurassic deposits in the Angostura Reservoir area, South Dakota. He also briefly revisited the Boysen and Canyon reservoir areas. That summer, his party also explored Lewis and Clark and Broadwater counties in Montana and established the presence of Oligocene and Miocene strata there. The Garrison Reservoir area in North Dakota was visited at which time the Paleocene Cannonball Formation was determined to be more

Figure 12.2 Theodore White (center) at camp during his Canyon Ferry Reservoir, Montana, paleontology survey.

extensively exposed than previously thought (Metcalf and White 1953). White's group collected Cretaceous Pierre Formation fossils from sites in the Fort Randall Reservoir area, South Dakota. In September, White, accompanied by Morris Skinner from the Frick Laboratory, American Museum of Natural History, surveyed 10 reservoirs in Nebraska. White finished his 1950 fieldwork in the Missouri River Basin prospecting for Pliocene fossils in the Bonny Reservoir area in Colorado.

White began his 1951 field season prospecting for Cretaceous fossils in Wyoming's Keyhole Reservoir area. A skeleton of a plesiosaur was discovered in the Graneros Formation, which was the first record of vertebrate fossils from that formation. White returned to the Canyon Ferry Reservoir, Montana, to continue studying Oligocene and Miocene fossils during which several new genera were added to the list of mammals from the Oligocene and Miocene of Montana. White went on to conduct paleontological investigations in 12 reservoir areas including Cretaceous sites in the Tiber Reservoir area, Montana; Paleocene Fort Union Group exposures in the Garrison Reservoir area, North Dakota; and Cretaceous Pierre Formation strata in the Oahe Reservoir area, South Dakota. Of particular interest was White's discovery of a complete skeleton of the mosasaur, *Clidastes pumulis*, from the Pierre Formation near the dam site for the Oahe Reservoir in South Dakota (Cooper 1955) and the skull and lower jaws of a large, black bear-size mammal from the Sentinel Butte Formation near Williston, McKenzie County, North Dakota. Gazin (1956) determined the later to be from the rare pantodont, *Titanoides primaevus* (Hartman and Kihm 1991). White completed the field season in the Fort Randall Reservoir area exploring for Cretaceous fossils.

White completed preliminary reconnaissance surveys of the Tuttle Creek and Lovewell reservoir areas in Kansas, the Ashton and Trenton reservoir areas in Nebraska, and the Gavins Point Reservoir area in South Dakota and Nebraska in May, 1952. For most of June, White investigated Cretaceous strata exposed in the Keyhole Reservoir area in Wyoming. From there, he returned to the Canyon Ferry Reservoir area to resume collection of Oligocene and Miocene fossils from important and productive fossil sites. In August, White was joined by George Metcalf to explore Paleocene Fort Union Formation outcrops in the Garrison Reservoir area near the soon-to-be inundated town of Elbowoods, North Dakota. A nearly complete skeleton of a crocodile-like *Champsosaurus* was collected. White completed his 1952 field season in the Oahe Dam area of South Dakota prospecting for fossils in the Cretaceous Pierre Formation, finding the remains of many marine reptiles although none were well-enough preserved to collect. At the request of the National Park Service, White briefly visited Theodore Roosevelt National Park in North Dakota to examine fossils that had been found there.

Again at the request of the National Park Service, White spent the early part of June, 1953, in South Dakota undertaking a paleontological inventory of Oligocene White River strata in peripheral areas of Badlands National Monument to help determine boundaries of the Monument (White 1953). He then returned to the Canyon Ferry Reservoir area to prospect for Oligocene and Miocene mammalian fossils, at times by boat because of the initial flooding of the reservoir. White's party also traveled to Fort Peck Reservoir in Montana to examine a plesiosaur skeleton from the Cretaceous Bear Paw Shale that had been collected by the U.S. Fish and Wildlife Service.

White's paleontological service with the RBS was terminated at the end of June, 1953, thereby ending paleontological surveys for the program. It is generally assumed that fossil surveys were discontinued because of lack of funding (R. Lee Lyman, personal communication 2012; Thiessen 1999). After leaving the Program, the National Park Service hired White as the first paleontologist at Dinosaur National Monument, and he retained that position until his retirement in 1973. The National Park Service "loaned" White to the Missouri Basin Project for three weeks in 1954, six weeks in 1957, and four weeks in 1961 to identify zooarchaeological specimens.

Significance of the Missouri River Basin Paleontology Surveys

The number of paleontological resource surveys that Theodore White undertook during his short tenure as RBS paleontologist is remarkable. He and his crew visited most of the 78 Missouri River Basin reservoir areas he first examined in 1947 (Figure 12.1, Table 12.1). The baseline data collected during those surveys allowed White to formulate a research plan for more indepth inventories in subsequent years of reservoir areas that potentially contained significant paleontological remains destined to be flooded. It should be kept in mind that during his time as the RBS paleontologist, White also did field studies outside the Missouri River Basin area. Paleontological surveys were undertaken in at least 121 reservoirs within the greater RBS project area, and it is assumed that White was involved with most or perhaps all of these (Roberts 1953). These were the first, and mostly the last, paleontological surveys that would be done in many of these reservoir areas. Unfortunately, in most cases the results of White's paleontological activities were not published and, if at all, are only summarized in a few internal reports by White (1947a, 1947b; 1948, 1953) and Metcalf and White 1953 or in the *Bureau of American Ethnology Annual Reports* from 1947–1953.

White published two professional papers describing the results of some of his Missouri River Basin paleontological studies. One concerned two reptilian taxa, one of which was a new species (*Glyptosaurus donohoei*), and 23 mammalian taxa, including the new species (*Cynodontomys lundeliusi*), collected from the Eocene Wind River Formation at the Boysen Reservoir area near Shoshoni, Wyoming (White 1952a) (Figure 12.1). Of particular importance is that this fossil mammalian fauna established the presence of the Lost Cabin faunal zone of the Wind River Formation at Boysen Reservoir. The other was a comprehensive publication in which he summarized his paleontological investigations in the Canyon Ferry Reservoir area, near Townsend, Montana (White 1954) (Figure 12.1). In that article, he described 22 species of mammals from Oligocene strata, four of which were new taxa, the nyctitheriid *Kentrogomphios strophensis* and the cricetids *Eumys cricetodontoides, Eumys latidens,* and *Eumys spokanensis*, and mentioned the occurrence of a few mammalian fossils from the Miocene. Significantly, White recognized paleobiogeographic differences between the Canyon Ferry Reservoir intermountain Oligocene mammalian fauna and mammalian faunas of the same age from localities in the Great Plains and suggested that the differences reflect climatic zoning of the faunas.

White discovered other important fossils during the Missouri River Basin surveys that were mentioned in the Bureau of American Ethnology Annual Reports but were not formally published. These fossils included the skeleton of a plesiosaur from the Cretaceous Graneros Formation in the Keyhole Reservoir area, Wyoming, in 1951; the skeleton of *Clidastes pumulis*, a mosasaur, from the Cretaceous Pierre Formation near the Oahe Dam site, South Dakota, in 1951; the skull and lower jaws of the pantodont mammal *Titanoides primaevus* from the Paleocene Sentinel Butte Formation near Williston, in the Garrison Reservoir area,

North Dakota, in 1951; and the skeleton of the crocodile-like reptile, *Champsosaurus*, from the Paleocene Sentinel Butte Formation collected near the now submerged town of Elbowoods, in the Garrison Reservoir area, North Dakota, in 1952.

White also made a contribution to archaeology. In addition to his paleontological duties, White also spent considerable time identifying, studying, and writing about faunal remains from archaeological sites during his time with the RBS. He was one of the first scientists to expound on the importance of faunal remains in archaeological sites and was one of the early founders of the field of zooarchaeology (R. Lee Lyman, personal communication 2012; Thiessen 1999; Wood, this volume).

Discussion and Reflections

Premature termination of paleontological surveys for the Missouri River Basin project is unfortunate. Many paleontologically sensitive areas were identified during the surveys, but only the Canyon Ferry Reservoir area received reasonably comprehensive attention. Other reservoir areas of great paleontological potential, such as Garrison and Oahe, received only cursory consideration. Deletion of paleontological investigations from the program is particularly puzzling because conducting these kinds of surveys was mandated in both the 1961 and 1965 Memoranda of Understanding between the National Park Service and Smithsonian Institution reaffirming continuation of the RBS. The fate of the paleontology program is foretold in a 1952 Smithsonian Institution archives report probably written by Frank H. H. Roberts, Jr., Director of the RBS at that time:

> Paleontological remains involved are not as significant as those of an archeological nature because bone deposits are not restricted to the banks of streams and in many cases similar material may be found at locations which will not be flooded. Also, such deposits as are exposed in reservoir basins are not as extensive nor as numerous as the camp and village sites left by the Indians (Smithsonian Institution Archives 1952).

Funding for paleontology was possibly redirected to the Missouri River Chronology Program initiated in 1958 to examine and detail the chronological sequences of prehistoric cultures in the Missouri River Basin through radiocarbon dating, dendrochronology, palynology, and other chronological indicators (Caldwell 1960).

The number of scientifically important paleontological sites and fossil specimens that were lost to dam construction and inundation of reservoirs will never be known and cannot even be adequately estimated.

It is certain, that many significant fossils now lost to science could have been retrieved before inundation if the Missouri River Basin paleontological surveys had continued. Shoreline and bank erosion, however, by impounded water have also exposed countless new fossil-bearing rocks, fossil sites, and specimens. Federal and many state resource management agencies recognize this and now either employ staff paleontologists or collaborate with other paleontologists to assess paleontological resources along waterways and reservoirs under their stewardship.

Acknowledgments

Documentation about the paleontological surveys done during the RBS is difficult to obtain because much of the information is only found in internal agency reports. I therefore thank the following people for providing access to many of these documents: Ann Jenks, Jim Davis, and Rachel White of the State Historical Society of North Dakota Archives; Tad Bennicoff of the Smithsonian Institution Archives; Rachel Benton, Megan Cherry, and Sylvia Cox of the National Park Service; and Ray Wood, Tom Thiessen, and R. Lee Lyman. I would particularly like to thank R. Lee Lyman for sharing with me his recently submitted Theodore White biography manuscript and Tom Thiessen and Ray Wood for providing wise counsel about the Missouri River Basin surveys. I thank Kimball Banks for providing pictures of Theodore White and his and Jon Czaplicki's insightful reviews of this manuscript.

CHAPTER 13

Women in Reservoir Salvage Archaeology

Ruthann Knudson

American women experienced a lot of changes from the 1940s through the 1960s, and this is reflected in their participation in the River Basin Surveys (RBS) and the Interagency Archeological Salvage Program (IASP). Women were consulting experts, field crew members, and occasionally even crew chiefs, laboratory managers and technicians, editors, office staff, and unsalaried working wives in the field and laboratory, as well as at their typewriters. Using a conservative ratio of 1:5 for field to laboratory work hours on a mid-1900s archaeological project, I estimate that women completed at least 75 percent of the RBS, IASP, and related archaeological work across the United States, most of which was in the laboratory: washing, labeling, describing, and reporting on the artifacts, site layout, and archaeological patterns. Despite this, there is still a public (and even scholarly?) perception that only those bronzed, lean, and usually shirtless male field surveyors and excavators were RBS archaeologists.

A number of women who were already established archaeologists were called in by RBS archaeologists as subject experts, but were rarely included as report coauthors. These experts included (Anonymous ca. 1946–1970a):

- Dorothy Cross Jensen (1906–1972), New Jersey archaeology;
- Frederica de Laguna (1906–2004), Athapaskan ethnologist, author of paper on salvage ethnology (de Laguna 1961);
- Sister Inez Marie Hilger (1891–1977), ethnologist working with Arapaho, Anishinabe (Chippewa), and Amskapi Pikuni (Blackfeet);
- Isabel Kelly (1906–1983), Southern Paiute ethnologist and linguist;
- Kate Peck Kent (1914–1987), textiles;
- Florence Lister (1920–), ceramics;
- Lucile Eleanor St. Hoyme (1924–2001), physical anthropology;

Dam Projects and the Growth of American Archaeology: The River Basin Surveys and the Interagency Archeological Salvage Program, edited by Kimball M. Banks and Jon Czaplicki, 180–201.

- Anna O. Sheppard (1903–1973), ceramic chronology;
- Ruth DeEtte "Dee" Simpson (1919–2000), California and Southwest archaeology;
- Erminie Wheeler Voegelin (1903–1988), ethnography and ethnohistory; and
- H. Marie Wormington (1914–1994), Paleoindian archaeologist.

The roles of American women changed significantly during World War II, as they had to take over many jobs traditionally assigned to men to keep farms, industry, the government, and even homes running. Most women born in the 1920s, 1930s, and early 1940s understood Depression-era needs for economy, hard work, and an education while still being trained in sewing, cooking, accounting, and other "household" jobs. New families in the post-war baby boom had to cope with changing roles, young children, income-producing work, and their own intellectual desires, often at the same time. The 1950s brought television and other new technologies, and a new emphasis on consumerism. The "youth era" of the 1960s brought birth control, new legal rights and privileges, and new social awareness.

While women had been able to vote since 1920, they did not have full civil rights in government employment until many decades later. Under the *Equal Pay Act of 1963* (Public Law [P.L.] 88-38, 77 Stat. 56), employers subject to the *Fair Labor Standards Act* could not pay different wages based on sex. This act did not make much difference if you were not hiring women for archaeological field positions. The *Civil Rights Act of 1964* (P.L. 88-352, 78 Stat. 241) outlawed major forms of discrimination against blacks and women, but its enforcement was initially weak. Executive Order (EO) 11246, signed in 1965, ensured equal employment opportunity for people based on race, color, religion, and national origin, but not sex. The EO 11375, signed in 1967, banned discrimination in the federal workforce and among its contractors based on sex. The Smithsonian Institution's (SI) Washington, DC office sent out a letter in March, 1968 (Kennedy 1968), notifying RBS directors that they needed to report on their installation's compliance with the SI's Equal Employment Opportunity program by April 30th of that year.

The inclusion of women in field, laboratory, office, and writing activities reflected these national patterns. In some places (e.g., the Southeast), these rights were still not available to participants until after the passage of the *Equal Employment Opportunity Act of 1972* (P.L. 92-261, 78 Stat. 253), which amended Section 701 (42 U.S.C. 20000e) of the Civil Rights Act of 1964 to include sex. Still, women's access to work on RBS projects was determined in large part by the personalities of the administrators who approved hiring decisions.

The following discussion of women known to have worked on RBS projects must be considered incomplete. For instance, the list of Missouri

River Basin participants (Anonymous ca. 1946–1970a) includes only those salaried women identified in SI-funded project reports; women who worked for cooperating institutions were not included. Information about activities of the Athens, Eugene, and Berkeley field offices was difficult to acquire. My apologies and appreciation to the women whose contributions are not remembered here.

Missouri Basin Project Office

The Missouri Basin Project (MBP) office, located in Lincoln, Nebraska, administered projects throughout the Missouri River Basin. Information about women on MBP projects was retrieved from a list in the RBS records (Anonymous ca. 1946–1970a) although some of the information in that list is inaccurate or incomplete. However, this list and other sources provide a picture of women involved in MBP activities.

One of the earliest projects was a survey of the Cherry Creek Reservoir in Denver that H. Marie Wormington ([Mrs. George D. "Pete" Volk] 1914–1994; Knudson 2004) undertook in 1948 (Wormington 1948). Wormington (1939, 1944) had completed two editions of *Ancient Man in North America* and was working on a third. The archives also contain correspondence between the RBS and Wormington from 1957 to 1964.

Carol Franke (dates unknown) worked at the Harlan County Reservoir in Nebraska and Kansas in 1946 (Anonymous ca. 1946–1970a; Kivett 1953). Her tasks there are unknown.

Joyce Wike (1918–) and her husband, Preston Holder, were employees of the University of Nebraska State Museum, which was a cooperating partner of the RBS. She and Preston excavated the Allen Site on Medicine Creek in southwestern Nebraska during the summer of 1948. They published their findings (Holder and Wike 1949, 1950), in which they defined the Frontier Complex in the Central Plains. Their work was the basis for Chapter VII (Allen Site) in E. Mott Davis's 1954 dissertation that outlined the culture history of the Central Plains.

Dorothy E. Fraser (dates unknown) was a student of William Duncan Strong at Columbia University when she worked on Works Project Administration field projects in 1939. She finished her Master's degree in 1941 on the Anton Rygh Site in the Oahe Reservoir in South Dakota (Fraser 1941). She was on the Fort Randall and Oahe Reservoir projects in 1947 (Cooper 1949) and 1948 (Wedel 1953). In 1948 she was a special consultant at Oahe with Robert Hall, Warren Wittry, and Gordon F. McKenzie (Hall 2006:646 [Hall appears to have the years mixed up]); she lived in the Tomas House while the men were in tents.

Carlyle Smith noted (2006:533) that he was the first member of the University of Kansas Museum of Anthropology staff to take a wife,

Judy Smith (dates unknown), into the field, which he did in 1949 and then throughout his career (*see* Reyman 1992:74 for a comment on archaeological "working" wives). Judy was with him at the Talking Crow Site in Fort Randall Reservoir at least in 1949, 1950, 1951, and 1952 (Smith 1977; Woolworth 2006:652). Smith noted that in the 1950s, women were not allowed to be in Kansas state-owned vehicles.

Mary Farrell (dates unknown) is illustrated in a picture of the 1951 South Dakota Archaeological Commission field crew (Breternitz 2006:723). University of Kansas student Holly Wagner (dates unknown) analyzed projectile points from Talking Crow (Wagner 1955). Wesley Hurt (2006) took his wife and infant son with the Smiths to work at the Swanson Site in 1950 and the Thomas Riggs excavation in 1952. The Swanson Site is in the Big Bend Reservoir and Thomas Riggs in the Oahe Reservoir, South Dakota. Alice Shroyer (dates unknown) worked on the Oahe Reservoir project in 1950–1951 (Lehmer 1954).

Anthropologist Dolores (Dee) Bellamy Gunnerson (1923–) was the wife of James H. Gunnerson, who noted that "rarely have I been in the field doing archaeology when she was not with me" (Gunnerson 2006:616; Gunnerson and Gunnerson 1952). The two worked in the Lincoln laboratory and provided field training there (Woolworth 2006:652). Ruth (Mrs. Carl) Miller (dates unknown) was on the RBS staff at Fort Randall in 1950 (Miller 1960b), Tiber Reservoir in 1955 (Miller 1956), and Oahe in 1956 (Miller 1964). Dorothy Thurlow Cooper (dates unknown) was Paul Cooper's wife and was employed at the Koehler Site in the Heart Butte Reservoir, North Dakota, in 1948 (Anonymous ca. 1946–1970b). Cooper (1958; Hall 2006:639) says she was a volunteer, but other sources say she was salaried.

Beth Ogden Davis (1918–2006), E. Mott Davis's wife, was in the field with Davis from the late 1940s into the 1950s at the Lime Creek and Red Smoke sites in Medicine Creek Reservoir, Nebraska (Knudson and Wyckoff 2013), and they usually had their young sons Jonathan and Hugh with them. Beth Davis often helped run the camp and kept the laboratory operating. She and Alice Rowe Bell (dates unknown) drew some of the figures in Davis's (1954) dissertation. Davis noted that Beatrice Beutel (dates unknown) and Joyce Lasse (dates unknown) were part of the project's laboratory crew and that several other unnamed women advised him on report style and form and typed the dissertation.

In the late 1950s, Gladys M. Bowers (1910–2005), wife of Alfred W. Bowers, maintained an artifact classification program while caring for two children when Bowers was excavating the Anton Rygh Site, South Dakota (Knudson et al. 1983). Flavia Waters Champe (1902–1992), John Champe's wife, was an excellent cook and went into the field each summer where she supervised food preparation at the camps (Woolworth

2006:651). In 1953, Caroline Kivett (1916–2003), along with her son Ronnie, was at her husband Marvin Kivett's excavation at 39LM26 in South Dakota (Woolworth 2006:655), and probably many other times. Roger T. Grange's wife, Jane Whitner (dates unknown), was at Talking Crow (Smith 1977) and on the RBS staff again in 1953 (Grange 2006:693); she took care of three children, did the shopping, cooked, and helped on survey.

Mildred Mott Wedel (1912–1995; Gradwohl 1995; Mason 1992:97) accompanied her husband Waldo Wedel with their children on the 1951 excavation at the Cheyenne River Site, South Dakota, and probably at other sites as well. Mildred completed her Master's thesis (Mott 1938) on Iowa ethnohistory while being a field excavation supervisor on a site in Hamilton County, Iowa. She married Waldo in 1939 and took their three children into the field on numerous RBS projects. Mildred contributed her own expertise to the projects and often served as the field crew cook. She continued her own research (Wedel 1959, 1961, 1976) on the relationship of Chiwere-speaking Siouan people (Ioway and Oto) to Oneota sites, and with the Wichita Indians and other Caddoan speakers, and was recognized for her methodological rigor (cf. Wedel 1976).

LaVerne Pendleton (personal communication July 16, 2010) was the Lincoln office manager from early 1951 to mid-1953, under the direction of Paul Cooper (Thiessen 1999:26). She was working in the Museum of Anthropology at the University of New Mexico (UNM) when she first applied for an RBS position in 1948 through the Civil Service Commission. She received a BA in archaeology from UNM in 1949 but was not able to acquire her Lincoln position until 1951. Her museum and anthropological training was useful in her RBS position as a museum aid, though she reports that she was effectively the office manager.

Evelyn B. Stewart (dates unknown) was another staffer in the MBP office and was probably also a museum aid. Missouri Basin Project Chief Robert Stephenson (1958) recommended Stewart for an outstanding performance rating for being "responsible for the maintenance and operation of the archeological records and references files" and for her "truly remarkable record for constant control of the files and the hundreds of thousands of individual site records that they contain." Stewart maintained all photographic record files, assigned and directed photographic work, and coordinated archaeologists' needs with photographer workloads.

Hester A. Davis (1930–; personal communication April 7, 2010; White 1999a; White et al. 1994:104) says that her claim to fame is that she was the first woman hired by the MBP office in 1952. She was photographer and, for the last part of the project, cook on Richard P. Wheeler's crew at Jamestown Reservoir, North Dakota (Wheeler 1953, 1963) and her brother E. Mott Davis's 1954 Fort Randall field school excavations (White 1999a:212 says 1955). She was cook and laboratory specialist on Reynold J. Ruppé's 1958 field school at the Helen Davis Site in Iowa (Henning 2006:787), and worked at the Utz Site field school in Missouri in 1960 (Bray 1991; McGimsey et al. 1991). She began teaching museum methods in the University of Arkansas, Department of Anthropology in 1960, was appointed assistant director of the University of Arkansas Museum in 1963, and became the Arkansas State Archeologist in 1967. She helped write *Stewards of the Past* (McGimsey et al. 1970), was involved in the Airlie House Seminars in 1974 (McGimsey and Davis 1977), and taught a University of Arkansas public archaeology course from 1974–1991. She has been president of the Society of Professional Archaeologists, the American Society for Conservation Archaeology, and a number of other organizations. She retired as Arkansas State Archeologist in 1999.

In 1952, Roger T. Grange hired Mary Kiehl (Rusco 1928–2005; Figure 13.1; Hattori 2005) and Shirley Lyon (dates unknown) for his crew excavating the Talking Crow Site with Carlyle Smith (Smith 1977). In 1953, Kiehl was hired (Anonymous ca. 1946–1970b) to work at the Spain Site in South Dakota (Smith and Grange 1958) and 14RY10 just below the Tuttle Creek Reservoir near Manhattan, Kansas (Cumming 1958). She presented a paper on the Glen Elder and White Rock sites at the Nebraska Academy of Sciences and Affiliated Societies that year (Kiehl 1953). Kiehl subsequently married Elmer Rusco and finished her Master's thesis at the University of Nebraska in 1960, which the university immediately published (Rusco 1960). Her thesis was on the White Rock Aspect of south-central Nebraska and north-central Kansas and included material from several RBS-excavated sites. She and her husband subsequently left Nebraska for California where she finished her coursework for a Ph.D. in anthropology at the University of California,

Figure 13.1 Mary Rusco from Gene Hattori.

Davis. She left her doctoral program to work with Catherine Sweeney Fowler (*see below*) on the ethnography of southern Nevada and eastern California and in 1967 became the records manager for the Nevada Archaeological Survey in Carson City. She directed the Survey between 1974 and 1983 and kept up her involvement with ethnography and archaeology until her death.

In 1952, Carling Malouf (1952) had two women on his Garrison Dam Project in North Dakota (Malouf 1963). Both had undergraduate training in archaeology and were over 21 years of age. Malouf considered their inclusion to be "something of an experiment;" there is no information about the success or failure of that experiment.

In 1953, Barbara Trotter (dates unknown) was on the crew at Fort Randall Reservoir area (Smith and Grange 1958). Among other things, the crew excavated the Spain Site.

In 1955, Dena Ferran (Dincauze 1934–), Ann Stofer (Figure 13.2; Johnson 1934–2002), and Alice Beck (Kehoe 1934–) were all Barnard College best friends testing the world of Plains archaeology that summer. Ferran and Stofer worked on Carlyle Smith's excavations at Two Teeth (Smith and Johnson 1968), Medicine Crow, and Fire Cloud. Beck was at Modoc Rock Shelter with Melvin Fowler (Fowler and Winters 1956) where she worked with Bettye J. Broyles (*see below* Davis 1999:120–122). The three graduated from Barnard in 1956. In the meantime, Waldo Wedel refused to take Ferran and Stofer on his dig after Smith's projects had closed. However, the two and Valerie Wheeler (Nammour Esainko 1940–) were accepted on Valerie's father Richard P. Wheeler's crew on the Oahe Reservoir for the last half of the summer (Dincauze 1992; Gradwohl 1997). Ferran's diary (personal communication July 8–14, 2010) documents that her lab and fieldwork was mixed with laundry, movies, hail storms, Johnny's bar, rodeos, swimming, movies in town, campfire singing

Figure 13.2 Ann Stofer Johnson, 1971.

with visiting David Gradwohl's guitar, Lakota dances on the Rosebud Reservation, South Dakota, and visits to other nearby RBS projects. Ferran applied for work on an RBS field crew the next summer, 1956, but received a letter from RBS Missouri Basin Project Director Robert Stephenson saying they did not hire women. Instead, she was a crew member at Modoc Rock Shelter that summer, where she worked with Bettye J. Broyles (*see below*), Phyllis Anderson (Morse dates unknown), and a woman bookkeeper (Davis 1999:123). Dincauze (1967) completed her doctorate in anthropology at Harvard University and was on the faculty of the University of Massachusetts at Amherst from then until her retirement. She was President of the Society for American Archaeology in 1987–1989.

Stofer married Alfred E. Johnson and completed a Master's degree at the University of Arizona with J. Charles Kelly, Emil W. Haury, and Ray H. Thompson (no thesis required, but see Johnson 1958). She subsequently handled field school logistics and supported her husband by working in the laboratory at the University of Kansas, Museum of Anthropology while she raised their two daughters.

Alice Beck Kehoe went on to become an important figure in Plains anthropology and archaeology (e.g., Kehoe 2010). While an intern at the Museum of the Plains Indian in Browning, Montana, she met Thomas A. Kehoe, who she married and with whom she raised two sons. She completed her Ph.D. at Harvard University (Kehoe 1964).

Valerie Wheeler (Nammour 1940–) was a regular crew member with her father in 1955. She was at the Leavenworth Site in 1961 and at Norton Reservoir with Emily Blasingham in 1962 (*see below*). She received a Master's degree in anthropology at the University of Nebraska in 1967, then a Ph.D. at the University of Oregon (Nammour 1974). She then taught at California State University, Sacramento (CSUS), for 40 years. She and her husband, anthropologist Peter Esainko, are currently working on completing a long-range study of organic farming in California, Ohio, and France.

Al Johnson (personal communication June 15, 2010) noted that Waldo Wedel employed Mary Elizabeth King (1926–2005) on a Missouri River project in the 1950s. King (1965a) subsequently reported on Hohokam textiles from the Painted Rocks project in Arizona and completed her doctorate on Peruvian textiles at the University of Arizona (King 1965b). She spent the rest of her career studying archaeological textiles and their curation at the Textile Museum in Washington, DC, Texas Tech University, University of Pennsylvania Museum, and Laboratory of Anthropology, Santa Fe.

Barbara Durbin (Anderson 1936–) was in the University of Iowa field school in 1955, and she and her husband, Adrian Anderson, were in the field in Missouri in 1959 before they were hired at the Comanche County (Oklahoma) Historical Society Museum and Archives in 1960 (Anderson 2006:801). In 1959, Barbara Richards (dates unknown) worked on the

Stricker Site under Carlyle Smith; she coauthored the site field journal with Bert Salwen (Salwen and Richards 1959). In the late 1950s, Amy Evelyn Harvey (Henning 1937–1998; Anonymous 1998) worked with then-husband Dale R. Henning and Carl Chapman's University of Missouri crews on sites in southwest Missouri (Chapman et al. ca. 1955–1960a, ca. 1955–1960b; Chapman et al. 1960; Harvey 1960) as did Bonnie B. Keller (dates unknown; Chapman et al. ca. 1955–1960a, ca. 1955–1960b). The Hennings and Kenneth Cole worked on the Joanna Reservoir project in Missouri in 1961–1962 (Henning 1964) when Amy reported on Starr Mound.

Richard A. Krause (personal communication June 6, 2012) recalls that he was working at the Leavenworth Site (Krause 1972) in 1960 when he realized that times were changing in Plains archaeology. He was sitting in the mess tent when he overheard Preston Holder and Robert Stephenson discussing how they were going to comply with a federal order requiring them to integrate their crews. They would need to provide separate camps and separate toilets for the new crew members. Krause had spent a lot of time in Georgia and Tennessee and to him "integration" meant Negroes but that didn't make much sense, because he was already sharing a field tent with Ernie Gunter, a black track star from Nebraska, and Ojo Arewa, an aspiring black anthropologist from Nigeria. Finally, Krause realized they were talking about including women on their crews! This was before passage of federal legislation mandating inclusion on field crews; apparently Holder and Stephenson were prescient!

In fact, Holder with Krause had four women on the Leavenworth field crew in 1961: DeLu Frederickson, Petra Sibylla Moore (Reyes), Astrida Blukis (Onat), and Valerie Wheeler (Figure 13.3). The women spent the summer skimming house floors and hiding out in cache pits (Onat personal communication, March 11, 2013). Frederickson and her husband had their small child with them in a relatively primitive camp. The crew took turns with mess duties; potable water was hauled for miles, and the crew bathed in the Missouri River.

Blukis Onat (1940–) received a Master's degree from the University of Kansas (Blukis Onat 1964) and taught anthropology/archaeology at Seattle Central Community College from 1966 to 1995, during which she directed more than 20 archaeology field schools. After completing her Ph.D. at Washington State University (Blukis Onat 1980) she founded Blukis Onat Anthropological Services (BOAS), Inc. which has conducted both archaeological and ethnographic research and management projects for a variety of clients, including local tribes.

Moore Reyes (1942–) received a Master's degree in anthropology from the University of Kansas, an MPH at the University of California, Berkeley, and a DSc at Johns Hopkins University in health administration and international health. She has spent a lifetime building institutional

Figure 13.3 Leavenworth Site 1961: Astrida Blukis, Petra Moore, Bruce Blinco, Valerie Wheeler, and Richard Krause.

and contract partnerships with local and international organizations in over 80 countries. She noted that "the River Basin experience was a good bootcamp for training in endurance, physical stamina, negotiation, tolerance for all manner of oddity and caprice, teamwork, for confronting fears and personal vulnerabilities—which turned out to be essential for managing my career" (personal communication, March 17, 2013).

Teresa Lynn Holder (1955–) is the daughter of Preston Holder and Joyce Wike and was also present during the 1961 field season. She went on to complete a Ph.D. in cultural anthropology at Ohio State University in 1992 and is now a social worker at the University of Washington Medical Center-Roosevelt. Several other women, including Dele Doke and Alice Benfer at the University of Missouri, were involved in the preparation of the Leavenworth report.

Robert L. Stephenson was Chief of the Lincoln MBP office from September, 1952, until October, 1963 (McNutt 1995; Thiessen 1999). He became the acting and then permanent RBS Director from 1963 until 1966 when he moved to the University of South Carolina. Stephenson had strong feelings about hiring women; comments from men on the crews ranged from "we'd have to wear clothes if there were women around" to "where would we pee?" (see Neuman 2003).

A typical letter from Stephenson to a female applicant for summer fieldwork included the statement "I regret to say, however, that we do not employ women in any capacity in the field and consequently would not be able to accept your application" (Stephenson 1964b). However, in going through Stephenson's correspondence in the SI's National Anthropological Archives, Kimball Banks (personal communication, July 12, 2010) noted that Stephenson routinely suggested that women apply to the state and university partners.

I finished a BA in anthropology at the University of Minnesota in 1963, working as a GS-4 Seasonal (Interpretive) Ranger at Bandelier National Monument in New Mexico that summer. Eldon Johnson, University of Minnesota Professor of Anthropology, hired me as a field assistant on the Shakopee Mounds Archaeological Field School the summer, 1964. He allowed me to take the field school for graduate credit because he said the RBS would not hire women for fieldwork, and he thought I needed fieldwork on my resume. At the 1990 Society for American Archaeology meeting, Stephenson apologized to me for his attitude toward women in the RBS.

Thomas A. Witty, Jr.'s wife Maryanne (dates unknown) was usually in the field with him and his crews from the Kansas Historical Society, another RBS partner. She was at Council Grove and John Redmond Reservoir in 1961 (Witty 1961a, 1961b); Elk City Reservoir, Cheney Reservoir, and Wilson Lake in 1962 (Witty 1962a, 1962b, 1962c, 1962d); Marion Reservoir and Milford Reservoir in 1963 (Witty 1963a, 1963b); Perry Reservoir in 1964 (Witty 1964); the Upper Verdegris Watershed in 1965 (Witty 1965); the Bogan Site in 1967 (Marshall and Witty 1967); and the Lower Salt and Plum Creek Valley in 1968 (Marshall and Witty 1968).

David M. Gradwohl (personal communication April 5, 2012; *see* Gradwohl, this volume) had three women on his 1964 Iowa State University crew, including a student's wife who was the cook. Woolworth (2006:651) noted that women students were common and well accepted at the University of Nebraska, so that having women on the field crews was never an issue for most of the males. The late 1960s MBP office staff had its usual complement of women typists, stenographers, and administrative clerks (Thiessen 1999:56).

In 1962, Emily Jane Blasingham (1926–2007) worked with Preston Holder on a University of Nebraska crew at the Norton Reservoir in Kansas (Blasingham 1963). She and her cat arrived at the excavation in an early 1960s model Corvair, which intrigued everyone. Emily had completed a Master's thesis on the Yankeetown Site at the University of Indiana in 1953 (Blasingham 1953) and a Ph.D. on Illinois Indian ethnohistory (Blasingham 1956). She taught at Loyola University in 1965 where she completed the Norton Reservoir report and again excavated at Yankeetown (Blasingham 1965). She also was on the faculty of the University of Alabama and Illinois State University and was a Head Curator of Anthropology at the Illinois State Museum.

Sally (Mrs. James O.) Marshall (dates unknown) was in the field with her husband in 1963 and 1964 on Preston Holder's Glen Elder Reservoir project (Marshall 1969a); at Big Hill Reservoir in 1966 (Marshall 1966); at the Bogan Site in 1967 (Marshall and Witty 1967); in the Lower Salt and Plum Creek Valley in 1968 (Marshall and Witty 1968); at the Grove Reservoir in 1969 (Marshall 1969b); and at the Elk City Reservoir in the early 1970s (Marshall 1972). Janet Goldenstein Ahler (1942–) was on Preston Holder's Solomon Valley project from 1963 to 1966; at least one of those years she was in charge of an all-girl crew of six or seven. She completed a Ph.D. (Ahler 1974) and taught Educational Anthropology and Native American Education at the University of North Dakota and the University of Missouri. Mary Ann Behlen (Hruska 1941–) was also on the Glen Elder crew in 1963.

Janie Botos (Koltermann 1944–), Susanne Hengeveld (Miller 1943–), Rose Myers (dates unknown), and Rosemary (Mrs. Darrell) Henning (1942–1991 and who was pregnant and shortly gave birth to daughter Sarah at the end of the field season) worked at the Fire Creek Site on the Oahe Reservoir in 1964 (Anonymous ca. 1946–1970b; Lehmer 1966). Ellen Starr (dates unknown) worked for Oscar Mallory at Pipestem Reservoir, North Dakota, in 1966 (Mallory 1966), and Shirley McCrary (dates unknown), Barbara Richards (dates unknown), Gwendolyn Hall (dates unknown), and Patricia Arrington (dates unknown) were at Clinton Reservoir, Kansas, that year (Johnson 1968). Nancy Osborn (dates unknown) worked on NPS-Iowa State University cooperative projects from 1966 to 1970 in Red Rock and Saylorville reservoirs along the Des Moines River (David Gradwohl personal communication April 5, 2012, and January 3, 2013). Wilfred M. Husted (2006:754–755) had at least three women—Donna Slatin, Mary Szmkowiak, and Cynthia Vann—on his Fort Union Trading Post crew in 1969, saying "the times had changed."

Jane Holden Kelley (1928–) completed work at Tuttle Creek Reservoir, Kansas, in the mid-1960s (Kelley 1966a). Kelley (1992) was the daughter of William Curry Holden and grew up around the Lubbock Lake Site in the Texas Panhandle. She received a Ph.D. from Harvard University (Kelley 1966b), raised her and David Kelley's children, and spent her career teaching in the Department of Archaeology at the University of Calgary doing research in New Mexico, El Salvador, and Chihuahua.

Women have been involved in analysis and interpretation of RBS archaeological collections since 1969. In 1973 I reanalyzed the material from the Anton Rygh Site (Knudson et al. 1983) and the Red Smoke Site (Knudson 2002). Amy Rosebrough (1969–) analyzed material from the Helb Site in the Oahe Reservoir (Rosebrough 1994, 1995). Ann Mary Johnson completed her dissertation on the John Ketchen, Durkin Village, and Hallam I and II sites in the Big Bend area of Lake Sharpe, South Dakota

(Johnson 1977, 1979). For her dissertation, Lynn Snyder used faunal material to examine the use of domestic dog as a food source in the Middle Missouri (Snyder 1995). While working for the U.S. Bureau of Reclamation, Bobbie Ferguson (?–2007) assessed the ownership of RBS archaeological collections from federal, tribal, state, and private lands acquired through a complex array of federal, state, and private agencies (Ferguson 2001a).

Berkeley Field Office

The Berkeley office was housed at the University of California. As far as the author can find, no women had significant roles in the field, laboratory, or reporting activities of the office. However, they may have been there, just unacknowledged and unrecognized.

Eugene Field Office

The Eugene office was housed at the University of Oregon. The office oversaw projects in the Columbia River drainage in Idaho, Washington, Oregon, and western Montana. Luther Cressman, Chair of the University of Oregon Department of Anthropology, was known to take women on his crews (White 1999a:210).

Douglas Osborne led several RBS projects out of the University of Washington. His wife Carolyn M. Osborne (1919–2008) coauthored a paper on an RBS-excavated rockshelter (Mills and Osborne 1952). Earl Swanson's report on his 1953–1954 excavations in the Wanapum area (Swanson 1962) noted that Carolyn had taught him the technical aspects of prehistoric textiles and cordage. Helen Branan (dates unknown), Alys Tree (dates unknown), and Vera Washburn (dates unknown) were on Swanson's crew in 1954. In his 1957 report on McNary Reservoir on the Columbia River, Osborne acknowledged the help of Mrs. R. D. Busey (dates unknown) and Mrs. J. F. Flock (dates unknown) of the Oregon State Button Society. Carolyn analyzed the perishables, did the photography, and provided "generally wholesale non-compartmented assistance" for that report (Osborne 1957); Maureen Connors (dates unknown) of the university did the drafting. Carolyn did drafting and darkroom work and sorted sediments in search of textiles from the Sheep Island Site (Osborne et al. 1961:271). She also did darkroom work and typing along with fabric analysis and editing for Osborne's (1967:iii–iv) report on the Lower Grand Coulee project. Jean Parsons (dates unknown) typed, proofread, and rearranged the manuscript for that report. Similarly, Maxine (Mrs. Joel) Shiner (dates unknown) also typed and proofread her husband's report on McNary Reservoir excavations (Shiner 1961).

The Eugene office 1942–1956 archives mention Martha R. Flahout (dates unknown). Osborne's September, 1956–April, 1964 archives have records of Annette Parent (dates unknown) (Anonymous ca. 1946–1970c). Those records have not been examined, so the two women's roles in RBS projects are unknown.

Richard Daugherty excavated the Lind Coulee Site in 1951 (Daugherty 1956a:223). Betty Joyce Enbysk (dates unknown), University of Washington Curator of Paleontology, analyzed the vertebrate and Mollusca collections (Enbysk 1956). Daugherty used Lind Coulee and other Columbia Basin archaeological sites for his dissertation (Daugherty 1956b), in which he acknowledged University of Washington faculty members Erna Gunther (1896–1982) and Catherine McClellan (1921–2009) for their helpful criticisms. Ann M. Irwin (1940–2006) and Ula Moody (1945–) reinvestigated Lind Coulee in the early 1970s under an NPS contract (Irwin and Moody 1976, 1977; Monseth et al. 1973). Irwin went on to a career in the environmental division of the Texas Department of Transportation, and Moody taught at Florida State College at Jacksonville, North Idaho College and other campuses.

In 1956, Ruth Gruhn (Bryan 1935–) worked on B. Robert Butler's (1960) University of Washington field crew excavating the Wakemap Site at the Dalles (Caldwell 1956), and in 1957 at sites behind the Priest Rapids Dam on the Columbia River (Greengo 1986). She received her Ph.D. from Harvard University (Bryan 1961) and, with her husband Alan Bryan, has spent her career at the University of Alberta, Edmonton, researching early Americans and teaching.

Margaret M. Lyneis (1938–) was on Robert Greengo's field school in the Priest Rapids Dam area in 1958 (Greengo 1982, 1986). She worked in the University of Washington archaeology laboratory during the summer of 1959, and in the summer of 1960 she was a graduate assistant at Greengo's Wanapum Dam project. Lyneis left to finish her Ph.D. at the University of California, Los Angeles (Lyneis 1968). She has had an extensive career teaching at the University of Nevada, Las Vegas, and doing research and cultural resource management across the intermontane West. She is currently a research associate at the Museum of Northern Arizona (MNA), and she is completing a monograph on excavations in the Lost City area of Southern Nevada.

Excavation of the Wildcat Canyon Site in the John Day Reservoir on the Columbia River began in 1958 and continued for over two decades (Dumond and Minor 1983). Don E. Dumond and Rick Minor reported that Karen Kuil (dates unknown) was responsible for field or laboratory work at Wildcat Canyon. Becky Saleeby (1949–) and Ruth Greenspan (1952–) addressed bird seasonality. Carol Steichen (Mrs. Don E. Dumond 1923–) drew the report cover art while Margaret Chodos (dates unknown) provided artifact drawings and Karen Anne Toepel (1953–) drew the base map. Becky Saleeby was an archaeologist in the Alaska Regional Office of the NPS and Ruth Greenspan is a Technical Management Section Manager with the Arizona Department of Transportation. Carol Dumond spent a career illustrating archaeological materials and illustrating and writing other publications. Margaret Chodos-Irvine also has had a successful career as an illustrator and artist. Kathryn Anne Toepel is president and owner of Heritage Research Associates in Eugene, Oregon, with her husband Rick Minor.

Charles M. Nelson (1969) excavated the Sunset Creek Site on the Columbia River south of Trinidad from 1957 to 1963. Carolyn Osborne (1969) analyzed and reported on perishable materials from rockshelters on the Quilomene Bar where the Sunset Creek Site was located.

In the early 1960s, Washington State University (WSU) began salvage excavations in the proposed Lower Monumental Reservoir area; these excavations lasted for more than a decade. The 1962 fieldwork focused on the Palus Village Site and Marmes Rockshelter. The field crew included Erika Fogg (dates unknown), Barbara Grater (*see below*), Elva Hammerstrom (dates unknown), Helen Horton (dates unknown), Barbara Ratcliff (dates unknown), and Rosemary von Elling (Bolaria, dates unknown) (Fryxell and Daugherty 1962:1). Sonja Solland (dates unknown) was the lab person on the Palus Burials project (David G. Rice, personal communication, June 29, 2012). She went on to receive her Master's degree and a Ph.D. in anthropology from the University of Washington (Solland 1974). She then taught at Shoreline Community College in Washington state until 2012.

Many women, whose names are not documented here, were active in the WSU field and laboratory projects throughout those years. During the 1962–1968 field seasons at the Marmes Site (Hicks 2004), David Rice (1969:xvii–xviii, 1972) noted that a number of women participated in the field and laboratory work. Madge Gordon Gleeson (1945–) did probably 95 percent of the stratigraphy during the 1968 field season, assisted by Joan Brodhead (dates unknown). Claudine Weatherford (1945–), Gail Thompson (1947–), Jeanne Riggs (dates unknown), Kathleen (Kay) MacKenzie (dates unknown), and Madilane Perry (1947–) were part of the field crew. Ann Monseth Irwin directed the Marmes field laboratory and Diana Ames Marshall (1948–; personal communication June 16, 2012) and Caren Ross (dates unknown) worked for her. Barbara Daniels (dates unknown) and Janet Nunn (dates unknown) also worked at Marmes during various field seasons. Shirley Twehus (dates unknown) illustrated artifacts, Tracy Youmans (dates unknown) and Judi Wutzke (dates unknown) provided secretarial services, and Rice's wife Signe Johnson (1952–) was supportive throughout the project. Ruth Kirk (1925–) wrote a lay book about the Early Holocene human remains from Marmes, and illustrates several unnamed female excavators (Kirk 1970).

Barbara Alice Grater (1938–) spent two summers on Frank Leonhardy's Lower Snake River WSU field projects in the mid-1960s, one summer with Roderick Sprague on the Palus Burial Site (Sprague 1965), and the next full year processing burial goods in the WSU lab. She finished a Master's degree on the Votaw Site (Grater 1966). Leonhardy's 1967 and 1968 field school crews at the Granite Point Locality included Grace Miller (dates unknown) in the field and Barbara Daniels, Donna Mitchell (dates unknown), Janet Nunn, and Caren Ross in the laboratory (Leonhardy 1970:iii). Jill Osborn (dates unknown) worked in the laboratory following the 1969 excavations at Wexpusnime.

Austin Field Office

Edward Jelks (2006:605) noted that in the early days of the Texas RBS all the field crews were male because it was thought that the language and behavior were too coarse and crude for women. However, his wife Judy (1924–) accompanied him in the field at Texarkana in 1952, Iron Bridge in 1958, and Whitney in 1969. She was never a paid employee, but helped with the digging and data recording, and sometimes cooked. Frances Irene (Pat) Moore Wheat (1916–1987) accompanied her husband Joe Ben Wheat on the Addicks Dam Basin Survey and "by her constant aid and encouragement, has in a very real sense made this work as much her own as mine" (Wheat 1953:150).

Dee Ann Suhm (Story 1931–2010; Anonymous 2011; Shafer 2011) was a part-time secretary in the Austin office between 1951 and 1954 (Jelks 2012). She was hired in 1952 to catalogue the artifacts, prepare the physical anthropology notes, and help interpret three sites excavated in the Texarkana Reservoir area on the Sulphur River (Jelks 1961:1). After doctoral work at the University of California, Los Angeles (Story 1963), and two years on the Glen Canyon Project (see below), she returned to Texas. In 1960, she became the Curator of Anthropology at the Texas Memorial Museum and, after marrying Hal Story in 1961, became an assistant to Jelks at the Texas Archeological Salvage Project (Jelks 2006, 2012). In 1964, Jelks had her run all-male crews at Waco Reservoir where she earned their respect. In addition to Suhm's unflappable personality, acceptance was probably also due to her regional expertise—she had published a typology of Texas points a decade before (Suhm 1954a, 1954b, and 1954c). Her experience on the Glen Canyon Project was also a significant asset. She became Director of the Texas Archeological Research Laboratory in 1965, a position she held until she retired in 1987.

Athens Field Office

The records of the Athens field office have not been searched except those in the National Anthropological Archives. In general, there were few archaeological field schools in the Southeast, and women were not allowed to do paid fieldwork until passage of the *Equal Employment Opportunity Act* in 1972 (Lynne Sullivan, personal communication June 5, 2010). However, Bennie C. Keel (personal communication March 22, 2013) notes that the Florida State University and University of Alabama's 1960 archaeological field schools in the Woodruff Reservoir area, an RBS project, contained a number of women. Also, see Carol Mason's (1992) discussion of women in Southeastern archaeology in the 1950s.

One of the earliest activities involving women was Lucile St. Hoyme's (1924–2001) analysis of human remains from Kerr Reservoir that were collected between 1947 and 1951. St. Hoyme, who was at the National Museum of Natural History, documented evidence of scalping (Hoyme and Bass 1962; Miller 1962).

Sheila Kelly Caldwell (dates unknown) (*see* Mason 1992:96) photographed the historic Vann House in 1951 with her husband Joseph R. Caldwell and authored a publication about the resource (Caldwell 1991).

Martha Ann Rolingson (1935–; personal communication, May 10, 2010) excavated the Henderson Site in 1961 with a crew of three men and two women, the first mixed-gender crew in Kentucky. The crew lived in a house of the U.S. Army Corps of Engineers which partially supported the project. In 1963, Rolingson surveyed the proposed Grayson Reservoir on the Little Sandy River in eastern Kentucky (Rolingson 1963; White 1999b). She walked small bottomland fields and pastures and narrow, rugged valleys by herself. Later that year, she worked for Douglas Schwartz on the Laurel River Reservoir Survey (Rolingson and Schwartz 1963). In 1967, Rolingson had a short field season with a crew of women and men on Eagle Creek north of Lexington (Rolingson 1968; White 1999b). She directed the University of Kentucky Archeology Program during the 1967–1968 academic year. Rolingson tested sites in the Cave Run Reservoir in the eastern Kentucky Mountains in June, 1968; again with a crew of women and men (Rolingson and Rodeffer 1968a, 1968b; White 1999b:238). They lived in a house with cold water, an outhouse, no electricity, and nearly constant rain. Stephanie Holschlag (Rodeffer 1947–), was a crew member. Holschlag completed a Ph.D. in anthropology at WSU (Holschlag 1975) and is the Museums Services Program Manager at the NPS Western Archeological and Conservation Center, Tucson.

Rolingson completed her Ph.D. in anthropology at the University of Michigan, working with James B. Griffin and James Fitting (Rolingson 1967). From 1968–1972 Rolingson was the Research Station Archeologist at the University of Arkansas Monticello research station, and from 1972–1977 she was in Fayetteville teaching, editing, and conducting contract work. She became the Toltec Station archaeologist in 1977 where she worked until she retired in 2005.

In 1964, Dorothy Cross (Jensen 1906–1972; Mason 1992:97) carried out excavations in the Tocks Island Reservoir in New Jersey (Cross 1964). In 1967 Mildred L. Fryman (dates unknown) worked with her husband Frank on the proposed Ulvah, Parke Branch, and Wolf Creek Dam Cargo Life Areas (Fryman and Fryman, 1967; Fryman et al. 1967a, 1967b).

NPS Southwest Regional Office, Santa Fe

The Glen Canyon Project (GCP) in Utah and Arizona was the largest IASP-sponsored project in the NPS Southwest Regional Office in Santa Fe. The office also oversaw projects at Flaming Gorge Reservoir in Wyoming and Utah, at Navajo Reservoir on the San Juan River in New Mexico and Colorado, and at Cochiti Reservoir on the Rio Grande River west of Santa Fe. Fowler (2011:273) noted that Jesse Jennings, director of the University of Utah Glen Canyon Project, and the MNA male staff were generally products of their time, believing that Glen Canyon was too dangerous for women and that women were generally best suited for washing and labeling artifacts and typing/editing reports.

Katherine Bartlett (1907–2001; Anonymous 2001; Babcock and Parezo 1988:170–173) was a founding staff member at MNA, and she served as its first Curator of Anthropology from 1930 to 1952. The Museum of Northern Arizona was the southern partner (with the University of Utah) on the GCP. Bartlett established an archaeological collection catalogue system that served throughout the GCP. Gene Field Foster (1917–1983; Olberding 2011) began MNA's involvement in Glen Canyon in 1952 when she started recording sites in the project area (Foster 1952, 1954). She was hired again in 1957, in advance of a contract, to continue surveying. Her crew included Bartlett (her housemate and volunteer), and three other women (Fowler 2011:317). Xerpha Mae Gaines (dates unknown), from WSU published her study of the plants there (Gaines 1957; *see also* Gaines ca.1960) and Inez Marie Haring (1876–1968) studied and published on the bryophytes (Haring 1961). Olberding (2011) presents the words and images of Bartlett and Foster on the GCP.

Despite the prevailing male attitude toward women in archaeology, the GCP had four graduates of Bertha Dutton's Girl Scout archaeological training schools ("Dirty Diggers"; Cohen 2006; Fowler 2010): Elizabeth (Nettie) Kessler (Adams 1934–), Elizabeth Galligan (1938–), Mary Anne Stein (1938–), and Catherine (Kay) Sweeney (Fowler 1940–). All came to GCP with significant training and *chutzpa*, and all contributed, as well as built successful careers within their own specializations. Catherine Sweeney was an undergraduate anthropology student at the University of Utah when she was hired as the anthropology department secretary in summer 1961. By that fall, she was working on GCP ethnohistory studies

(Sweeney 1961) and a Southern Paiute bibliography. Sweeney worked in the field with Robert C. Euler in 1962 (Sweeney and Euler 1963) and married Don Fowler in June, 1963. She and Don finished doctorates in anthropology at the University of Pittsburgh (Fowler 1972), had distinguished careers, and are both retired professors at the University of Nevada (Fowler 2011:312–313).

Elizabeth Galligan directed the MNA GCP archaeology laboratory until fall 1961 when Mary Anne Stein replaced her, at Galligan's recommendation. In 1961, Stein worked on the GCP excavations at Segazlin Mesa with Alexander (Lex) Lindsay, Richard Ambler, and Paul Long. At the Pottery Pueblo excavations in 1962 (Lindsay et al. 1969) Stein again ran the Paiute Mesa field laboratory, with the help of Ada Hatch (1941–; married Jim Merrill) and Dorothy (Dottie) Diehl (Carnine 1941–). Fowler (2011:326) noted that the women washed their hair first and then the sherds. That was the first year Lindsay had a field laboratory to support the excavations, probably with encouragement from his wife Jane (dates unknown). The four women running the laboratory felt they were the first blow for women's inclusion in GCP fieldwork. Hatch (personal communication June 10, 2012) recalls that the Navajos built a ramada for the crew and that the girls slept on the north side and the men slept on the south; there was no latrine. They had a cook for a short time, but after she left, the crew prepared meals together. A Navajo man often passed by on the way to his cornfields, and Hatch remembers him singing as he passed. The crew also went to a traditional Navajo wedding that summer. Fannie Bitsinnie (Navajo, dates unknown) of the bride's family also worked in the field lab, as did other family members.

Mary Ann Lively (dates unknown) worked with Stein on a survey of Paiute Mesa, which was the basis of Stein's University of Oklahoma Master's thesis (Stein 1966). Her Southern Methodist University Ph.D. dissertation (Stein 1984) was on Pottery Pueblo. Subsequently, Stein became a wildlife lawyer in 1987 (personal communication May 25, 2012).

Galligan (personal communication June 1, 2012) received a Master's degree in anthropology, raised two sons, and then in her 50s completed a Ph.D. in Curriculum and Instruction in Multi-Cultural Teacher Education at the University of New Mexico. She taught at Menaul School, Eastern New Mexico University, and Adams State University and recently published *Secrets of the Plumed Saint* (Galligan 2012), a mystery novel set in northern New Mexico. Hatch completed a degree in anthropology and a Master's degree, then taught in Page, Arizona, for most of her life and continues to volunteer there. Because she could speak some Navajo, she spent (and spends) a lot of time with Navajo children at the school. Diehl (personal communication June 1, 2012) completed a BA in dietetics and, after marrying a rancher in western Nebraska, raised children and taught dietetics in the area.

Figure 13.4 Nettie Adams and Floyd Laughter cooking breakfast.

While working for MNA in 1957, Nettie Adams (Figure 13.4), William (Bill) Y. Adams, and Maurice Cooley, a geologist, surveyed the lower San Juan Canyon within the GCP. Their job was to locate access points to the project area, working overland to save money. They hired a Navajo, Floyd Laughter, to provide horses and pack stock for a week's survey of Paiute Canyon (Nettie Adams, personal communication, June 16, 2012). The winter of 1957–1958, Nettie Adams washed artifacts while her husband did the analysis. The Adams surveyed the canyons of the San Juan River in the spring, 1958, and were supported by Frank Wright, an experienced boatman from Blanding, Utah, and in the fall they did overland-accessible excavations (Adams and Adams 1959). Christy G. Turner II (personal communication, June 17, 2012) and his wife Jacqueline (Jackie) Loy Adams Turner (1934–1996) joined the GCP in mid-1958. Jackie and her husband worked sites at the mouth of Rainbow Bridge Canyon, in West Canyon Creek, and in Nokai Canyon, a tributary of the San Juan River, where she was an excavator and also helped prepare meals. Nettie and Jackie did lab work in Flagstaff during the spring and summer of 1959 while their husbands did river-accessible excavations in the Canyon. The Adams left the GCP at the end of 1959 to pursue careers in the Nile Valley and elsewhere. Jackie raised their daughter, who was born in December, 1959. Throughout her life, she worked with her husband collecting comparative osteological and dental information in collections around the world and helped excavate sites in Alaska. She and her husband published a study of cannibalism in the American Southwest (Turner and Turner 1998).

Dee Ann Suhm was a doctoral candidate at University of California, Los Angeles, when Jesse D. Jennings hired her in May, 1958, as the GCP laboratory supervisor, which she ran for over two years (Fowler 2011:273; Gunnerson et al. 1959). Mary Mulroy (dates unknown), Susan

Figure 13.5 Carol Condie Stout.

Clark (dates unknown), and several other women worked in the laboratory (Fowler 2011:311). Suhm usually had an all-female crew (Ohlson 2011–2012), and provided significant guidance in artifact cataloguing (Suhm 1960a, 1960b). Suhm also spent two weeks directing the excavation of three sites in the lower Red Canyon of Utah (Suhm 1959a). She left the GCP to return to Texas in 1960.

Carol Condie (Figure 13.5; Stout 1931–) and Norma Mikkelsen (1927–2010; Anonymous 2010; Fowler 2011:273) worked as GCP editors in the early 1960s, as did June Finley (Lipe 1935–2005; Matson and Kohler 2006:4–6, 11, 16) Condie received her Ph.D. in anthropology from the University of New Mexico (Stout 1972) and owned and operated Quivira Research Center/Associates from 1983 until 2011. When asked why she went on to get her Ph.D. and undertake lifelong fieldwork after being around the restricted (for females) GCP, she commented that men were having all the fun and she wanted that too (personal communication, March 30, 2012)! Mikkelsen was head of the University of Utah Anthropology Laboratory for many years and later lead editor for the University of Utah Press anthropological publications. June Lipe did archaeological illustration for most of her life, as she moved with her husband. Florence Lister (1920–) completed a study of regional ceramics in 1960 (Lister 1964) and concluded that the Kayenta-Virgin ceramic dichotomy was spurious (Jennings 1966:86, 88–89). This assessment impacted many of the subsequent GCP interpretations.

Women did not have much involvement in the Flaming Gorge Reservoir project, but Dee Ann Suhm (1959b) reported on two sites there. Polly Dix Schaafsma (1935–) studied the rock art in the Navajo Reservoir area in 1961–1962 (Schaafsma 1962, 1963), and Beth Dickey (dates unknown) worked in the project laboratory. In 1965, Schaafsma completed a study of the kiva murals from LA 70 in Cochiti Reservoir

(Schaafsma 1965) and wrote up the rock art for the project in 1967 (Schaafsma 1975). Eleanor Jane Nettle (dates unknown) was a coauthor on Jim Schoenwetter and Frank Eddy's alluvial and palynological reconstruction at Navajo Reservoir (Schoenwetter et al. 1964).

Women in the RBS and IASP Workplace

The above discussion cannot be considered all-encompassing with regard to the participation of women in the RBS and IAPS. The discussion does demonstrate that women were involved in a various ways in RBS and IASP projects between 1946 and 1969—but most of them remain invisible. Of the women involved in RBS field and laboratory work, many of them went on to successful careers that involved archaeology in one form or another. They endured significant and enabling changes in the federal legal structure, usually worked for pittances, and with their predecessors from the first half of the twentieth century, set the stage for subsequent noteworthy contributions by women to American archaeology. The men who hired and appreciated the RBS/IASP women were also part of setting the stage for participation of women in the field today. They all truly are heroes!

Dincauze (1992) has noted factors that contributed to women's invisibility in American archaeology from the late 1940s through the 1960s: cultural styles, professional values, competitive exclusion, personal arrogance, institutional depreciation, and deeply held social habits of self-abnegation by women. All these factors affected the involvement of women in RBS projects, whether government or partner-run. Many of these factors still hold today, despite changes in federal law. The RBS program was truly *cultural resource management* before that term was coined (Lipe and Lindsay 1974). It is notable that men and women who had been part of the RBS created this term—and women now form a major portion of today's practitioners.

Acknowledgments

The inventory of all the people who helped me compile this paper is too extensive to list. Many of the women cited here have talked about their RBS lives with me, as have several of their surviving spouses and children. Many of those bronzed, lean RBS men have also been invaluable sources of women's names and often guidance on how I could find the women today. Thank you, everyone!

CHAPTER 14

The Missouri River: The Backbone to Survival

Gerard Baker—Yellow Wolf

In the mid-nineteenth century, when she was a child, Buffalo Bird Women of the Hidatsa tribe lived along a bend of the Missouri River named "Like a Fishhook." As an old woman, she looked back on those faraway times:

> I am an old woman now. The buffaloes and black-tail deer are gone, and our Indian ways are almost gone. Sometimes I find it hard to believe that I ever lived them.
>
> My little son grew up in the white man's school. He can read books, and he owns cattle and has a farm. He is a leader among our Hidatsa people, helping teach them to follow the white man's road.
>
> He is kind to me. We no longer live in an earth lodge, but in a house with chimneys; and my son's wife cooks by a stove.
>
> But for me I cannot forget our old ways.
>
> Often in the summer I rise at daybreak and steal out to the cornfields; and as I hoe the corn I sing to it, as we did when I was young. No one cares for our corn songs now.
>
> Sometimes at evening I sit, looking out on the Big Missouri. The sun sets, and dusk steals over the water. In the shadows I seem again to see our Indian village, with smoke curling upward from the earth lodges; and in the River's roar I hear the yells of the warriors, the laughter of little children as of old. It is but an old woman's dream. Again I see but shadows and hear only the roar of the River: and tears come into my eyes. Our Indian life, I know is gone forever. (Nabokov 1992)

When I first read this passage, it was one of the saddest "messages" I had ever read. I could feel the heartbreak that is still felt by those of us that never even had the chance to see our beloved Missouri River before

Dam Projects and the Growth of American Archaeology: The River Basin Surveys and the Interagency Archeological Salvage Program, edited by Kimball M. Banks and Jon Czaplicki, 202–213.

its backbone was broken by the dams that the U.S. Government built on the River. This passage takes me back to our dining table in our little log house on the Fort Berthold Reservation, listening to my relatives, my elders tell of times when we lived.

I am a member of the first generation not born and raised within the valleys of the Big Missouri, but in the "uplands" of the Northern Plains away from the River and most of our history as a Mandan Hidatsa Indian. It was my ancestors that lived in those earthlodge villages, and it was my father and mother's generation that the U.S. Government moved out when it built Garrison Dam and flooded the River forming Lake Sakakawea. In fact, my family moved, as all did, from the bottom lands with a small Ford tractor, a team and wagon (full of our belongings), and my oldest brother Frederick on horseback chasing our range and milk cows and saddle horses.

I remember that my mother and father did not talk much about that move or too much about how they lived in the early part of my life, probably because it was too painful. Later as I got older, they would speak more about that life, and it seems my mother, Cora Young Bird Baker, would always cry. I could never read my father too well, but I knew he was angry. He was so angry, my mother would say after we lost him to cancer, that he moved his family as far away as he could from the River, to the west side of the Fort Berthold Reservation, and it was there that he and my mother built our ranch. It was here that I learned from many elders that our "home" was truly along the Big Missouri, where the Creator helped us live, where our women learned how to grow our crops, such as corn, beans, melons, sunflowers, squash, tobacco, etc. Since time began, according to our oral histories, we were the middle voice and management in the trade that linked the tribes in the Northern Plains; and our presence would increase during the time of the early white fur traders of the Hudson's Bay Company and the American Fur Company.

Various authors and researchers have written much about life-ways of our people during this period, and archaeologists have done much to help us understand or verify our oral histories. The waters of the Big Missouri, though, rose so fast following the construction of Garrison Dam that many of our sites were lost, and to us Hidatsa and Mandan, much of our way of life was drowned, never to come back. Professional archaeologists were able to excavate the major sites before the bottom lands were flooded. They did the best they could, given the deadlines they were up against, but many of our historic sites and sacred sites were lost to the water.

I do remember the elders talking about this, saying that "lots of white people came to dig, and look for stuff" (personal contacts of elders). Most of the people, it seemed had no idea what these white people were doing. A lot of the elders, including my parents, would state that the U.S. Government did not tell them what was going on about the rising waters, and I remember many would say that they just did not believe that the great Missouri River could have a dam on it. They could not understand that these rising waters would cover up all they knew—many would not move until the waters came to their doorsteps, and they would move with only what they could put in their wagons. And as my father told me, they would leave their houses when the horses had water up to their ankles.

Most tribal people did not understand what the archaeologists were doing, nor did they assist much, if at all. It was not proper to go back to the villages and ceremonial sites and dig; that was improper and the belief was, if you did this, that action could bring bad luck to your family and sometimes your tribe. Today, we do see the benefit of this scientific research; many Indians look upon this research as a verification of our oral history. Most of the elders, many now gone including my parents, would say that the people that dug those sites would not talk to them nor would we talk to the archaeologists, although some did. We look at that archaeological work today and understand that those excavations are part of our history and the tribal members born after the dam see these reports as a partial view into our past, one that will be complete with the addition of our oral history.

As a young man, I enjoyed hearing stories told by some of the elders and my folks about the times they knew, right before the dam was built. They talked of a time that most of us today do not know, a time when each neighbor helped one another, and as my father would say, "we did not know what welfare was." He talked about all having cattle, farms, and everything that went with it. We still had great gardens, that was so much of our life. The gardens were still tended by the female side and their pride, according to my mother, was very high. Our people still had tough times in that this was a period of a lot of cultural change and upheaval, the construction of the dam and flooding of the River, new "white-man" schools and religions, and we were still fighting the battles that were introduced by the traders: booze!

My mother grew up in the Hidatsa community of Lucky Mound by Independence Hill. All you see now of the very tall hill she described is nothing more than a piece of land sticking out of Lake Sakakawea. It seems every generation had a major obstacle to contend with. In the early days, it was small-pox, then came the rest of modern civilization and all of its vices. In my folk's time, along the River, they say it was the River itself. The River that was given to us during the creation of time as the backbone to our nations, an area where we were to live forever, as long as we learned how to take care of this River.

My mother lived by this River with my grandparents, Joseph and Suzie Young Bird. My mother would tell me many stories of the River, how it could be a friend to the people, but you had to respect it. She told of the mighty River under ice; not a silence that you might expect, but a rumbling and cracking of one that was alive. She told stories of crossing that River with my father and oldest brother, Frederick, where in the middle of the River the "wood parts" of the wagon would start to float, meaning the wagon would be floating a bit downstream with the current, as the horses pulled straight ahead. She said that this was probably the scariest for her, and carrying a baby made it worse. She also told of the people that came up the River, remembering how her mother made her and her sisters hide when the last of the steamships came up the River with "bad men" on board.

She also described the community of the River and how even in the dead of winter, folks would follow the River to a community building where they would sing and dance, and have give-a-ways and gatherings. In her stories, she told of how her parents would get their best team of horses and fill the wagon with some hay so they could put up with the sometimes 40-below temperature. She would remember sitting in the back of the wagon, seeing the smooth snow-covered River on a day so bright that everyone had to shield their eyes and watching other travelers come along that River in that same type of transportation and getting together as they had done since the time of Creation.

My father grew up in the Mandan community of Red Butte, along Short River of the Fort Berthold Indian Reservation. He would tell of chasing cattle and horses across this River, most of the time swimming alongside his horse. To me, these and a host of other stories that I learned from my parents that lived along the Missouri River told me of a time of "peace" and of working together as neighbors and, once again, not knowing what welfare was.

After the move out of Like-a-Fishhook Village (Figure 14.1) in the 1880s, the people settled in various parts of the Fort Berthold Reservation that would become new communities. Communities were established where various groups lived:

> The demography of the reservation was also undergoing changes. From the time the people began to scatter over their allotments (established by the allotment act, which encouraged the Indians to move family into separate tracts of land called allotments), for the Mandans to locate south and west of the Missouri, and for the Hidatsas to concentrate near Elbowoods and Shell Creek and to locate in smaller numbers elsewhere in company with the Mandans. Although the allotment system prevented the establishment of closely grouped population centers, people tended to settle in loosely defined communities. (Meyer 1977)

The Arikara people, a branch of the Skidee Pawnees, moved to the west side of the reservation. Besides Elbowoods, considered by the U.S. Government as the "center" of business for the reservation, there were other communities that were established: Independence, Shell Creek (Figure 14.2), Red Butte, Lucky Mound, to name a few. My parents came from the communities of Red Butte (Mandan) and Independence

Figure 14.1 Like-a-Fishhook Village.

(Hidatsa and Mandan). They and my oldest brother Frederick would tell me that the people that lived in these communities never really "saw" Elbowoods as being the "center" of the reservation, although my mother talked about the big fairs there and it being where the Bureau of Indian Affairs (BIA) agency offices were. Each community had its social area, or "Round Hall" where the community would come together for dances and other social doings. These communities were important as they were where tribal members did business and went to day schools. More importantly, these communities were where families and the clans relocated. Almost in a way, these communities were reminiscent of the historic villages in the Knife River area: still close enough to each other to help one another and maintain that extended family way of life.

We saw many hardships along that River, but the all time worst at least for me, is to go to the "River," now a lake formed behind one of the biggest earthen dams in the world, and watch people recreate on it, not knowing what the land under several hundred feet of water means and that many, many of our "creation" sites and other important archaeology sites are lost forever. The loss of our River was devastating and, as

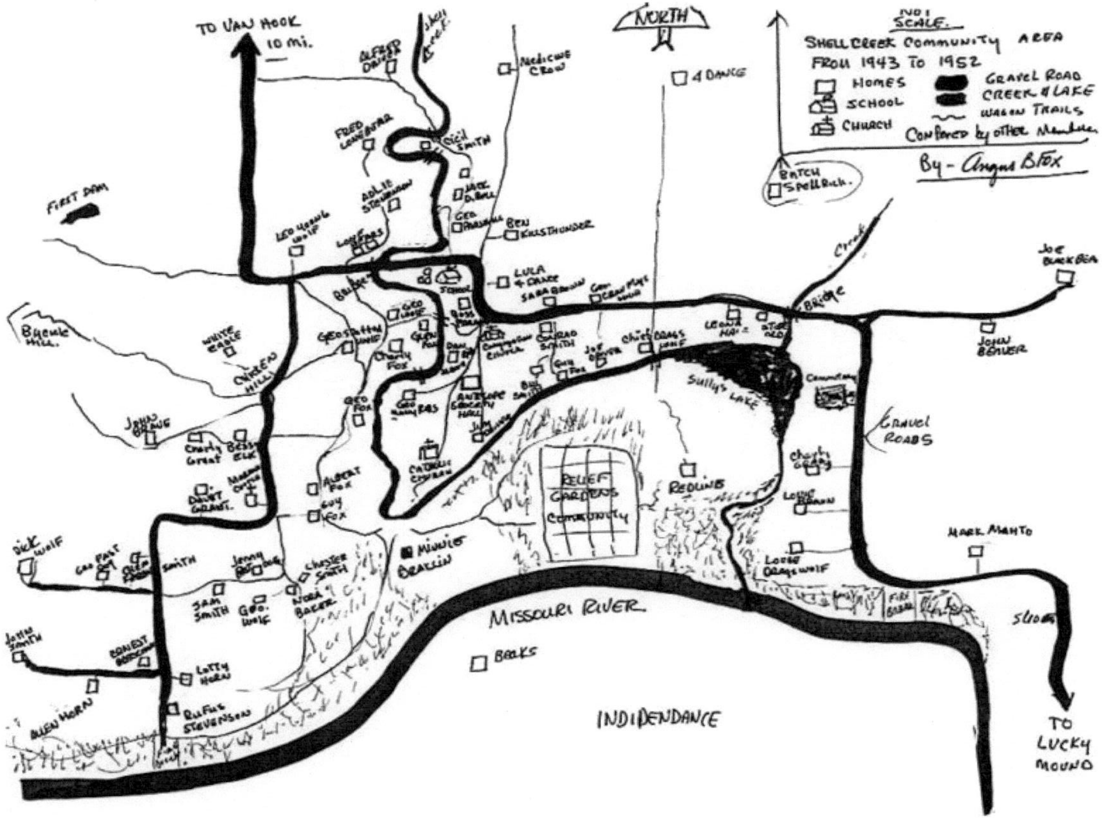

Figure 14.2 Shell Creek Map.

mentioned, changed our way of life. This change was started by the Pick-Sloan plan to dam the Big Missouri, primarily for flood control. But the impact was far reaching. According to my father, the people just did not understand this plan. Many of the older people simply did not believe that one could dam a River such as the Big Missouri, and when they finally realized the U.S. Government would and could, it was beyond belief. One of my grandfathers, Mr. Joe Black Bear, who gave me the name Yellow Wolf when I was a young man, stated:

> We lived in a fertile valley; we didn't need irrigation; had good crops, berries and this was our way of way of life since the first time we begin living in the River bottoms.
>
> When we first heard of the Garrison Dam there was some people on the reservation that agreed with the idea and then there was some that didn't. This was the beginning of our controversy on the reservation over the building of the Dam. When this was going on about the building of the Garrison Dam, once again the Army took over in the form of the Army Engineers [Editors' note: U.S. Army Corps of Engineers]. The Army Engineers were hard to deal with and then the government made promises. They promised us electricity; they promised us free electricity when the Garrison Dam was completed—the power from the dam—but today there is no such thing as free electricity here. Then again, what they paid us for our bottomlands wasn't near enough. It should have been three times that amount what we got.
>
> Before the coming of the Garrison Dam, every family had a few head of cattle, gardens and maybe a field of some sort—wheat, oats, or some type of field. They also had timber where they could get their firewood for the winter and coal mines. Wood and coal that they could get in abundance at no cost to us whatsoever except labor.
>
> I'd like to reminisce a little bit about the cost of living in them days. Coal was cheap; groceries was very cheap. With a $10 bill you could come home with a large box of groceries.
>
> I'd like to reminisce about that I used to hear years ago. The old Indians, they didn't even understand English, would use an interpreter to talk to the Government, and at that time they would be able to get some claims for our land that we have lost and this today with the young people being able to talk English fairly well don't seem to accomplish what the older used to.
>
> Going into education a little bit, it seems that in my day we weren't really encouraged to go to school and today in my old age realize how important it is to get an education and so it is today even some of us older Indians don't encourage our young ones to get an education. There is a few that are going on to get a higher education and you can see they are getting along in life real good.
>
> Now going back to the Garrison Dam relocation, we all had to move out of our big valley. Move on to our lands some of us had to move to inherited lands which was no comparison to our rich bottomlands near the Missouri River. Even in the rich bottomland I used to so [do] some farming which was a lot of my income. Today that type of life does not exist for me on higher

ground. When the Garrison Dam was planned years ago, it was supposed to come up to a certain line and now certain times of they [the] year when the water comes high it comes way over that imaginary boundary. This as far as I know we don't get paid for another of our so called treaties broken. As far as I know, we have never been paid for the water gone over this imaginary line. Most of these people have returned to our reservation; from the way I see things this Garrison Dam as [has] ruin our way of life. With the coming of the Garrison Dam the chairman of the Tribal Council was George Gillette and as I stated before we lost a lot of lands, coal, wood, berries, way of life which will never be repaid by our government. Thank you.
(This is a recording that I personally have from Mr. Black Bear; no indication of when it was recorded.)

The Garrison Dam, like the small-pox epidemics, forever changed our way of life. The people of Fort Berthold did try to fight the construction as pointed out by a resolution that was unanimously adopted on May 21, 1945, at Elbowoods, North Dakota, now many feet under water. The resolution says in part:

WHEREAS; We the Mandan, Gros Ventre (Hidatsa), and Arikara people known as the Three Affiliated Tribes, residing on the Fort Berthold Indian Reservation, State of North Dakota, parties to the Fort Laramie, Wyoming, treaty in 1851; and

WHEREAS; The present Tribal Business Council of said Indians appointed Martin Cross, chairman; Douglas Standish, Earl Bateman, and Floyd Montclair under date of May 10, 1945 to study the proposed Garrison Dam embodied in the Rivers and harbors bill H.R. 3961, and the flood-control bill H.R. 4485 and

WHEREAS; The said subcommittee, after careful consideration of the proposed legislation, have made their report hereto follows;

THEREFORE BE IT RESOLVED; That the proposed Garrison Dam is designed to the detriment of Fort Berthold Indians, and the judgment of the Three Affiliated Tribes is to comply with the recommendation and report of the Subcommittee named herein, and we the people hereby voice our expression in unison opposing the Garrison Dam. We further tender the following points as hereinafter set forth, which is construed to mean by us as not to be to our best welfare and interest to the end that we may secure to ourselves economic independence.

Proposed Garrison Dam

First, It will destroy the Three Affiliated Tribes of the incentive to build substantial homes for their families in the future days to come.

Second, It will explicitly create more intricate problems in determining the individual equities in the tribal estates of the Fort Berthold Indians.

Third, It will repeal legislative enactments of vested rights on Indian allotment lands, and treaty stipulations of the Federal Government.

Whatever the good intentions of the proposed Garrison Dam are, they will be lost in the maze of red tape, and we respectfully request the

Figure 14.3 George Gillette.

> exemption of the Fort Berthold Reservation from the provisions of H.R. 3961 and H.R. 4485 respectively.
>
> Final Conclusion
>
> We, the Subcommittee, by Memorial of a recent session sent copies to officials and interested people, and expressed what we believe are the sound and basic principles be it.
>
> FURTHER RESOLVED; That copies of this resolution be sent to officials and interested people in the United States of America.
>
> This was signed by Chairman Martin T. Cross.

Of course the answer the U.S. Government gave can be seen in the picture of tribal Chairman George Gillette weeping at the forced signing of the final document to give our land away for the construction of Garrison Dam (Figure 14.3) and the creation of Lake Sakakawea. Looking on in what seems to be an unbelieving stare are some of the Elders that resisted the construction of the dam. Various letters that have more personal messages of tribal members not wanting to give up their land also talk of some political splits and understandings or misunderstandings among these folks. One example is a letter written on May 23, 1949. The letter is in the form of a statement:

> STATEMENT OF MRS. LILLIE WOLF MADE VOLUNTARIALLY (*sic*) BEFORE A PUBLIC MEETING HELD AT ELBOWOODS COMMUNITY HALL, ELBOWOODS, NO. DAK.

My deceased husband Daniel Wolf was very much against the building of the Garrison Dam. So, therefore, I feel bound that as long as I am alive, I will carry on the fight to oppose the building of the Garrison Dam and all the evil things that entail.

His last fervent wish before he died was that Martin Cross (then chairman) might carry on the work that he would have done if he were living yet. He is gone to a place where there is no return.

My reasons to make this statement is based on the fact that I still own and hold all my lands and property, although, I am poor and need money badly, I never sold any part of my land and if I had sold or lost my land through taxation or otherwise, I would be ashamed to look towards any deal that will give me some land at the sacrifice of others. Sometime ago James Hall a member of the tribal Council brought a whiteman to my place, Mr. Hall said to me that this whiteman has some good words to say to you. I never answered or acknowledged Mr. Hall address to me. Instead I talked to the whiteman (Mr. Hunter) in my own poor English, this is what I said. This is my own house, it is mine and I like it very much, I will not part with it for anything, and I will not answer any of your questions regarding your paper work, then I started to cry, because even our Tribal Council Chairman crys [*sic*] before the public, even though they never held any guns to his head. This whiteman (Mr. Hunter) said I agree with you, all you have said is true, this valley and bottom lands are the best lands and no equal lands can be found nowhere else. He further said that all Indian people on the western end of the reservation feel the same way as you do. He told me it was all right to form one's own opinion. He did not try to scare me or force me to sign any papers or do anything for him. The whiteman (Mr. Hunter) explained that he was working for other people and he started to fold up his papers and left my place. When outsiders feel that way about our problem, why should we spend any of our time listening to the words of Jim Hall, Tribal Council men do not have any worries or even felt any sorry of us Indian people. So why should we respect or have any fears for them. Nobody asked the council to go to Washington DC and further they did not have our permission to use our tribal funds at $20.00 a day to live in Washington, and use some of this money to buy half dozen silk dresses for their wives. One that you call Blackie has just now growed (*sic*) up, he has no experience or any wisdom, but others have influenced him. We know who I mean. Blackie is a tool, I say again he has no experience in matters of this big. As for our Supt. Ben Reifel, I want to know if he is our real superintendent or not. If he is in charge of us. He should tend to business and stay in the office to look after our needs and to sign our checks and so forth. That is what he is hired to do. Since he does not own any lands or property on our reservation he should keep his hands off the Garrison Dam business. Also, there is our missionary Reverend Mr. Case, he has no business to get mixed up in this matter and to work with the Tribal Council. We have our own Catholic priest here, he refused to go to Washington, DC with the Tribal Council, he told them that he was only interested in his church work among the Indian people, and had no wish to interfere with our business. I used to feel critical of our priest once but now I realize he is a better person that these people I have mentioned.

I heard that the Tribal Council has picked some outside delegates to be used as representing people at large, if that is so, it is fraud, because, we have never authorize them or even sanction such thing.

I heard that Mr. Case brought some men back with him, what is their business here, they are not invited by us. I suppose Rev. Case told them there was a gold mine here. I am very much disappointed today because our Washington delegates did not come to our meeting. I put the notice of the meeting on Thursday, we also invited the delegates in person to make their reports to us. Surely they have seen the Notice of the meeting. I am sure that if the tribes have any funds posted they would be quick to notice it. Joe Packineau claims he did not [hear] anything about the meetings being held—the Chairman also said he did not know anything about it. We asked Floyd Montclair to notify them about the meeting and he did. They used some excuse to go somewhere else in order to avoid attending our meeting. If these delegates were in the right they would have no fear to attend our meeting or go any place, it is because they feel guilty that they do not want to come here. They said that Davis Paint asked them to come to Nishu church, I asked Mr. Paint about it, he did not know we want them here, he said he never begged them to come to Nishu. That was their excuse. It seems to me that they went to Nishu to rehearse their act to fool us people. Why do you men remain silent, we are in dire emergency, let's do something about it. If those 17 men delegation came in here, I would not be pleased to see them, even though they are Superintendent's, Preacher's and Tribal Councilmen's. I wish Frank Crow's Breast was here I would tell him something.

Our council men are not with us poor people, otherwise we would not be feeling as, let's do as we do now. All the Councilmen are interested in is using our tribal funds, they are not to fill up their pockets full of our tribal money . . .

I do not like to hold meetings in the School houses, because most of us seemed to be afraid to talk in such places . . . all discussions should not be one sided.

(This letter was signed by Lillie Wolf, and the Witness was John Brave.)

I found this letter among some of the many papers I got from my father, and it is just one of many that tell of a people losing their land . . . forever. This letter also tells, as stated, of the people facing a challenge so massive that they had nowhere to turn, and as only one answer came from the U.S. Government, their only response was to argue among themselves sometimes and distrust leadership of all kinds: tribe, government, church, and individuals.

My folks told us kids of a terrible time, a time of forced movement of a people that lived along a River they believed to be alive and put there for a purpose, the purpose being to help our people live and prosper.

Our parents, grandparents and elders, all those that had to leave that beautiful River, left and I believe left a large part of themselves and their spirit there on the banks of the River, left our sacred sites, left our dead in some cases and moved to higher ground. I wonder how many would go back and look down at the dam now and, like all of our elders, cry for what was lost, the life, the feeling, the spirits, and a hope for a future.

The message that comes out from all this, though, is that we survived and are still proud members of the Three Tribes of the Fort Berthold Indian Reservation. Those of us that never had the opportunity to experience the lifestyle of our elders along the River, have had to learn how to live in the uplands. We never realized many of the promises that the U.S. Government bestowed upon us (and what people think . . . free!)—free electricity, free houses, free health facilities, free education . . . none of this (and more) never did come through. I and many others also need to have people understand, that none of these were FREE, it was paid for by the land we had no choice but to give up. They were paid for by the hearts that were broken by this removal, never to be fixed.

We have survived, learning how to enjoy that dam, swimming, fishing, boating, picnics, and gatherings at the shore of Lake Sakakawea. Today, we are on the brink of another big change, this time I hope we are ready, only time will tell. We are at the beginning of the development of a large oil find on our Reservation. I wonder how our long gone elders would feel, what advice they would give, how they would tell us to be ready.

And I wonder if anybody will once again stand on the banks of this (now) lake, and once again see the earthlodges and hear our warriors and our children in this midst of our dreams, and wonder what is to happen once again to our River and our People that lived by and on our backbone, the Missouri River?

Today, we still exist, but the sadness of this all, is that we are losing the generation that lived and knew our "Backbone" of our Culture, the Big Missouri. I hope that all people can learn about this lost life, for all the people that got misplaced and moved, and learn also about our sites and people, so maybe one day you too can stand on the banks of the lake and "see" in your mind our villages, and "hear" our people the Mandan-Hidatsa in our villages once again.

Aho.

Yellow Wolf

Author's note: As mentioned I never got to experience the "bottom Lands," nor did I ever get to see the landscape of the communities that now lay underwater. But I got to listen to the elders as a young boy, only to see them in my mind and through their stories. Most of my recall is from the stories that came from my father, Paige Baker Sr., my mother Cora Young Bird Baker, and a host of clan grandpas and grandmothers and other relatives including Sam Smith, Joe Black Bear, Mercy Walker, John Yellow Wolf, and many others.

CHAPTER 15

Flapping Tents, Outhouses, Hail Storms, and Running for Water: An Insider's View of Life in Interagency Archeological Salvage Program Field Camps, 1950s–1970s

David Mayer Gradwohl

The initial contact from Kimball Banks and Jon Czaplicki inviting me to join this review of the early days of the Interagency Archeological Salvage Program (IASP) some 60 years ago was not only surprising but a bit daunting. Before I could articulate a polite negative response to what I perceived as a gargantuan research project, Banks clarified that he wanted me to present an insider's view of what life was like in the IASP field camps. He said he was looking for someone who was old enough to have experienced those times, but not so ancient as to have lost his memory—in other words, antiquity without senescence. My Nebraska-registered birth certificate documents my age, and so far, it has not been contested by any "birthers" casting aspersions on my parents or nationality. Only Hanna, my wife of 55 years, can testify as to my mental decrepitude; and her lips are sealed.

Like many other Plains anthropologists of my generation, I began my career as a "shovel bum" on IASP field parties (Gradwohl 1997). Without running water and electricity, we lived in tent camps that were sometimes flattened by summer storms. The weather was sizzling. The work was strenuous. The pay was pitiful. Adding insult to injury, the Missouri Basin Project (MBP) crews, and those of most cooperating agencies, were restricted to males. Despite the near monastic ambiance and "no frills" accommodations, we had the opportunity to train under some outstanding professional mentors, explore some dramatic and significant archaeological sites, and establish productive and enjoyable

Dam Projects and the Growth of American Archaeology: The River Basin Surveys and the Interagency Archeological Salvage Program, edited by Kimball M. Banks and Jon Czaplicki, 214–227.

professional associations and personal friendships with colleagues spanning the decades since those summers of our youth.

As a matter of fact, the camp life component of IASP field projects has not been published in any detail. Site reports may contain a sentence or two along with a revealing photograph, but (rightly so) they emphasize historical and scientific aspects of the investigations. For the most part, the folklore of camp life in those days has been more the domain of stories around the campfire, bar-room reveries with old buddies and neophyte students at professional meetings, and occasional mentions in autobiographical essays. Perhaps the most informative brief summary of IASP field camp life was published by Donald J. Lehmer in his *Introduction of Middle Missouri Archeology* (1971). There, in five succinct paragraphs and five wonderful photographs, Lehmer outlines the story. Additional photographs in a report by Thomas Thiessen (1999) help in imagining what the IASP field camps were like.

My entrada into Plains archaeology in 1952 was the result of looking for a summer job during the spring of my freshman year at the University of Nebraska. My friend, James Bailey, an engineering major from Lincoln, had just obtained a job on a "salvage" archaeological crew of the Nebraska State Historical Society (NSHS). For several years, the NSHS had been a cooperating agency in the IASP. Since Bailey was the first to apply for the NSHS crew, he had been made the field assistant. The NSHS, however, still needed more field workers. I was curious about the possibility of getting out of my hometown for the summer and hanging out with several of my buddies. So, at Bailey's suggestion, I sought out Mr. Marvin F. ("Gus") Kivett at his office, then in the Nebraska state capitol building. Kivett explained that the NSHS would be working along the Missouri River, near Chamberlain, South Dakota. Crew members would live in a tent camp with no running water, spend all day digging, and be paid a grand wage of $.75 per hour. The latter item was bad news, since I was a card-carrying member of the International Hod Carriers and Common Laborers Union of America Local #540 and had earned the princely sum of $1.37 1/2 per hour doing construction work the previous summer. Kivett, however, in his quiet way, stressed that significant protohistoric material remains of the Arikara Indians (about whom I had never heard) would be discovered, and that these priceless antiquities would be saved from irretrievable loss by the construction and operation of the Fort Randall Dam. On the one hand, I was disappointed that poverty was apparently a vow archaeologists had to take; on the other hand, silence and chastity were not also required. So I joined the crew consisting of seven teenaged high school and college men. Little did I know the serendipity of this decision. What started out to be a summer job became a life-long career; and Gus Kivett became my first archaeological mentor.

The NSHS 1952 field camp at the Oacoma Site (39LM26/27) was everything Kivett had promised. The site and our encampment were west of Chamberlain, right on the main highway that used to stretch across South Dakota from Sioux Falls to the Black Hills (roughly the route of the extant Interstate 90 highway). We had old, heavy, army surplus tents for sleeping quarters. A large and very unwieldy surplus bi-pyramidal tent was erected to provide space for two kitchens, the dining hall, and laboratory. A.T. Hill's old trailer served as the summer home for Gus, Caroline, and Ronnie Kivett along with their dog Rocky.

My friend Bailey and I shared a dark, olive drab canvas wall tent measuring roughly six feet wide, 10 feet long, and six feet high with no mosquito netting and no floor. The gabled canvas cover was supported by a horizontal wooden ridgepole with metal ferules into which fitted the metal spikes on top of the two vertical wooden poles at the front and back of the tent. Outside, ropes angled down from the support pole spikes to the ground and were fastened to wooden or metal stakes. On each side of the tent, ropes extended from grommets (located along the seam where sloping roof and vertical wall canvas met) down to the ground where they were likewise fastened to stakes. In theory, this whole system was supposed to result in a solid, stable structure, especially if the ropes were maintained at a proper tightness. In practice, due to South Dakota's wind, however, the ropes tended to loosen and the tents flapped incessantly. The ridgepoles swayed back and forth, occasionally torqued, and then snapped. To make matters worse, all the stakes and ropes provided easy obstacles on which to trip as you walked around the camp at night.

Inside our tent, the furnishings were sparse and Spartan: army cots, orange crates that served as shelves, a rustic table, aluminum washbasin, and Coleman lantern. Ah, yes, Coleman lanterns were our only means of lighting since we had no electricity. I have many memories of pumping up those lanterns to do lab work and read at night, replacing the delicate mantles often, and driving into town to obtain "white gas" to fuel the devices. The lanterns hissed as the gas burned, but eventually we became oblivious to that sound. Our few changes of work clothes and other personal belongings were kept in suitcases underneath our cots. We put bricks or blocks of wood beneath the suitcases so they would not absorb ground moisture and cause our clothes to mildew. Since Al Rea, on whose land we were camping, was a junk dealer of sorts, Bailey and I borrowed some scrap marble slabs to cover most of the ground within our tent. All in all, I guess you could call our tent décor "Squirrelly American!"

The aforementioned large bi-pyramidal tent may have led some locals to think that the circus had come to town. Unfortunately, after we had been on the site for a short time, a very mild breeze came up and the bi-pyramidal tent blew down and ripped. We quickly regrouped, salvaged what canvas we could, bought some two-by-fours, and built a rectangular box-shaped tent to carry us through the season. A long table and benches took up the center of the tent. On one side in the back of the tent was a kitchen area where food for the crew was prepared on a Coleman stove. That area had stacked orange crates that were scrounged, under my supervision, from grocery stores in town. There we kept our pots and pans, dishes, glasses, tableware, and so on. We also had an old style icebox to keep small amounts of food refrigerated in between frequent forays for groceries in Chamberlain.

Kivett had arranged for Bea Rea, wife of the landowner, to cook for the crew. The food was simple, but adequate for teenagers who had managed to survive meals at Boy Scout or YMCA camps. On the other side of the tent was a second kitchen area where Caroline Kivett prepared food for her family. The crew members had KP duties consisting of setting and clearing the table, washing dishes, and making sure there was enough water for the camp. In the evenings, the crew usually sat around the mess tent, washing selected artifacts, gluing potsherds together, or reading. The Coleman lanterns attracted millions of bugs, so we had plenty of company.

We needed to run into town in the NSHS carryall truck to obtain water at gas stations. We hauled water in five-gallon "jerry" cans acquired through army surplus. Cans for drinking water were stored in the kitchen tent, and remained reasonably cool. Other cans were left in the hot sun all day and thus provided hot water for field "showers," that is to say, pouring a basin or two of water over your body, quickly soaping, and then rinsing off with more water. Clothes washing generally occurred on Saturday or Sunday afternoons (Figure 15.1). The NSHS provided metal washtubs

Figure 15.1 Doing laundry at the Oacoma field camp, 1952. Left to right: Peter Schmitt, Dennis Johnson, and David Gradwohl. (Photo from the collection of David Gradwohl.)

and several corrugated metal washboards for this chore. Clotheslines stretched across the camp gave our little habitation area a festive appearance, as our jeans, t-shirts, and underwear flapped in the wind.

Frequent runs also were made into Chamberlain to obtain block ice for the kitchen's small icebox. Normally, the ice was shipped in from Mitchell. One morning, however, when we went to pick up ice, the gas station attendant said that the ice from Mitchell had not been delivered, but if we came back during the afternoon, they would have "Huron ice." He meant, of course, ice from Huron, South Dakota, but we had understood "urine" ice and were not too excited about the possibilities.

Upon our arrival at Oacoma, Kivett assigned me the task of digging a latrine pit, perhaps a sort of doo-doo detail saved for the greenest of greenhorns. Directly west of our encampment was a small, raised, abandoned railroad spur, which extended south from the main railroad, then located north of the highway. Beyond the railroad spur was the portion of 39LM27 we would be excavating. So Kivett directed me to dig the latrine pit at an intermediate location so it would be readily accessible from both our living and work areas. Having never excavated a latrine pit before, I had absolutely no idea how large it should be except from the dimensions of a wooden two-seater platform that the NSHS museum carpenters had built in Lincoln as part of the field equipment to be hauled to South Dakota. With only a vague idea of how much "fill" might be generated by the field crew over an eight-week period, I decided to dig a nice deep pit that would definitely last as a "kaibo" for the whole field season. But, alas, I dug the pit a bit too wide for the wooden platform. So we jury-rigged a partial pit cover with army surplus canvas. This worked fine and dandy for our field party, but one uninformed visitor from the University of Nebraska crew slipped down into the upper stratigraphic levels of the pit before he could extricate himself from the situation.

A sort of semaphore flag system erected on top of the railroad spur assisted traffic control to the kaibo from the encampment. Red signaled "occupied," green announced that there was immediate seating available. The Kivetts maintained their own latrine tent near their trailer. While these facilities were admittedly a bit primitive, they seemed more advanced than the concrete culvert that Bea and Al Rea apparently used below the highway near their house. Their privy was flushed whenever it rained. Our kaibo was not flushed, but maintained by each user applying a cup of lime as a consequence of doing business there.

The summer's work was hard, but I learned a lot about the Arikara Indians and their neighbors along the Missouri River (Kivett 1958); and I acquired some knowledge concerning how to excavate and map circular earthlodge floors, dig cache pits, and identify some artifacts diagnostic of different times and cultures. At the end of the field season, I was right in the middle of digging a very rich cache pit, filled with all sorts of pottery, projectile points, scrapers, awls, and catlinite pendants. At my request, Kivett allowed me to finish excavating the pit after nightfall with the assistance of a Coleman lantern. The next morning as we left Oacoma, power machinery was backfilling the site. I had a lump in my throat. The Oacoma Site is now covered by the waters of Fort Randall Reservoir, officially known as Lake Francis Case.

My activities during the summer of 1953 further illustrated my lack of monetary acumen. Upon returning to Lincoln in the fall of 1952, I registered for an Introduction to Anthropology course taught by Dr. E. Mott Davis. I found the subject matter fascinating and the professor charismatic (or, as I later told him, charisMOTTic). Davis pressured me to sign up for his summer field school in archaeology. I grumbled that I already had a season's field experience and could not afford to go to summer school. Davis shot back with his characteristic lack of subtlety, "If you are going into anthropology as a career, you can't afford NOT to go on the field school." Thus, caving into a negative cash flow problem, I ended up *paying* money for the privilege of doing archaeology!

Davis was conducting excavations at the Red Smoke Site (25FT42), a stratified, multicomponent, Paleoindian site in the proposed Medicine Creek Reservoir, now called Harry Strunk Lake, near Cambridge, Nebraska (Davis 1953). Salvage archaeology at Red Smoke was another IASP project, in this case with the University of Nebraska State Museum (UNSM). The University of Nebraska field crew was billeted a short distance from the Red Smoke Site in a little, abandoned rural schoolhouse on the prairie, a place considerably less exciting than those described by Laura Ingalls Wilder or, for that matter, Willa Cather. But we did have the luxury of a wooden outhouse instead of a canvas kaibo.

Toward the end of July, however, Davis turned to another cooperative contract the UNSM had within the IASP, namely, an archaeological survey of the proposed Big Sandy Reservoir in southwestern Wyoming (Davis 1956). With all our field equipment packed into a sedan and a carryall truck, we drove to Farson, Wyoming, the quintessential wide place in the middle of the road. Our field party consisted of Mott Davis, two graduate students, and myself. Two tents comprised our field camp near the location of the proposed dam (Figure 15.2). We had no kaibo, but we did have shovels and miles of desert floor with no other humans around. Our refrigerator consisted of a stainless steel milk can submerged in the cool water of Big Sandy Creek. The Big Sandy survey project extended two weeks beyond the summer field school semester, so Mott offered to pay me for the time beyond my indentured servitude. He asked me how much money I needed for my summer's goal of purchasing a Weston-2 light meter. I answered $45.00. Davis checked his budget and graciously offered me $15.00.

Figure 15.2 Writing up field notes at night in the tent after a day of reconnaissance survey, Big Sandy Reservoir, Farson, Wyoming. Note Coleman lantern, upper right; Coleman stove, lower left. Left to right: Tom Newman, Jim Christensen, and David Gradwohl. (Photo by E. Mott Davis; River Basin Surveys and University of Wyoming Archaeological Laboratory, courtesy of Danny Walker.)

My financial position and proto-professional status improved during the summers of 1954 and 1955. Kivett advanced my rank to field assistant and increased my hourly wage to $1.75 or $2.00 per hour, as I recall. During those summers, our base camp was at the Crow Creek Site (39BF11) on the Crow Creek Reservation, south of Fort Thompson. This site, dramatically positioned on a high bluff overlooking the broad Missouri River flood plain, is a multicomponent locality with Woodland mounds, several Plains Village tradition components with fortification ditches, and some historic Sioux manifestations (Kivett and Jensen 1976).

For the seasons at Crow Creek, the NSHS significantly upgraded its field equipment and accommodations for camp life (Figure 15.3). Of particular delight to me was the purchase of new white canvas wall tents with separate canvas flies. These tents were larger, much cooler, lighter, and more pleasant than the army surplus olive drab tents, though they still flapped in the seemingly unending winds. Several sleeping tents

Figure 15.3 NSHS field camp at the Crow Creek Site, 1955. (Photo by David Gradwohl.)

housed crew members, and a larger one contained the kitchens, dining area, and laboratory. Two old army surplus tents were used to store field equipment and supplies and to stack bags of excavated artifacts until they could be hauled back to Lincoln. As before, the Kivett family lived in A.T. Hill's old trailer, which was fast becoming an historic artifact in itself. At Crow Creek, we had the luxury of a propane gas refrigerator that ran most days unless the wind extinguished the flame.

The closest potable water was the public pump at Fort Thompson. Water was hauled back to the field camp in large galvanized garbage cans, as well as jerry cans. Despite the fiasco of the latrine pit I prepared at Oacoma, Kivett appointed me as the official Sanitary Engineer at Crow Creek. Although I had not yet read Charles Sales' classic book, *The Specialist,* I was aware that aesthetics, as well as function, are important in latrine design. As a consequence, the kaibo at Crow Creek was positioned just over the edge of the steep bluff, out of sight from the field camp, but looking down some 60 feet to Wolf Creek. From a sitting position, one had an absolutely spectacular view out over several miles of the Missouri River bottoms. We assumed there was nobody out there to have spectacular views of our bottoms. The crew affectionately referred to the latrine as "the dapper crapper."

In addition to the field party from Lincoln, the crew at Crow Creek included young Yankton and Yanktonai men from Fort Thompson (Figure 15.4). Wayne Shields, Mark Shields, Billy Picotte, and a couple of others drove down to the site each day to assist in the excavations. Those of us of European descent normally worked in short pants and without shirts. Our goal was to get as complete and deep brown suntans as possible. At the time, this was considered "cool." One Native American

Figure 15.4 Native American crew members at the Crow Creek Site, 1955. Left to right: Wayne Shields, Mark Shields, and Billy Picotte. (Photo by David Gradwohl.)

coworker joked that he had a better suntan on his butt in the middle of winter than we had on our shoulders in the middle of summer. In fact, the Native Americans worked in long pants and long-sleeved shirts. They did not want darker skin colors, especially given the fact that there was rampant discrimination against Native Americans off the reservation at that time. For example, in Chamberlain, certain restaurants would not allow Indians to come inside and eat—they could only buy food at the take-out windows; and the municipal swimming pool did not welcome Indians. In retrospect, the long pants and long-sleeved shirts were probably part of Native American wisdom: most of us who sought deep suntans in our youth now suffer pre-cancerous skin lesions if not melanomas.

In Lehmer's words, "the weather is a matter of immediate concern to the field archaeologist, both on the dig and in the field camp" (Lehmer 1971:13). I could speak *ad infinitum* on this point, but a few examples will suffice beyond the destruction of the NSHS bi-pyramidal cook tent by wind at the Oacoma site in 1953. The slumping or collapse of tents due to high winds was not uncommon. Normally the tent poles held up and the canvas did not rip, but it was discouraging to have to spend time maintaining our sleeping and eating quarters. It was necessary to put a ditch around the tents to keep the ground inside the shelters from flooding when it rained. Sudden rain and hail storms could, and often did, ruin stratigraphic columns or horizontal surfaces that had been prepared for photographic documentation. The worst such instance occurred in 1955 at Crow Creek where a large area including an earthlodge, external cache pits, and postholes representing a palisade had been meticulously cleaned up so it could be photographed the next day along toward noon for optimal lighting. Not long after the crew quit work that day, a late afternoon hail storm (the likes of which I have rarely seen) hit the site with golf ball-sized chunks of ice and water; postholes and cache pits were filled with water, and the earthlodge floor looked like "shell-torn France." It was a long and discouraging process waiting for the site to dry out so we could clean up the excavation again and succeed with the photography. Scorching hot weather also impacted our work schedule. Frequently, we started work at 6:30 or 7:00 a.m. to avoid some of the heat of the day; and if the temperature soared over 100 degrees, we were usually allowed to quit work for the day.

Recreational outlets were minimal at these field camps. Driving to Fort Thompson to replenish our drinking water supply was, of course, an excuse to buy popsicles at the small store there or visit John Saul, a Yanktonai elder who carved wooden figurines for his grandchildren and for those of us archaeologists who commissioned him to do so. Occasionally, we foraged for berries or chokecherries, some of which were later made into jam; or we hunted for rattlesnakes, whose meat we fried up to eat and whose skins were fashioned into hatbands by one crew member. On most evenings, the crew assembled in the mess tent to engage in various activities by the light of Coleman lanterns: reading, writing up notes or letters, or gluing potsherds together. We found the latter pastime more interesting than jigsaw puzzles. There also were, of course, the proverbial evening campfires associated with burned marshmallows and sing-alongs. I made my musical contributions—not proficient to be sure—on a ukulele or guitar.

On Saturday nights, the crews were permitted to take the state carryall into town for grocery shopping after which we might take in a movie. Since most of the crew members were under the legal drinking age of 21, bars were not really an option. Usually, we just did what the locals did: walk around main street in a clockwise direction, then walk around main street in a counter-clockwise direction, and finally just stand around and look at everybody else.

On most days after work, we went over to Lake Bedashosha, an artificial impoundment along Crow Creek, for bathing and swimming (Figure 15.5). On Saturday and Sunday afternoons, we would lounge on the beach there (the better to get a darker suntan), write letters, or hike around the eroded banks of the lake looking for selenite crystals and other interesting minerals. The lake also offered a place to meet some of the local people. On one occasion, we met an Episcopal minister from Fort Thompson. He invited us to attend Sunday morning services at his church. A couple of the crew members went to take Holy Communion, and I tagged along as an "ethnographic observer." I was intrigued with the bilingual prayer books and hymnals, written in English and Dakota. As I compared the texts, I wondered (and still do) if the term *Wakantanka*

Figure 15.5 NSHS crew members relaxing after bathing and swimming in Lake Bedashosha. Left to right: Peter Schmitt, Bernie Leonard, Jim Bailey, and David Gradwohl (with ukulele).

really does translate specifically as *God* in the Judeo-Christian worldview. Regardless, we learned to sing "For The Beauty of the Earth" and several other hymns in the Dakota language.

Leisure activities additionally included opportunities to visit other archaeological field camps and sites. Particularly popular was Carlyle Smith's University of Kansas (KU) field crew, which was also excavating near Fort Thompson. Smith's excavations were, of course, interesting. Even more exciting, however, was the fact that women were included on the Kansas field parties, giving real meaning to the word "party." Specifically, I recall Dena Ferran [Dincauze] and Ann Stofer [Johnson] who graced the KU crew. Our crew also visited the MBP excavations at the Cheyenne River site (39ST1), supervised by Waldo Wedel, who was accompanied by his children and wife, Mildred Mott Wedel. Mildred, of course, provided an impressive role model for younger women who were hoping to have careers in anthropology.

Despite the rugged living conditions and hard work, the crew's *esprit de corps* was normally high. There were the usual in-group jokes and teasing which, as structural anthropologists would observe, contributed to the "social solidarity" of the group. This also was manifested in a song we put together about our project. The chorus went, "On the Crow Creek Reservation/On the Crow Creek excavation/Lookin' for the stratification/Cache pits and fortification." As far as I know, nobody wrote down the words to the verses, and I have long since forgotten them.

A sense of larger involvement in the IASP salvage operations was facilitated by the weekly field reports disseminated by the MBP. The Smithsonian parties were obliged, and cooperating agencies invited, to submit short progress reports each week. The MBP office then reproduced the reports on a ditto machine, and sent each field party the purple-printed news of the important finds from the previous week. Kivett shared those reports with the crew, and we all found them instructive and interesting. Additionally, an inspection group from the National Park Service (NPS) made annual visits to all crews working within the MBP. I recall, in particular, visits from John Corbett (Chief Archeologist for the NPS), Frank H. H. Roberts, Jr. (Director of the Smithsonian River Basin Surveys), Paul Beaubien and Wilfred D. Logan (successive Regional Archeologists for the Midwest Region of the NPS), and Jackson W. "Smokey" Moore, Jr. (NPS archaeologist). Needless to say, our crews anticipated these occasions by special efforts to tidy up the site and by plans for a field dinner of fancier than normal fare to impress the visiting dignitaries. After all, our next year's field budget was in the hands of these men! The annual inspection tours added to our general sense of engagement in the IASP enterprise and heightened an awareness of its purpose beyond our hourly wages.

In the fall of 1955, I went off to graduate school, first at the University of Edinburgh, Scotland, and then Harvard University. Thanks to the draft, I also served two years in the United States Army. In 1962, I accepted a teaching and research position at Iowa State University (ISU) in Ames. My wife and I wanted to return to the midlands, and I was anxious about the possibilities of continuing my research in Prairie-Plains archaeology. For better or worse, I was the sole anthropologist in ISU's Department of Economics and Sociology, that is agricultural economics and rural sociology (Gradwohl 2003). Prior to arriving in Ames, however, I had received a communication from Wilfred D. Logan at the NPS Midwest Regional Office in Omaha. He congratulated me on snagging a tenure-track job and said his office had money for salvage archaeology along the Des Moines River. Thus it was, via IASP cooperative agreements between ISU and NPS, that I was able to establish the Iowa State University Archaeological Laboratory (ISUAL), initiate the ISU summer field school in archaeology, and run salvage archaeological projects in Red Rock Reservoir and Saylorville Reservoir from 1964 through 1970.

During those years, ISU was able to arrange with the NPS and U.S. Army Corps of Engineers for the archaeological crews to use vacant farmhouses and other structures (prior to their destruction by the reservoirs) as field headquarters. Perhaps our cushiest field headquarters consisted of a cluster of summer cabins along the river near Madrid, Iowa, in 1968. A pseudo-log cabin provided space for the classroom, dining room, kitchen, girls dorm, and field laboratory; a separate house was the boy's dorm; a small two-room cabin with a screened porch provided the residence for Hanna, me, and our three children; and another structure, dubbed "the kibbutz," housed married couples (legal or common law) on the crew. Although we were able to hook up to electricity at these field camps, we still had to acquire outhouses, and truck in potable water.

The biggest change was that I decided that women, as well as men, could apply for the field crews (Figure 15.6). ISUAL's first archaeological crew in 1964 created a bit of a stir in that regard. More than one of my professional colleagues asked, "won't you have trouble with women in the field?" My answer was, "none that we don't already have on campus!" Several farm men commented, "boy, I wish I could get my wife to work like that," apparently oblivious to the fact that milking cows, slopping hogs, driving tractors, preserving garden produce, raising children, and so on constituted work for their wives. The presence of women on archaeological crews was still rare enough that the *Des Moines Register* sent a woman reporter (in high heels) down to our Dunreath field camp to write an article, headlined "The Women Who Work at Red Rock" which appeared on what used to be called the women's "society page" (*Des Moines Register*, July 26, 1964). I think this may have been the

Figure 15.6 Women and men on ISU's first archaeological field crew, 1964, at Milo's Silo site, near Dunreath, Iowa. Left to right: Georgina Granberg, Ronald Seastrand, Patricia Trow Mericle, Ronald Kivett, Peter Brennan behind Michael Wonio, John Rosenberg, Mark Peterson, Lynn Swanson, Robert Knowles, Donald Graybill, Allan Sharlin, and Steve Polley. (Photo by David Gradwohl.)

first (and also the last) time Plains archaeology has been featured on a newspaper's society page.

The reporter did not ask many, if any, questions about the archaeology per se. She interviewed five women: three on the excavation crew, the cook (a male crew member's wife), and Hanna. She recorded the kinds and amounts of food we ate, and asked the female crew members what it was like working for Gradwohl in the hot sun. One answered, "he's absolutely relentless." The newspaper article included a photo of Hanna and our two oldest children. Since the reporter neglected to note that my wife has a first name, the photo caption reads, "Mrs. David Gradwohl, whose husband is director of the expedition seeking ancient objects at Red Rock Reservoir area, manages to look fresh and pretty, although her family is living under crowded conditions in [an] old farmhouse serving as headquarters for the crew and living quarters for some. She is shown with her daughter Jane, 11 months, and son Steven 2 1/2." We all laughed at the article because that day was the only time Hanna wore a dress all summer. All of the subsequent ISUAL crews included women as well as men. In due course, women served as field assistants and supervisors; and one, Nancy Osborn Johnsen, became my long-time research associate and co-principal investigator. To wit, there was none of the "trouble" some of my colleagues had predicted. Female crew members always performed their tasks as well as or better than the men.

In conclusion, I hope I have given at least a glimpse into what life was like in IASP field camps, in addition to my good fortune in excavating at some significant archaeological sites, helping to recover some important and finite cultural remains before they were destroyed, training under some impressive mentors, working with a number of supportive colleagues, and teaching many outstanding students. If it were not for the IASP, I would not have become a professional anthropologist. I would have missed my fascinating journey from the era of salvage archaeology to that of cultural resource management—an expedition from *peon* to *P.I.* (Principal Investigator). For that I am grateful.

CHAPTER 16

From RBS to CRM: Late Twentieth-Century Developments in American Archaeology

Francis P. McManamon

Introduction

With five notable events, 1974 was a banner year for the development of cultural resource management (CRM) in the United States. The first was the publication of the Advisory Council on Historic Preservation's (ACHP) "Procedures for the Protection of Historic and Cultural Properties" in the January 25th issue of the *Federal Register*. These procedures are designated formally as Title 36 (Parks, Forests, and Public Property), Chapter VIII (Advisory Council on Historic Preservation), Part 800 of the United States Code of Federal Regulations (36 CFR Part 800). They are more commonly and regularly referred to as the "Section 106 procedures," named after the section of the *National Historic Preservation Act of 1966* (NHPA) that they implement. By whatever name, these regulations have substantially impacted how most archaeological investigations have been conducted in the U.S. for nearly 40 years.

The second and third events were the cultural resource management conference held at the Federal Center in Denver (Lipe and Lindsay 1974) and the Airlie House Archeological Resource Management Seminars held at the conference center of the same name in northern Virginia (McGimsey and Davis 1977). These meetings focused on how to interpret new laws, regulations, policies, and directives affecting archaeological resources and the anticipated changes in the practice of archaeology that would be necessary to address these new legal and regulatory requirements. The Denver CRM conference and the Airlie House seminars were attended by many of the individuals in academia, public agency, and museum organizations who were trying to solve the problems associated

---

Dam Projects and the Growth of American Archaeology: The River Basin Surveys and the Interagency Archeological Salvage Program, edited by Kimball M. Banks and Jon Czaplicki, 228–252.

with the new legal, policy, and regulatory situations that influenced the management and preservation of archaeological resources in the U.S. Gathered at these meetings were archaeologists who began the practice of CRM. They initiated the development of concepts, methods, and procedures that still guide CRM.

The fourth event was the publication of Lipe's (1974a) article "A Conservation Model for American Archaeology," in which he forcefully and effectively pointed out that the archaeological record is a "non-renewable" resource. He rightly noted that while highway construction, mining, dam and reservoir construction, looting and vandalism, and natural processes destroy archaeological sites, archaeologists also exploit the resource. Archaeological fieldwork, specifically testing and excavation, preserves data about and from sites but, in the process, destroys the in situ resources. One of Lipe's important points in this landmark article was that archaeologists must work at conserving the in situ archaeological record and rescuing archaeological data threatened by development. He advocated an approach that would "avoid our getting to last-ditch, emergency salvage situations" (Lipe 1974a:215). Along with McGimsey's (1972) *Public Archeology*, Lipe's article established the philosophical foundation for a conservation-oriented CRM.

The fifth event was the passage of *The Archaeological and Historic Preservation Act* (AHPA). Known popularly as the Moss-Bennett bill (named for the primary sponsors of the bill) the act amended the *Reservoir Salvage Act of 1960*. The AHPA broadened the obligations of federal agencies to take into account archaeological sites affected by "any federal construction project or federally licensed project, activity or program" (Section 1; 16 U.S.C. 469).

In this essay, my use of the term "cultural resource management" focuses on archaeological resources. In its broadest sense though, the term refers to ways of managing a wide range of cultural resource types including archives, historic structures, cultural landscapes, ethnographic resources, museum collections, traditional cultural properties, in addition to archaeological resources. These other types of cultural resources present management challenges similar to those for archaeological resources with respect to identification, evaluation, treatment, preservation, and management (King et al. 1977:7–10; Knudson 1986; McGimsey and Davis 1977:27).

In the U.S., CRM developed within the discipline of archaeology during the early 1970s and was influenced by the laws, events, ideas, and regulations mentioned above and others. Fowler (1982:1) attributes the first use of the term "cultural resources" to specialists within the National Park Service (NPS) in 1971 or 1972. "Management" became linked with cultural resources shortly after this, as exemplified in the titles of the

Denver conference (Lipe and Lindsay 1974) and Airlie House seminars (McGimsey and Davis 1977). McGimsey (1991, 2004:3–7) concluded that, while the term came into use during the early 1970s, "as a coherent, identified concept applicable to . . . archaeology, it crystallized in the minds of a few archaeologists at the Airlie House conference in 1974 and was given formal birth, or at least christened, with the publication of that report" (McGimsey 1991:xvii).

American Archaeology before 1974

Although it was a new term and acronym in the early 1970s, CRM is related to concerns about the preservation and protection of archaeological sites that developed in the late nineteenth and expanded throughout the twentieth century. In the late 1800s, archaeologists and others concerned with preserving archaeological artifacts and sites persuaded the federal government to preserve and protect a few sites from destruction and deterioration, such as Casa Grande Ruins near Coolidge, Arizona (Clemensen 1992; Lee 2000:205–209). In 1906, Congress passed and Theodore Roosevelt signed the *Antiquities Act* to protect archaeological sites on public lands and regulate how they were investigated (Lee 2000; McManamon 2006a; Thompson 2000). During the first third of the twentieth century, the Department of the Interior and, after its creation in 1916, the NPS attempted to preserve archaeological sites but allocated few resources to these efforts (Browning 2003:4–11). In 1935, Congress enacted the *Historic Sites Act*, which provided additional authorization for federal involvement in preserving archaeological resources (McManamon 2001:259–261, 2006a:168–169). Also in the 1930s, the back-to-work programs of Franklin Roosevelt's New Deal included several federal agencies that sponsored archaeological investigations (e.g., Lyon 1996).

In the late 1940s, concern about destruction of and need to preserve archaeological sites, artifacts, other remains, data, and information about sites resulting from projects authorized by the *Flood Control Act of 1944* led to the authorization of the River Basin Surveys (RBS) program. Archaeologists got wind of the planning for this massive public works program scheduled for implementation following the conclusion of World War II. The act authorized the construction of dams, reservoirs, and irrigation projects in major drainage basins throughout the U.S. Concerned individuals and scientific organizations lobbied government agencies to take archaeology into account when planning these projects (Brew 1947; Jennings 1985; Johnson 1947, 1951a; Roberts 1948; Wendorf and Thompson 2002). The RBS was the outcome and the archaeology practiced came to be called "emergency" or "salvage" archaeology.

"Emergency/salvage archaeology" is used here to refer to the methodological approach the RBS and related programs adopted. By the late 1970s, many archaeologists derided the term "salvage archaeology" and held it in low regard. Even today, the term is derisive and has a negative connotation to many. The term "emergency" is linked with salvage archaeology in deference to J. O. Brew (1961, 1968:1), who used this term, and Jesse Jennings (1986:59), who preferred the term. Brew spent much of his career involved in and in promoting the preservation of data from sites that would be destroyed by development. He is the archaeologist most often associated with development of the emergency/salvage approach (Brew 1961, 1968; Brew et al. 1947). Jennings created and administered an exemplary emergency/salvage program in Utah (Fowler 2011; Jennings 1963; Jennings and Sharrock 1965; Lipe 2012) and wrote clearly and forcefully about both the benefits and failings of emergency/salvage archaeology (Jennings 1963, 1985; Jennings and Sharrock 1965).

The RBS program was a cooperative effort involving the NPS, Smithsonian Institution (SI), U.S. Army Corps of Engineers, Bureau of Reclamation, national archaeological and scientific organizations, universities, museums, and historical societies throughout the country. Numerous publications document the development of this program (Brew 1968; Committee for the Recovery of Archaeological Remains 1958; Corbett 1961; Johnson 1947, 1951a; Kohler 1953; Thiessen 1999). The SI, through the Bureau of American Ethnology, published a series of reports of investigations (River Basin Surveys Papers, 1–39). A second series of reports (Publications in Salvage Archeology, No. 1–13) was published by the RBS itself. Petsche (1968) produced a book-length listing of publications resulting from or associated with RBS investigations. Summary reports (e.g., Roberts 1952b, 1961b) and at least two books for the general public (Baldwin 1966; Silverberg 1967) also were published.

The NPS developed and administered the program, building upon its mandate in the 1935 *Historic Sites Act* to undertake surveys of archaeological and historic sites and provide technical assistance in archaeology and historic preservation. The passage of the *Reservoir Salvage Act* (RSA) in 1960 strengthened this mandate, as the act authorized emergency/salvage archaeology as part of dam and reservoir construction. RSA applied mainly to the U.S. Army Corps of Engineers and the Bureau of Reclamation, the principal agencies involved in dam and reservoir construction; it did not address other kinds of federal actions (Davis 1972).

While the SI coordinated RBS investigations and some studies were conducted by SI staff, typically investigations were carried out through contracts or cooperative agreements with universities and state museum that the NPS or SI administered. Most work was done just prior to or during reservoir construction, not as part of project planning. This "just ahead of the bulldozers" approach became one of the chief criticisms of RBS archaeology. In the early 1950s, however, archaeological investigations associated with other kinds of development also had to "salvage" data in the face of construction. Pipeline and highway emergency/salvage programs were developed in some regions of the U.S. Nusbaum (1956) was instrumental in the development of pipeline salvage archaeology in the Southwest, while Wendorf (1962) established a similar program with the New Mexico State Department of Transportation for highway projects.

By the late 1960s and 1970s, heightened concern about environmental issues resulted in the enactment of laws to protect aspects of the human and natural environment. Prominent among these laws were the NHPA and the *National Environment Policy Act of 1969* (NEPA). Both statutes affected the development of archaeology in the U.S. Both required federal agencies and their state counterparts to take cultural resources, defined broadly and including archaeological sites, into account as they planned, reviewed, or undertook projects or activities. Executive Order (EO) 11593, signed in 1971, required federal agencies to identify, evaluate, and take into account when planning or undertaking projects' cultural resources "under their jurisdiction or control that appear to qualify for the National Register of Historic Places" (Section 2 [a]). This requirement, which was interpreted broadly, subsequently was incorporated into the NHPA by amendment.

These changes occurred during a time of cultural upheaval and social change in the U.S. Challenges to authority, the environmental and civil rights movements, and protests against government actions, especially the Vietnam War, were hallmarks of these decades. Within archaeology, an explicitly scientific approach in theory, method, and technique, the "new archaeology," was growing and affecting how archaeological investigations were conceived and conducted (Binford 1962, 1964; Binford and Binford 1968; Watson et al. 1971).

CRM Develops

On both local and regional fronts, criticism of and dissatisfaction with the emergency/salvage approach increased during the 1960s and early 1970s. Wendorf (1962:76–84) spoke to this criticism in an attempt to define salvage archaeology and its potential for contributing to the discipline. Jennings (1963, 1985) also addressed this criticism, which he turned on its head in a convincing justification and advocacy of emergency/salvage archaeology, as least as it was practiced in Utah under his direction. Also, there was recurrent criticism of the lack of dissemination of data, results, and interpretations from emergency/salvage projects. Lehmer (1965) criticized the publication of results from the Missouri River Basin salvage program, estimating that only 25 percent of the data was published or distributed as of 1964, nearly 20 years after the program began.

In the southwestern U.S., the Arizona State Museum, Arizona State University, NPS Arizona Archeology Center, and Museum of Northern Arizona sponsored a series of conferences, meetings, and workshops in the early 1970s (Lindsay and Lipe 1974:viii–xi). These meetings focused on the changing perspective and practice of emergency/salvage archaeology and the emergence of CRM and led to the Denver Federal Center Cultural Resource Management Conference. Lindsay and Lipe (1974:ix–xi) described how NEPA, with its emphasis on environmental impact studies and planning, and EO 11593 injected greater concern about conservation and management into the treatment of threatened archaeological resources. Some archaeologists, particularly in the western states, observed that federal agencies were beginning to hire resource specialists to address wildlife concerns, watershed quality, and so on. These archaeologists began to suggest that land and resource management agencies also ought to have professional archaeologists on staff to manage the archaeological resources for which agencies were responsible (Lipe, personal communication, 2012).

The 1974 Cultural Resource Management Conference

The Bureau of Land Management hosted the conference at the Denver Federal Center. Topics included institutional responsibilities for conserving archaeological resources, how contracting for archaeological studies should be organized, standards for archaeological investigations, and how to address the anticipated increase in archaeological investigations that new laws and regulations required. Five months after the conference ended, Lipe and Lindsay (1974) published the proceedings, which included the presentations, formal comments, and transcripts of discussions related to the topics. The publication provides a detailed picture of the topics and concerns associated with the early development of CRM and, along with McGimsey's *Public Archeology* (1972), was one of the few books on the topic at the time. Table 16.1 is a compilation of presentation titles and their authors, as well as the formal discussants and others who made comments during the discussions that followed each presentation (Lipe and Lindsay 1974). The conference also spawned the American Society for Conservation Archaeology (ASCA), which focused on administrative, legal, methodological, policy, and regulatory issues associated with managing and conserving archaeological resources (Mayer-Oakes 1974). The ASCA, which continued to operate into the 1980s, organized and shared information on issues important to CRM.

In their introduction to the proceedings, Lindsay and Lipe (1974:vii) contrast the new and developing perspective of "conserving" the archaeological

Table 16.1 Information on topics, presentations, authors, commenters, and discussants 1974 Denver CRM conference

Paper Title and Author(s)	Commenters and Discussants
Introduction—Alexander J. Lindsay, Jr. and William D. Lipe	
History of Archaeological Conservation Policy and the Moss-Bennett Bill—Douglas H. Scovill	
Institutional Responsibilities in Conservation Archaeology—Raymond H. Thompson	Jesse D. Jennings, Charles R. McGimsey, III, Bob Butler, James Judge, William D. Lipe, Tom Layton
Archaeological Contractual Agreements—James I. Carpenter	Jesse D. Jennings, Clayton Hagen, Jack Rudy, Douglas H. Scovill, Hester Davis, Cal Cummings, James Judge, Dave Madsen, Larry Banks, Donald S. Miller, Jerry Nash, Earl Swanson, Jim Hester
Gearing Up for Contract Archaeology—Douglas H. Scovill	Thomas F. King, S. Alan Skinner, William D. Lipe, Jim Hester, Stan Bussey, James Judge, Hester Davis, Bob Butler, Raymond H. Thompson, Dee Green, Dave Rice, Cal Cummings, Lloyd Pierson, Donald S. Miller, Dave Breternitz, Thomas F. King, Larry Banks, Charles R. McGimsey, III, Garland Gordon
Preparation of Archaeological Inputs to Environmental Reports and Environmental Impact Statement—Roberto Costales	Larry Aten, Louis S. Wall, William D. Lipe, Tom Layton, Larry Aten, Louis S. Wall, Garland Gordon, Jim Hester, Dee Green, Roberto Costales, Donald S. Miller, Hester Davis
Contract Standards for Archaeological Studies—Keith Anderson	Leslie E. Wildesen, David G. Rice, David Cole, Larry Aten, Hester Davis, Adrian Anderson, Floyd Sharrock, Bob Butler, Roberto Costales, William D. Lipe
Certification for Archaeologists and Amateurs—Donald S. Miller	Hester Davis, William Mayer-Oakes, Alexander J. Lindsay, Jr., William D. Lipe, Bob Euler, Alvin Briggs, Charles R. McGimsey, Jr., J. O. Brew, Mark Grady, Don Fowler
The Restructuring of a Profession—Charles R. McGimsey, III	R. Gwinn Vivian, James Judge, Larry Aten, William D. Lipe, Stan Bussey, Tom Layton, Jim Sears, R. Gwinn Vivian, Thomas F. King, Hester Davis, Bob Butler, Don Fowler
General Comments on the Conference—J. O. Brew	

From: *Proceedings of the 1974 Cultural Resource Management Conference, Federal Center, Denver, Colorado*. Museum of Northern Arizona Technical Series No. 14, 1974 record with the "exploitative" manner of emergency/salvage archaeology. In his presentation, Thompson (1974:15–16), then the Director of the Arizona State Museum, emphasized and elaborated on this contrast in perspective:

> I draw a contrast between salvage and conservation archaeology solely for the purpose of emphasizing that conservation programs are not just overgrown salvage programs. I have no desire to denigrate salvage programs. In fact, I think that we can look back upon the salvage period in American archaeology with some justifiable pride. However, this laudable progress has been achieved largely at the expense of the resource base, that is, archaeological sites. I say this not because the archaeology done on salvage projects has been substandard, but because the decisions having to do with the use of the resource have not been made in most cases by archaeologists. Very often decisions are made that have a major impact on the archaeological resources without any consultation with the archaeologist.

Scovill, then the newly appointed NPS Chief Anthropologist, how new laws and regulations might allow greater emphasis on conservation. He noted (Scovill 1974a:3):

> Our task at this conference can be to begin to work out the specifics of how, at the operational level, we can achieve a rational balance between development to meet our industrial society's needs and conservation of historic resources so that our society and its future generations can reflect upon the past.

Scovill reviewed the new legal and regulatory landscape. He noted that the ACHP's "Procedures for the Protection of Historic and Cultural Properties" took as it's statutory authorities Section 106 of NHPA, EO 11593, and NEPA. He correctly interpreted the procedures as requiring a planning and review process that federal agencies and their state counterparts would have to follow. He further noted that this process would precede any archaeological data recovery that might be authorized under the new Moss-Bennett law (Scovill 1974a:4–8). Such a process to identify important archaeological resources that might be impacted by public works, assess whether the impacts would adversely affect the resources, and, if so, mitigate the resources by recovering data systematically prior to destruction would make a conservation approach possible.

Among the participants in the Denver conference were two archaeologists very familiar with the emergency/salvage approach. One was J. O. Brew, and the other was Jesse Jennings. Brew was dedicated to the preservation of antiquities and worked on both a national and international level to preserve archaeological monuments and sites. On the national level, Brew, along with others, was primarily responsible for the formation of the Committee for the Recovery of Archaeological Remains (CRAR), which provided informal professional oversight of federal emergency/salvage programs from the 1950s to the 1970s (Wendorf and Thompson 2002). On the international level, Brew chaired the UNESCO Committee for Monuments, which was responsible for such emergency/salvage projects as that associated with archaeological sites threatened by the Aswan Dam on the Nile River.

Brew was invited to make "General Comments on the Conference." In his introduction of Brew, Lipe spoke of his achievements and career focus: "I would like this time to recognize one man in particular who has played a formative role in this effort to save our archaeological heritage and who is continuing to work for archaeological conservation" (Lipe 1974b:137). It is hard to imagine a more symbolic "passing of the baton" between the emergency/salvage approach and the new conservation approach and Brew made fitting comments. He observed that the new approach had the benefit of federal land management, permitting, and construction agencies funding archaeological investigations, which was something that had been impossible, or at least difficult, in his experience. His conclusion emphasized continuity by noting that McGimsey and Thompson had joined CRAR and that Thompson would soon succeed him as chairman (Brew 1974).

The Airlie House Archaeological Resource Management Seminars

Later in 1974, the Airlie House seminars were organized as six one-week-long workshops involving six to eight experts. Each seminar included analysis, discussion, and reporting on key topics facing American archaeology at the time. Topics included certification of archaeologists and accreditation of institutions and programs; guidance for preparing and evaluating archaeological reports; management of archaeological resources; communication in and about archaeology; archaeology and the law; and archaeology and Native Americans (McGimsey and Davis 1977:2–4).

Most of the topics had a clear relationship with the changes in how archaeologists regarded archaeological resources at the time. In particular, the topics related to the new conservation approach to managing and treating archaeological sites. Many of the participants also attended the Denver conference and four of the six topics—archaeology and the law, management of resources, guidelines for reports, and certification and accreditation—were the same or similar to topics discussed at the Denver conference. Table 16.2 lists the participants in and the titles of the final reports from each seminar.

Table 16.2 The management of archeological resources: The Airlie House report

Seminar Title	Seminar Participants
A Consideration of Law in Archaeology	Lawrence Aten*, Vernon Bellecourt, Joe Brecher, Michael Moratto, Charles R. McGimsey III*, Marvin Woolf
Cultural Resource Management	Adrian Anderson*, Hester Davis*, Mark Grady William D. Lipe*, Bruce McMillan, Lloyd Pierson* Margaret Weide
Guidelines for the Preparation and Evaluation of Archaeological Reports	Keith Anderson*, Hester Davis*, Rob Edwards Michael B. Schiffer, Stanley South, Gwinn Vivian*
The Crisis in Communication	Louis Brennan, Brian Fagan, Frank Hole, Alice Kehoe. Tom King*, Charles R. McGimsey III*, Nathalie Woodbury
Archaeology and Native Americans	Vernon Bellecourt, Hester Davis*, Cynthia Irwin-Williams, Elden Johnson, Clydia Nahwooksy, Emory Sakequaptewa, Marion White
Certification and Accreditation	Edward B. Jelks, James Judge*, Charles R. McGimsey III*, Stuart Struever, Raymond H. Thompson*, Fred Wendorf

*Following name indicates also participated in the Denver CRM conference.

Participants in the seminar about the management of archaeological resources were asked to consider their subject broadly. They should "plan with the total resource base in mind . . . there also must be established a set of archeological problem-oriented priorities, as well as a set of priorities oriented toward the total public good" (McGimsey and Davis 1977:3). They further were directed to develop a "philosophy of archeological conservation and of what constitutes mitigation, rather than the more restricted concept of salvage" (McGimsey and Davis 1977:3).

The CRM seminar affirmed the perspective that destruction of the archaeological record was a serious problem that needed to be addressed. The participants further noted that the emergency/salvage approach as most commonly practiced was not sufficient. A new approach was needed, and it should be guided by two principles:

1. Archaeologists . . . [must be] involved with the total planning process so they have an active voice in determining what must be destroyed.
2. Work in a project area—of whatever size—must be placed in a broad topical and regional context if there is to be any possibility of recovering maximum useful data. This awareness necessitates regional overviews and well-thought through research designs (McGimsey and Davis 1977:26).

In setting the stage for describing this new approach, the seminar participants briefly summarized recent developments in U.S. laws and public policies. In particular, NHPA, NEPA, AHPA, and EO 11593 were highlighted. They suggested that "this increasing concern by American archeologists for the total resource base paralleled and is perhaps related to a growing concern about resource conservation and environmental quality within American society at large" (McGimsey and Davis 1977:28).

The resource management chapter in the Airlie House report presents an extended discussion of three topics that continue to be relevant in CRM and managing archaeological resources: mitigation, significance, and compliance (McGimsey and Davis 1977:28–40). The section on the mitigation of adverse effects describes three types: avoidance, preservation (e.g., physical maintenance or protection of an impacted resource), and investigation (e.g., data recovery). The section also describes three kinds of impacts: direct, indirect, and potential. These terms and concepts are now standard in CRM as is how significance and compliance are considered.

In their introduction to the Airlie House report, McGimsey and Davis (1977:7) conclude that the report "attempts to pull together a great deal of current thinking about the archeological profession and its problems and thereby give everyone a solid point of departure for development and growth." Looking back, the report records the status of thought at an important point in the history of American archaeology and provides a solid foundation for consideration of how to deal with current challenges.

By the time the seminars were published in 1977, other publications covering CRM had appeared: *Conservation Archaeology: A Guide for Cultural Resource Management* (Schiffer and Gumerman 1977), *Anthropology in Historic Preservation: Caring for Culture's Clutter* (King et al. 1977), and the third section of *Social Archeology: Beyond Subsistence and Dating* (Redman et al. 1978). Publications of CRM project reports or report summaries also began to appear and circulate widely (e.g., Breternitz 1983; Schiffer and House 1975, 1977; Teague and Crown 1982). The NPS and the short-lived Heritage Conservation and Recreation Service also published reports on CRM topics (Benchley 1976; Dincauze and Meyer 1977; King 1978a; Portnoy 1978, 1979; Talmage and Chesler 1977). As a rapidly developing aspect of the profession, CRM quickly attracted professional attention and comment.

A Conservation Model for Archaeology

Lipe's 1974 publication perhaps had the most significant impact. Lipe began by reviewing the current state of archaeology. He recognized, as did others, that development was destroying archaeological sites at an alarming rate and that the federal government directly or indirectly supported much of this development. Lipe's insight was to turn the problem of site destruction on its head. Rather than developing better and more efficient ways to salvage archaeological data from sites about to be destroyed, Lipe recommended that archaeologists find ways to avoid site destruction. Focus on conserving sites; only if avoidance was not possible, should sites be excavated and the data recovered and preserved (Lipe 1974a:214).

To achieve this perspective, Lipe (1974a:215) noted that "archaeological resources must be accorded a higher value by society than they are now, so that more projects will be designed to avoid sites." He identified public education and outreach about archaeology and archaeological resources as an important means to social change. Another way of increasing regard for archaeological sites was to "greatly expand our efforts to gain institutional and regular access to the planning process with respect to land alteration schemes forthcoming from society" (Lipe 1974a:215–216). In her review of this article, Sebastian (2006:124–125) concludes that Lipe was prescient regarding many of the developments in CRM and had a broad perspective on how to achieve the goals he outlined.

The National Historic Preservation Act

In the 1960s, historians, historical architects, and other historic preservationists raised concerns about the destruction of historic buildings and structures by federal urban redevelopment and highway construction projects (Advisory Council on Historic Preservation 1986:15–24; Glass 1990; Hosmer 1981:809–1076; Murtagh 1988:39–77; U.S. Conference of Mayors 1966). These concerns led to the enactment of the NHPA in 1966 and its regulations, 36 CR Part 800, in 1974. The NHPA represents a major expansion of the basic policy of the *Antiquities Act of 1906*; there is a public interest in preserving archaeological and historic resources (McManamon 1996, 2006a:166–174; Rogers 2006). NHPA is a broad statute and has been expanded through substantial amendments in 1976, 1980, and 1992. The act embraces a wider range of historic property types than the *Antiquities Act* or the *Historic Sites Act of 1935* and is broader and more inclusive because its provisions encompass historic properties of local and state significance.

Like the earlier statutes, the central focus of NHPA is the public and noncommercial values of historic properties and how these properties are treated. The policy calls for considering historic properties within the context of modern development and the economy. The law and its procedures require involvement of cultural resource experts and consideration of impacts to historic properties in planning public projects and programs. Early consideration of how historic properties may be affected provides opportunities to avoid or mitigate potential adverse effects, thereby conserving these properties. Including or enhancing preservation of historic properties can have economic, aesthetic, associative, and historic benefits, as is the case with historic structures successfully rehabilitated for modern commercial uses.

The archaeological community was not noticeably active in the discussions, conferences, and planning that led to the passage of NHPA or in the development of the initial regulations to implement the new law (Davis 1996; King and Lyneis 1978:874–875; King et al. 1977:30–38; Townsend 1994). Only after the NHPA was enacted did some archaeologists begin to pay attention to it. Ironically, Brew, who was deeply involved with the development and expansion of emergency/salvage archaeology in the U.S. and other countries, wrote of the importance and potential benefits of NHPA after its enactment (Brew 1968:4–6). At the time, Brew was working with the NPS through CRAR, in particular with NPS Chif Historian Robert F. Lee and Chief archaeologist John Corbett. Along with Lee, Brew was a member of a committee charged by NPS Director George Hartzog with developing an internal plan for implementing the NHPA (Lee et al. 1966). Their report provided a general plan for how the NPS would address its archaeological and historic preservation responsibilities within and outside the national park system. The report acknowledged the existing Interagency Archeological Salvage Program of the NPS as a good example of how archaeological and historical assistance to other public agencies could be carried out. Lipe (personal communication, 2011) provides a personal recollection on this topic:

> In the early 1970s, those of us involved in salvage archaeology in the SW were not much aware of the NHPA and its implications. We tended to think more about NEPA as the master planning law under which we would be working, and thought that the Council on Environmental Quality might be the source of regulations that would more fully spell out how archaeological sites were to be handled.

The NHPA established the ACHP, the State Historic Preservation Officer (SHPO), the National Register of Historic Places (NRHP), and more recently, the Tribal Historic Preservation Officer (THPO) programs, all of which are important parts of the national historic preservation program. For CRM, the most important procedures are the ones in the Section 106 of the law, especially 36 CFR Part 800, as this regulation references EO 11593 and NEPA. As the importance of Section 106 review procedures became more apparent and were promoted by the NPS, archaeologists raised concerns that the procedures only addressed archaeological sites listed in the NRHP (Grady and Lipe 1977). In 1976, NHPA was amended to require that federal agencies also take account of properties that may be considered "eligible" for the NRHP. The 1976 amendments essentially codified the part of EO 11593 that required federal agencies to take account of cultural resources that might be eligible for the NRHP, as well as properties already listed on it. Regarding archaeological resources, it "effectively require[d] that [archaeological] surveys be done in advance of all federal undertakings" (King et al. 1977:42).

During the second half of the 1970s, there was considerable debate and disagreement among archaeologists about the need for and value of NHPA, the NRHP, and related policies and procedures (Davis 2009:27–32; Grady and Lipe 1977; King 1978b). King et al. 1977:41–44. Lipe (1978) was concerned that the NRHP focused on sites associated with events important in American prehistory or history; it was not the best tool for preserving the larger number of sites about which little if any information was known and which may have potential for furthering research. He objected to the NRHP serving as the main resource for identifying significant historic resources that public projects may affect as required by NHPA and NEPA. "It is this key role I object to because of the redundancy of the National Register and because of what I think are inherent limitations in its assumptions as viewed from a research and conservation orientation" (Lipe 1978:144).

During this period, the NPS Office of Archeology and Historic Preservation (OAHP), which was tasked with implementing the national historic preservation program outside the national park system, championed the use of the Section 106 procedures and the NRHP eligibility criteria. Archaeologists working in CRM, in particular those who worked in the SHPO offices and programs, received a flurry of guidelines and regulations suggesting or directing how Section 106 and the NRHP were to be used for archaeological properties (Advisory Council on Historic Preservation 1976; Interagency Archeological Services 1976a, 1976b; National Park Service 1977a, 1977b, 1977c). This network of archaeologists in SHPO offices, which received a substantial part of their funding from the federal program administered by the NPS, was instrumental in spreading the use of the NRHP and Section 106 procedures.

One measure of the centrality that the Section 106 procedures have taken in CRM today is the number of pages and text devoted to discussions of the procedures in current books on CRM. King's (2004) text on cultural resource laws and practice devotes over 100 pages (pp. 81–190) out of 400 pages to the procedures; Neumann and Sanford (2001:27–54) devote over half of their chapter on "Laws, Regulations, and Protocols" to Section 106. It must be remembered that the Section 106 procedures are just that, a set of steps linked to planning federal projects or programs. The procedures do not dictate nor guarantee a particular outcome, but aim to ensure that important historic properties, including archaeological sites, are appropriately considered during planning and implementation. Ideally, the outcome is that important archaeological sites are either not adversely effected by the project or program, or if they are, the effect(s) is mitigated in some way.

Plenty of critiques of the procedures and how they have been misapplied exist, but so too do examples of and suggestions for their effective use. Regional planning for cultural resources rather than project-specific reviews have been used effectively (e.g., Barker 2009). Archaeologists have long discussed alternative approaches to evaluating the significance of archaeological sites (e.g., Sebastian 2009). A wide range of ways to mitigate adverse effects have been used, and even more have been suggested (e.g., Chandler 2009). As a key aspect in implementing archaeological and historic preservation laws in the U.S., these procedures will continue to be utilized, hopefully in increasingly effective, economical, and productive ways.

The Archeological and Historic Preservation Act

In the late 1960s and early 1970s, archaeologists worked with Congress to address more broadly the adverse impacts to archaeological resources from a wide array of federally sponsored projects and programs (Davis 1972). The result was passage of the AHPA, which amended the RSA (King et al. 1977:38–40; McGimsey 1981, 1985:327). The AHPA requires federal agencies to take archaeological resources into account when executing, funding, or licensing projects. The act recognizes that archaeological investigations are a legitimate project cost. The AHPA followed the tradition of emergency/salvage archaeology, as might be expected, since the act was essentially an amendment to the RSA, the primary statutory authority for this approach.

The impetus for AHPA was to mitigate the destruction of archaeological sites as the result of actions funded or otherwise supported by federal agencies. The AHPA was intended to establish the same kind of archaeological emergency/salvage requirement of RSA for other kinds of federal activities. The expectation of the act seemed to be that excavation would continue to be, as in the existing emergency/salvage programs, the appropriate form of mitigation, rather than the broader palette of mitigation options that NHPA allowed, including in situ conservation.

The statute assigns to the Secretary of the Interior substantial authority to preserve historical and archaeological data and remains. The secretary can assist other federal agencies and even private organizations or individuals in historical and archaeological preservation. The ACHP authorizes the Secretary, upon notification that significant historical or archaeological data may be irrevocably lost or destroyed, to undertake necessary studies independent of, although in consultation with, the federal agency responsible for undertaking, funding, or licensing the project. The secretary is authorized to consult about the ownership of and appropriate repositories for artifacts and other remains recovered by investigations conducted under the statute. The ACHP is one of the statutory authorities for the federal regulations governing the curation and care of federal archaeological collections and associated records (36 CFR Part 79). Two other requirements are for the Secretary to submit an annual report to Congress on archaeological survey and recovery activities authorized under the statute and for federal agencies responsible for projects to transfer to the secretary funds to assist them in meeting their responsibilities, up to one percent of the total amount authorized for the project. This last continued the practice of having NPS (as delegated by the Secretary of the Interior) oversee the archaeology done for other agencies.

Developing regulations and procedures to implement aspects of the AHPA proved to be problematic. One example is differences in interpretations about Congressional intent regarding the one percent of project costs for archaeological studies. Some viewed this requirement as a strict limit on archaeological expenditures; others interpreted it as a general guideline. Another problem was that the statute did not provide for systematic reviews of federal projects to determine whether they would affect archaeological sites. The procedures for project review under Section 106 (36 CFR Part 800), on the other hand, did provide for a structured review of federal actions.

There were disagreements about how funding for archaeological investigations authorized by the act should be distributed. Some archaeologists (e.g., Jennings 1963, 1985:286–287; Scovill 1974b) believed that cooperative agreements between educational institutions and government agencies, like NPS, that were administering emergency/salvage archaeology were effective and efficient means of transferring funds. They argued that these agreements allowed for flexibility and economy in the use of funding while encouraging university professionals to develop research expertise related to the archaeological sites investigated. Others (e.g., King 1977; Wilson 1978) criticized such arrangements as "an old boy's network" that funneled funds to organizations that were not necessarily the best at doing the research.

Proposed guidelines to implement reporting requirements for archaeological data recovery projects were published in draft for public comment as 36 CFR Part 66 (National Park Service 1977b). However, the guidelines were never completed published as final regulations. Much of the guidance in the draft was incorporated in or superseded by sections of the *Secretary of the Interior's Standard and Guidelines for Archeology and Historic Preservation* (National Park Service 1983) and other official guidance.

Most archaeologists agree that the lack of explicit linkage between the AHPA and the NHPA, and NEPA was a serious oversight that should have been avoided (Davis 1996; King 1978b; King et al. 1977:30–44; McGimsey 1985:330). The drafters and proponents of AHPA, however, did not explicitly relate this legislation with the then-developing approach to archaeological resources that was part of the wider historic preservation movement. This preservation approach, based upon NHPA, came to emphasize planning, using the NRHP criteria in assessing site significance and the need for preservation, project review under Section 106, and preservation of sites in situ when possible and feasible.

The AHPA contributed importantly to the development of CRM (Davis 1996; McGimsey and Davis 1977; McGimsey 1981, 1985, 2004 [various chapters]). The act states clearly that federal agencies are authorized to fund archaeological investigations, reports, and other activities to mitigate the impacts of their undertakings on important archaeological sites. Furthermore, lobbying for passage of the act pressured agency managers to meet their archaeological responsibilities. The interactions with archaeologists supporting the bill "primed the pump" and pushed agencies to comply once the law was enacted. Finally, lobbying efforts by archaeologists and archaeological organizations between when the bill was first drafted in 1969 and its passage in 1974 alerted much of the archaeological community to the impacts of federal actions on archaeological resources and the importance of tracking government programs and statutes, regulations, policies, and new legislation.

There are more regulations, statutes, and professional developments since the early 1970s that have had substantial impacts on how CRM is practiced in the U.S. today. The *Archaeological Resources Protection Act*, the *Abandoned Shipwreck Act*, and the *Native American Graves Protection and Repatriation Act* are examples of key statutes that, with their related regulations, affect contemporary CRM. Archaeological ethics, the curation of archaeological collections and data, public education and outreach, and community and descendent group involvement all have come to play important roles in the conduct of CRM. But, available space requires that further consideration of how these laws and topics fit into the contemporary practice of CRM must be left for another essay.

CRM: A Better or Just a Different Approach to Preserving Archaeological Resources?

Was the development of CRM a reaction in opposition to the tradition of emergency/salvage archaeology or a continuation of that tradition improved by a greater focus on conservation and preservation? Although some CRM proponents draw a clear line demarcating the practice of emergency/salvage archaeology from the conservation ethic of CRM, there are clear links. One is that some archaeologists and archaeological organizations were involved in both. Perhaps the most prominent examples are Bill Lipe and Don Fowler. Both of them at the beginning of their careers in the late 1950s and early 1960s had important roles in the Glen Canyon project, a major emergency/salvage project (Fowler 1986:148–150, 2011:242–326; Lipe 2012). Both also have been influential in the development of CRM through their publications, projects, and training of graduate students. Likewise, Hester Davis, Tom King, and Bob McGimsey are prominent examples of archaeologists who participated in emergency/salvage programs early in their careers but moved into CRM, thereby influencing its development in their career trajectories.

Connections between emergency/salvage archaeology and CRM have been noted on other levels. Jennings offered his perspective regarding philosophical links between the two. He noted that the RBS marked

> a nationwide acceptance by archaeologists of a leading role in the conservationist movement . . . one can reasonably argue that Cultural Resource Management (CRM), currently the strongest force in shaping Americanist archaeology, is a direct legacy of the 1940s, 50s, and 60s. It is merely a more self-conscious posture, one more restrictively and complexly defined. (Jennings 1985:281)

Jennings (1985:293) believed that emergency/salvage archaeology "contributed heavily to the philosophy that shaped the course of CRM; total excavation simply became unacceptable . . . for both CRM and the salvage concept, preservation was the shared goal."

Wendorf (1962) and Jennings (1963, 1985) described aspects of emergency/salvage archaeology that also are basic to CRM—logistics, research strategies, method, and technique:

1) The locations of investigations were and are determined by the location of projects, not by the research interests of a particular researcher.
2) Both require a complex administrative framework, although that for CRM differs from the institutional relationships developed in emergency/salvage archaeology.

3) The substantive results of both reflect the diversity of the archaeological record, at least in part because investigations were and continue to have a wide geographical distribution; they are not limited to areas archaeologists might have selected because of known sites.

Jennings (1985:293) also noted three additional improvements that emergency/salvage archaeology developed and/or improved and that CRM has further developed and widely utilized:

1) The effective use of heavy equipment for archaeological excavation;
2) The development and use of permanent central laboratories for large archaeology projects, such as the RBS office in Lincoln, Nebraska; and
3) The use of multidisciplinary research teams, such as the one Jennings assembled for the Glen Canyon project (Fowler 2011:242–326; Jennings and Sharrock 1965).

Despite these similarities, the conservation perspective and the emergency/salvage approach differ in important ways. In their collection of articles on conservation archaeology, Schiffer and Gumerman (1977) took pains to describe what they saw as the differences:

> Cultural resource management is more than just new jargon or a subterfuge for salvage archaeology. It is the realization of a new social philosophy for the treatment of the all too ephemeral materials that contribute to our understanding of our cultural heritage . . . [In] compliance with its legislative underpinnings, cultural resource management studies entail sophisticated research planning, execution, and results—consistent also with the highest standards of modern archaeology (Schiffer and Gumerman 1977:1–2).

Further, because of the NHPA and NEPA, CRM archaeology is more directly involved during planning to identify and assess archaeological resources that a public project might affect. Under the NHPA, only those archaeological resources eligible for listing or included in the NRHP that may be affected by federal undertakings must be considered. Further, CRM also involves a full range of treatments for endangered resources rather than the limited focus on data recovery.

Finally, the growth of CRM has been accompanied by a transformation in the size and structure of the profession since the days of emergency/salvage archaeology. One response to meeting the requirements of the NHPA, NEPA, EO 11593, and the AHPA was that a wider range of public agencies, particularly at the federal and state levels, employed professional archaeologists. This was especially so of land managing agencies, such as the Bureau of Land Management, the U.S. Forest Service and U.S. Fish and Wildlife Service, and the NPS. Another response was among agencies, such as the Federal Highway Administration and the

Environmental Protection Agency, that did not manage land but funded or licensed projects. These agencies tended not to employ archaeologists, but they transferred their CRM responsibilities to state agencies or private firms that carried out their projects and, in turn, hired professional archaeologists. A third response was employment of professional archaeologists by multidisciplinary consulting firms in the private sector. A fourth was the appearance of consulting firms dedicated solely to CRM.

By the end of the 1970s, federal and state agencies had developed a network of professional archaeologists in headquarters, regional, and local offices. These archaeologists undertook the activities necessary to implement CRM laws, regulations, policies, and guidelines. At the state level, the NPS required SHPOs to have a professionally-qualified archaeologist on their staff for the state to qualify for annual funding from the Historic Preservation Fund authorized under the NHPA. In turn, many SHPOs required that state offices have professionally qualified archaeologists on their staff. There has been a corresponding growth of employment in the private sector, especially among consulting firms. Such firms range in size from large national or international multidisciplinary firms that undertake CRM activities associated contracted public projects to firms that specialize in CRM.

During the early years of CRM, the archaeological profession experienced a theoretical and methodological upheaval referred to as "new archaeology" (Redman 1991). King (1971:255–258) suggested that a conflict existed between the theoretical and methodological ambitions of the new archaeology and the requirements of emergency/salvage projects. One might surmise that his juxtaposition of the two perspectives is rhetorical because by the end of the article King proposed incorporating "problem oriented" methodologies and research designs into emergency/salvage projects. King suggested that coordinating and planning archaeological studies at a regional scale, rather than focusing on individual projects or sites, would permit a new archaeological approach to emergency/salvage projects. He cited studies in the North Coast Range of California to exemplify this approach (King 1971:259–261).

King's article raised concerns among some archaeologists that those advocating the new archaeological approach would abandon the archaeological record to the destructive forces of modern development and looting. Ruth Gruhn (1972:354) declared:

> Given the present self-righteous, more-scientific-than-thou attitude of new archeologists, which is so clearly expressed throughout King's article . . . it is doubtful that cooperation between academics and salvage program administrators in the face of the immediate destruction described by Davis can be achieved in time.

The debate about the new archaeology and emergency/salvage archaeology continued. Longacre and Vivian (1972) objected to Gruhn's characterization of the situation among archaeologists and her comments about the new archaeology. They argued that emergency/salvage projects did not preclude "theory-oriented" and problem-oriented research that the new archaeologists advocated. As examples of emergency/salvage projects that incorporated the methods the new archaeology advocated, they cited Prescott College's work on Black Mesa for Peabody Coal, studies by the Arizona State Museum for the Arizona Department of Transportation, and hypothesis-testing research by the University of Arizona using data collected as part of a development at Carter Ranch (Longacre and Vivian 1972:812).

Binford's investigations at the Hatchery West site near Carlyle, Illinois, most clearly debunked the notion that emergency/salvage projects could not incorporate aspects of the new archaeology (Binford et al. 1970). Binford was at the center of the new archaeology movement. The investigation at Hatchery West was part of an emergency/salvage project undertaken in 1962 and 1963. The analysis by Binford and his collaborators demonstrates that problem-oriented research based on an explicit and scientific research design, key aspects of the new archaeology, could be done as part of an emergency/salvage archaeology project.

Many archaeologists in the 1970s who began their careers in the public and private sectors were trained in graduate programs where aspects of the new archaeology were taught. Fred Plog, among the earliest advocates of the new archaeology, argued that "the intellectual traditions new archeologists have introduced . . . should form the basis for training for archeologists concerned with cultural resource management and the issues that cultural resource management raises . . . specifically . . . goals, 'regional archeology,' areal survey, and excavation" (Plog 1978:424). While some components of the methodological and theoretical orientations of the new archaeology were not directly applicable to CRM projects, explicit and detailed research designs were widely used to guide field and laboratory activities. Other scientific methods advocated by the new archaeology, such as hypothesis testing and statistical sampling, also were adopted. Schiffer and Gumerman (1977) cite many examples of early CRM projects that apply the methods and techniques of the new archaeology to site estimation and evaluation, impact assessment, and data recovery. Projects come from all areas of the U.S.: Arkansas, California, Missouri, New York, South Carolina, and the Southwest.

The connections between the new archaeology and CRM are both chronological and functional. Both involved changes to the status quo and occurred at about the same time. Some of the aspects of the new archaeology directly related to issues and topics at the center of CRM, such as the use of explicit research designs, links between analytical testing and interpretive outcomes, and site discovery and evaluation methods and techniques.

CRM: Critiques and Challenges

The focus of this essay is on the events, publications, individuals, laws, and regulations that affected the development of CRM as a new and important sector of archaeology. Additionally, CRM is a vital part of archaeology, as it is now conducted in much of the world. In the United States, a high percentage of professional archaeologists are engaged in CRM. And CRM impacts how archaeology is perceived and archaeological resources are preserved. Yet, the rapid and substantial changes within the archaeological community during the 1970s sparked discussions, debates, and disagreements regarding the benefits of CRM and the quality of archaeological work done in its name. Changes in the proportions of professional archaeologists employed in the public, private, and academic sectors and variations in and new kinds of professional responsibilities and duties caused debates about ethics, relevance, and values. In an essay reflecting his 50 years of archaeology, Jennings had some adverse commentary on CRM:

> I mention the last convulsive event merely in passing. . . . It is the traumatic refocusing and unfocusing of archaeological effort in the early 1970s on what is called Cultural Resource Management. This I can conservatively label as a mixed blessing, which has often led to grave mistakes and has had costs far beyond its scientific rewards; while there are a few bright spots in the record, on balance CRM has generally harmed our discipline (Jennings 1986:60).

Jennings (1985:281) was similarly critical of exorbitant costs attributed to some CRM projects: "one can also suggest that CRM is extraordinarily wasteful and inefficient if one measures funds expended against useful data recovered and disseminated." He also expressed concern about the professional standing of archaeologists who were in the CRM sector of the discipline:

> CRM and its inventories require a work force of specialists far larger than the salvage operations ever mustered, but the contributions to archaeological knowledge remain scandalously small for the funds expended. Even worse, and far more serious are the losses the CRM contract archaeologists themselves suffer. Those include reduced self-esteem, loss of autonomy and independence of action, and eroded scientific integrity, as bureaucrats and contracting officers reduce archaeology to numbers and even prescribe the procedures to be used in the field (Jennings 1985:293).

In general the debates and disagreements have moderated from vitriolic to civil. However, not all of the issues raised in the turmoil over CRM during its early days have been resolved. Much of the fieldwork done in the U.S. today is associated with CRM. So topics related to CRM, such as ethics, professionalism, quality of work, reporting and preserving information, costs, and so on continue to be important to American archaeologists.

Many, perhaps most, professional archaeologists support conservation as the appropriate treatment of the in situ archaeological record. Operationally, this approach has the goal of managing archaeological resources for future generations, yet allows for prudent, justifiable use of the resource for research. There is general agreement that professional archaeologists in the public, private, and academic sectors and the statutes, regulations, policies, and guidelines that protect archaeological resources are important in maintaining and, preferably, strengthening the preservation of archaeological sites.

Numerous challenges face archaeologists involved in contemporary CRM. In the conclusion of this essay, I mention four of the most pressing. One is the struggle to provide easy and wide access to archaeological information from and about past and current investigations so that the information can be used and inform future work. This challenge is not new, it has been with CRM since its early days (e.g., Longacre 1981:487; Schiffer and Gumerman 1977:85). The problem also was one of the critiques of the salvage/emergency archaeology. Inadequate access to and preservation of archaeological data continued to be a problem for CRM as it developed (e.g., Lipe 1996:25; McManamon and Wendorf 2000:45). The explosive growth in the number of archaeological investigations and production of data has increased the size of the challenge. It seems that digital technology offers a way to meet this challenge and there are exciting prospects here (e.g., Kintigh 2006; McManamon and Kintigh 2010; Snow et al. 2006). However, the comprehensive changes to CRM practice to meet the challenge are substantial.

A second challenge is how to maintain the professional archaeological workforce in the face of inevitable demographic changes and shrinking government services. In public agencies and CRM firms, retirements among the cohort of archaeologists hired at or near the beginning of CRM have begun and will accelerate in the next decades (e.g., Departmental Consulting Archeologist 2010:79–81). Programs and policies that make it easier to retain older, experienced archaeologists and permit flexible retirement arrangements should be developed. Older archaeologists, who are retained as younger professionals are hired as ultimate replacements for them, can pass along what they have learned during their careers and reduce the loss of "organizational memory" when they retire fully.

A third challenge is managing archaeological physical collections for effective access, preservation, and use is another continuing subject for CRM (e.g., Childs 2010). Archaeological collections held and managed by federal agencies, which of course are only a part of the collections from CRM investigations, are reported to total nearly 46 million artifacts, plus over 100,000 cubic feet of artifacts for which no counts are available (Departmental Consulting Archeologist 2010:48–51). Certainly progress in this area has been made since the era of salvage/ emergency archaeology and the early days of CRM, but a great deal remains to be accomplished to solve the problem.

Finally, there is the challenge of agreeing on what are the appropriate goals of archaeological site conservation. In a 1996 article, Lipe warned that a strictly "hands off" approach to site conservation would prohibit appropriate, focused investigations of archaeological resources to learn about the past. Site preservation programs, whether overseen by public agencies or private organizations, need to develop policies and procedures that allow for consumptive use of portions of sites for well-justified and well-designed archaeological research.

In the U.S., no single organization provides comprehensive public archaeological expertise and services. The NPS might have assumed such a role in the 1970s. Instead, NPS staff developed standards and guidelines for the identification, evaluation, planning, documentation, preservation, protection, and treatment of cultural resources for use in the public, private, and academic sectors (National Park Service 1983). How public archaeology in the U.S. came to be organized and the roles of various individuals and organizations in that process will continue to be debated and explored (Lipe 1996:23; McGimsey 2006a, 2006b; McManamon 2006b:7–8; Wendorf and Thompson 2002:327).

Whatever the specific interpretations might be, a primary point is that a broadly organized discipline without a strict hierarchical power structure needs strong, effective leadership and coordination. Due to the decentralized activities and responsibilities of many public agencies at various levels of government, the existence of private sector firms that play a vital role, and an academic sector that provides professional training along with general archaeology courses for non-professionals, there is no single obvious organization or office to provide effective leadership. Yet, such leadership is needed to ensure professional and consistent conduct, integration of activities, and accessibility of results. The four challenges identified in this section require concerted, coherent, and broad-based approaches and activities to overcome the problems that are faced. None of these challenges are insurmountable, but coordinated action will be most effective, and is needed to overcome them efficiently and economically. Leadership, whether from the public agencies, the

CRM industry, the university sector, museums, professional organizations, or some combination of these will enable solutions that are most effective and economical.

Acknowledgments

The initial idea for this essay was to expand an entry on "Cultural Resource Management" I had prepared for the *Archaeological Method and Theory: An Encyclopedia* (McManamon 2000). As I delved into the history of CRM and its foundational documents, I changed more and more of the original text. At this point, a forensic literary analyst would be hard pressed to find the snippets of the original text that remain. I appreciate the generosity of Jon Czaplicki and Kimball Banks in asking me to be part of the team that has created this collection and their patience regarding the preparation of this essay. I appreciate the assistance of Grant Snitker, Center for Digital Antiquity, School of Human Evolution and Social Change, ASU, in compiling the lists of topics and participants in the 1974 Denver CRM conference and the Airlie House seminar. Don Fowler and Bill Lipe read and commented on an earlier version of this essay. I appreciate very much that they took the time to share their perspectives and make comments. I hope they are satisfied with the way in which I have addressed their suggestions.

References

Adams, William Y. 2009 *The Road from Frijoles Canyon: Anthropological Adventures on Four Continents.* University of New Mexico Press, Albuquerque.

Adams, William Y. and Nettie K. Adams 1959 *Inventory of Prehistoric Sites on the Lower San Juan River, Utah.* Museum of Northern Arizona Bulletin 31, Glen Canyon Series 1. Flagstaff.

Adams, William Y., Dennis P. Van Gerven, and Richard S. Levy 1978 The Retreat from Migrationism. *Annual Review of Anthropology* 7:483–582.

Advisory Council on Historic Preservation 1976 *Guidelines for Making Determinations of "Adverse Effect" and "No Adverse Effect" for Archeological Resources in Accordance with 36 CFR Part 800.* Advisory Council on Historic Preservation, Washington, DC.

——— 1986 *Twenty Years of the National Historic Preservation Act.* Advisory Council on Historic Preservation, Washington, DC.

Ahler, Janet Goldenstein 1974 A Cross-Cultural Comparison of Meaning Ascribed to Social Studies Concepts. Unpublished Ph.D. dissertation, University of Missouri, Columbia.

Albert, Richard C. 2005 *Damming the Delaware: The Rise and Fall of Tocks Island Dam.* 2nd ed. Pennsylvania State University, State College, Pennsylvania.

American Antiquity 1936 News and Notes. *American Antiquity* 1:227–236.

——— 1937 News and Notes. *American Antiquity* 2:217–232.

——— 1938a News and Notes. *American Antiquity* 3:271–284.

——— 1938b News and Notes. *American Antiquity* 4:164–171.

——— 1939 News and Notes. *American Antiquity* 5:157–175.

Ames, Kenneth M., Don E. Dumond, Jerry R. Galm, and Rick Minor 1998 Prehistory of the Southern Plateau. In *Plateau,* edited by Deward E. Walker, Jr., pp. 103–119. Handbook of North American Indians, Vol. 12, William C. Sturtevant, general editor, Smithsonian Institution, Washington, DC.

Anderson, Adrian D. 2006 Recollections. In *Plains Archaeology's Past: A Collection of Personal Narratives,* edited by M. F. Hawley and V. A. Wulfkuhle. *Plains Anthropologist* Vol. 51, No. 200, pp. 797–808.

Anonymous ca. 1946–1970a River Basin Surveys Correspondents. Manuscript on file, River Basin Surveys, National Anthropological Archives, National Museum of Natural History, Smithsonian Institution, Washington, DC.

——— ca. 1946–1970b Missouri River Basin Table of Participants 1940s, 1950s, 1960s. Manuscript on file, River Basin Surveys. National Anthropological Archives, National Museum of Natural History, Smithsonian Institution, Washington, DC.

——— ca. 1946–1970c River Basin Surveys. National Anthropological Archives, National Museum of Natural History, Smithsonian Institution, Washington, DC.

——— 1998 Amy Evelyn Harvey. *Plains Anthropologist* 43(165):274.

Anonymous 2001 Katharine Bartlett, 1907–2001. *The SAA Archaeological Record* 1(5):32.

——— 2010 Norma Mikkelsen. *Salt Lake Tribune*, August 29, 2010.

——— 2011 Dee Ann Story 1931–2010. *The SAA Archaeological Record* 11(3):5.

Antevs, Ernst 1948 *Climatic Changes and Pre-White Man*. University of Utah Bulletin 38 (20):168–191. Salt Lake City.

Armelagos, George J. and Dennis P. Van Gerven 2003 A Century of Skeletal Biology and Paleopathology: Contrasts, Contradictions, and Conflicts. *American Anthropologist* 105(1):53–64.

Ayers, Howard G. 1965 An Appraisal of the Archeological Resources of the Blue Ridge Project: Grayson County, Virginia, and Alleghany and Ashe Counties, North Carolina. Interagency Archaeological Salvage Program. River Basin Surveys. Smithsonian Institution. Manuscript on file, National Anthropological Archives, Smithsonian Museum Support Center, Suitland, Maryland.

Babcock, Barbara A. and Nancy J. Parezo 1988 *Daughters of the Desert: Women Anthropologists and the Native American Southwest, 1880–1980. An Illustrated Catalogue*. University of New Mexico Press, Albuquerque.

Baby, Raymond S. and Martha P. Potter 1963a *Archaeological Survey of Proposed Reservoir Areas in Ohio*. Ohio Historical Society, Ohio State Museum, Columbus, Ohio.

——— 1963b *Archaeological Survey of Big Darby Reservoir Area*. Ohio Historical Society, Ohio State Museum, Columbus, Ohio.

Baby, Raymond S., Martha A. Potter, and Asa Mays 1966 *Explorations of the O.C. Voss Mound, Big Darby Reservoir Area, Franklin County, Ohio*. Papers in Archaeology of the Ohio Historical Society 3, Columbus, Ohio.

Baldwin, Gordon C. 1966 *Race Against Time: The Story of Salvage Archaeology*. G.P. Putnam's Sons. New York.

Banks, Kimball M. (editor) 1994 40 Something: The Smithsonian Institution River Basin Surveys. *North Dakota Archaeology: Journal of the North Dakota Archaeological Association*, 5.

Banks, Kimball M., Jon S. Czaplicki, and J. Signe Snortland 2012 The Price We've Paid: From Salvage Archaeology to Cultural Resource Management and Beyond in the Missouri River Basin. *Plains Anthropologist* 56(220):364–370.

Barker, Pat 2009 The Process Made Me Do It: Or, Would a Reasonably Intelligent Person Agree that CRM is Reasonably Intelligent? In *Archaeology and Cultural Resource Management: Visions for the Future*, edited by Lynne Sebastian and William D. Lipe, pp. 65–90. School for Advanced Research Press, Santa Fe.

Barton, Benjamin S. 1787 *Observations on Some Parts of Natural History*. C. Dilly, London.

——— 1797 *New Views of the Origin of the Tribes and Nations of America*. John Bioren, Philadelphia.

Basgall, Mark E. 1993 Chronological Sequences in the Southern North Coast Ranges, California. In *There Grows a Green Tree: Papers in Honor of David A. Fredrickson*, edited by Greg White, Pat Mikkelsen, William R. Hildebrandt, and Mark E. Basgall, pp. 167–195. Center for Archaeological Research at Davis, Publication 11. University of California, Davis.

Bass, William M. 1961 The Variation of Physical Types of the Prehistoric Plains Indians. Unpublished Ph.D. dissertation, University of Pennsylvania, Philadelphia. On file, National Park Service, Midwest Archaeological Center, Lincoln.

——— 1963 The Use of Heavy Power Equipment in the Excavation of Human Skeletal Material. *Plains Anthropologist* 8(20):122–123.

——— 1964 *The Variation in Physical Types of the Prehistoric Plains Indians*. Plains Anthropologist Memoir No. 1, pp. 65–145.

Bass, William M. 1995 *Human Osteology: A Laboratory and Field Manual*. 4th ed. Special Publications No. 2, Missouri Archaeological Society, Columbia.

Bass, William M. and Jon Jefferson 2003 *Death's Acre: Inside the Legendary Forensic Lab the Body Farm Where the Dead Do Tell Tales*. Berkley Books, New York.

Baumhoff, Martin A. 1951 Preliminary Statement on the Excavations at the Cachuma Dam Site, California, in 1951. Smithsonian Institution, River Basin Surveys, West Coast Projects. Manuscript on file, Tozzer Library, Harvard University, Cambridge.

Baumhoff, Martin A. and Robert Orlins 1979 *An Archaeological Assay on Dry Creek, Sonoma County, California*. Contributions of the University of California Archaeological Research Facility 40. University of California, Berkeley.

Beck, Charlotte and George T. Jones 2010 Clovis and Western Stemmed: Population Migration and the Meeting of Two Technologies in the Intermountain West. *American Antiquity* 75:81–116.

Bell, Robert E., Edward B. Jelks, and William W. Newcomb (assemblers) 1967 *A Pilot Study of Wichita Indian Archeology and Ethnohistory*. Final report for National Science Foundation Grant GS-964. Southern Methodist University, Dallas.

Benchley, Elizabeth 1976 *An Overview of the Prehistoric Resources of the Metropolitan St. Louis Area*. Cultural Resource Management Studies, Office of Archeology and Historic Preservation, National Park Service, Department of the Interior, Washington, DC.

Benedict, Jeff 2003 *No Bone Unturned: The Adventures of a Top Smithsonian Forensic Scientist and the Legal Battle for America's Oldest Skeletons*. Harper Collins, New York.

Berryman, Lorin E. and Albert B. Elsasser 1966 *Terminus Reservoir: Geology, Paleontology, Flora and Fauna, Archeology, and History*. Sacramento District, U.S. Army Corps of Engineers, Sacramento, in cooperation with the Western Region Office, National Park Service, San Francisco.

Binford, Lewis R. 1962 Archaeology as Anthropology. American Antiquity 28:217–255.

——— 1964 A Consideration of Archaeological Research Design. *American Antiquity* 29:425–441.

Binford, Lewis R., Sally R. Binford, Robert Whallon, and Margaret Ann Hardin 1970 *Archaeology at Hatchery West*. Memoirs No. 24. Society for American Archaeology, Washington, DC.

Binford, Sally R. and Lewis R. Binford (editors) 1968 *New Perspectives in Archaeology*. Aldine, Chicago.

Blasingham, Emily 1953 Temporal and Spatial Distribution of the Yankeetown Cultural Manifestation. Unpublished Master's thesis, Department of Anthropology, Indiana University, Bloomington.

——— 1956 The Illinois Indians, 1634–1800: A Study in Depopulation. Unpublished Ph.D. dissertation, Indiana University, Bloomington.

——— 1963 Appraisal of the Archaeological Resources of the Norton Reservoir. Report submitted by Loyola University, Chicago, to the Inter-Agency Archeological and Paleontological Salvage Program, National Park Service. Manuscript on file, National Park Service, Midwest Archeological Center, Lincoln, Nebraska.

——— 1965 Excavation of Yankeetown (12W1) 1965. Manuscript on file, National Park Service, Midwest Archeological Center, Lincoln, Nebraska.

Blukis Onat, Astrida R. 1964 Prehistoric Contacts Between Mesoamerica and Southeastern North America: A Comparative Study of Art Elements. Unpublished Master's thesis, University of Kansas, Lawrence.

——— 1980 Skwikwikwab: A Methodological Study of a Prehistoric Puget Sound Settlement. Unpublished Ph.D. dissertation, Washington State University, Pullman.

Bray, Robert T. 1991 The Utz Site: An Oneota Village in Central Missouri. *Missouri Archaeologist* 52.

Breternitz, David A. 1983 *Dolores Archaeological Program: Field Investigations and Analysis-1978*; Bureau of Reclamation, Engineering. Denver, Colorado. (tDAR ID:376704; doi: 10.6067/XCV8V1242M).

——— 1993 The Dolores Archaeological Program: In Memoriam. *American Antiquity* 58:118–125.

——— 2006 On the Western Periphery of the Plains, for ±50 Years. In *Plains Archaeology's Past: A Collection of Personal Narratives*, edited by M. F. Hawley and V. A. Wulfkuhle. *Plains Anthropologist* 51(200):721–732.

Brew, John O. 1947 Symposium on River Valley Archaeology: Introduction. *American Antiquity* 12:209–210.

——— 1961 Emergency Salvage Archaeology: Salvage in Advance of Technological Progress. *Proceedings of the American Philosophical Society* 105:1–10. Philadelphia.

——— 1968 Foreword. In *Bibliography of Salvage Archaeology in the United States*, compiled by Jerome E. Petsche, pp. 1–11. Smithsonian Institution, River Basin Surveys, Publications in Salvage Archeology 10. Lincoln, Nebraska.

——— 1974 General Comments on the Conference. In *Proceedings of the 1974 Cultural Resource Management Conference, Federal Center, Denver, Colorado*, edited by William D. Lipe and Alexander J. Lindsay, Jr., pp. 137–139. Museum of Northern Arizona, Technical Series 14. Flagstaff.

Brew, J. O., Wm. Duncan Strong, Frederick Johnson, Herbert E. Kahler, Frank H.H. Roberts, Jr., Waldo R. Wedel, John L. Champe, and Joseph R. Caldwell 1947 Symposium on River Valley Archaeology. *American Antiquity* 12:209–225.

Browman, David L. 1966 Contributions to the Prehistory of the Columbia Plateau: The Indian Dan Site. Unpublished Master's thesis, Department of Anthropology, University of Washington, Seattle.

Browman, David L. and David A. Munsell 1969 Columbia Plateau Prehistory: Cultural Development and Impinging Influences. *American Antiquity* 34:249–264.

Brown, James A. 1979 Charnel Houses and Mortuary Crypts: Disposal of the Dead in the Middle Woodland Period. In *Hopewell Archaeology: The Chillicothe Conference*, edited by David Brose and Neil Greber, pp. 211–219, Kent State University Press. Kent, Ohio.

Browning, Kathleen D. 2003 *Implementing the Antiquities Act: A Survey of Archeological Permits 1906–1935*. Studies in Archeology and Ethnography 2. Archeology Program, National Center for Cultural Resources, National Park Service, Washington, DC (tDAR document #376924; doi 10.6067/XCV8P84B6G; at http://core.tdar.org/document/376924.)

Bryan, Ruth Gruhn 1961 The Archaeology of Wilson Butte Cave, South-central Idaho. Unpublished Ph.D. dissertation, Harvard University, Cambridge, Massachusetts.

Burkett, Carl K. and Richard K. Cunningham 1997 The McFate Site and Late Woodland Settlements and Subsistence in French Creek Valley, Northwestern Pennsylvania. *Pennsylvania Archaeologist* 67(1):4–20.

Butler, B. Robert 1960 The Physical Stratigraphy of Wakemap Mound: A New Interpretation. Unpublished Master's thesis, University of Washington, Seattle.

——— 1961 *The Old Cordilleran Culture in the Pacific Northwest*. Occasional Papers of the Idaho State College Museum No. 5. Pocatello.

Butler, Mary 1938 *Three Archaeological Sites in Somerset County, Pennsylvania*. Pennsylvania Historical Commission Bulletin 753. Pennsylvania Historic Commission, Harrisburg.

Butler, Virginia L. 1993 Natural versus Cultural Salmonid Remains: Origin of The Dalles Roadcut Bones, Columbia River, Oregon, U.S.A. *Journal of Archaeological Science* 20(1):1–24.

Butler, Virginia L. and Jim E. O'Connor 2004 9000 Years of Salmon Fishing on the Columbia River, North America. *Quaternary Research* 62(1):1–8.

Butler, William B. 1995 Franklin Fenenga: 1917–1994. *Plains Anthropologist* 40(153): 271–282.

Byrne, Kevin B. and David C. Parris 1987 Reconstruction of the Diet of the Middle Woodland Amerindian Population at Abbott Farm by Bone Trace-Element Analysis. *American Journal of Physical Anthropology* 74(3):373–384.

Caldwell, Joseph R. 1953 *The Rembert Mounds, Elbert County, Georgia*. Smithsonian Institution, Bureau of American Ethnology Bulletin 154, River Basin Survey Papers 6:303–320. Washington, DC.

——— 1954 The Old Quartz Industry of Piedmont Georgia and South Carolina. *Southern Indian Studies* 5:37–39. Chapel Hill, North Carolina.

——— 1958 *Trend and Tradition in the Prehistory of the Eastern United States*. American Anthropological Association Memoirs 88, Menasha, Wisconsin.

——— 1974 Appraisal of the Archaeological Resources of Hartwell Reservoir, South Carolina and Georgia. *The Notebook* 6(2):35–44. Columbia, South Carolina.

Caldwell, Sheila K. 1991 Preliminary Report on the Vann House (9FO18). Manuscript on file, Georgia Department of Natural Resources, Historic Preservation Section, Atlanta.

Caldwell, Warren W. 1956 The Archaeology of Wakemap: A Stratified Site Near The Dalles of the Columbia. Unpublished Ph.D. dissertation, University of Washington, Seattle.

——— 1960 Tree-Ring Dating in the Missouri Basin Chronology Program. *Tree-Ring Bulletin* 23(1–4):14–17.

Cavallo, John A. 1983 Fish, Fires, and Foresight: Middle Woodland Economic Adaptations in the Abbott Farm National Landmark. *North American Archaeologist* 5(2):111–138.

Caywood, Louis R. 1954 *Archaeological Excavations at Fort Spokane 1951, 1952, and 1953*. U.S. Department of the Interior, National Park Service. San Francisco.

——— 1955 *Final Report, Fort Vancouver Excavations*. U.S. Department of the Interior, National Park Service. San Francisco.

Champe, John L. 1949 A is for Apple. *Plains Archeological Conference News Letter* [reprint] 1(3):45–48.

Chandler, Susan M. 2009 Innovative Approaches to Mitigation. In *Archaeology and Cultural Resource Management: Visions for the Future*, edited by Lynne Sebastian and William D. Lipe, pp. 115–140. School for Advanced Research Press, Santa Fe.

Chapman, Carl H., Richard A. Marshall, Robert T. Bray, W. Raymond Wood, Dale R. Henning, Rolland E. Pangborn, and Bonnie B. Keller ca. 1955–1960a Archaeological Investigations in the Table Rock Reservoir Area, Missouri. Part II. University of Missouri. Manuscript on file, National Park Service, Midwest Archeological Center, Lincoln, Nebraska.

——— ca. 1955–1960b Archaeological Investigations in the Table Rock Reservoir Area, Missouri. Part I. University of Missouri. Manuscript on file, National Park Service, Midwest Archeological Center, Lincoln, Nebraska.

Chapman, Carl H., Amy Evelyn Harvey, Dale R. Henning, Richard A. Marshall, Rolland E. Pangborn, and John R. Vincent 1960 Archaeological Investigations in the Table Rock Reservoir Area. Report to National Park Service on Contract 14–10–333–96 by the University of Missouri. Manuscript on file, National Park Service, Midwest Archeological Center, Lincoln, Nebraska.

Chiarulli, Beverly A., Douglas C. Kellogg, Robert G. Kingsley, William J. Meyer, Jr., Patricia E. Miller, Philip A. Perazio, and Peter E. Siegel 2001 Prehistoric Settlement Patterns in Upland Settings: An Analysis of Site Data in a Sample of Exempted Watersheds. Prepared for the Pennsylvania Historical and Museum Commission. Pennsylvania Archaeological Council. Harrisburg.

Chiarulli, Beverly A. and Sarah W. Neusius 2009 Burying the Past: Observations on Unintentional Site Reburial at the Johnston Site, Indiana County, Pennsylvania. Paper presented at the 88th Meeting of the Transportation Research Board, Washington, DC.

Childs, S. Terry 2010 The Dollars and Sense of Managing Archaeological Collections. *Heritage Management* 3(2):155–289.

Classen, Cheryl 1999 Madeline Black and White Women at Irene Mound. In *Grit-Tempered: Early Women Archaeologists in the Southeastern United States*, edited by Nancy M. White, Lynn P. Sullivan, and Rochelle Marrinan, pp. 92–114. Florida Museum of Natural History, Ripley P. Bulletin Series, University Press of Florida, Gainesville.

Cleland, Charles E. and James E. Fitting 1968 The Crisis of Identity: Theory in Historic Sites Archeology. *The Conference on Historic Site Archeology Papers* 2(2):124–138.

Clemensen, A. Berle 1992 *Casa Grande Ruins National Monument, a Centennial History of the First Prehistoric Reserve, 1892–1992*. Casa Grande Ruins National Monument, National Park Service, Coolidge, Arizona. (http://core.tdar.org/document/368791; DOI: 10.6067/XCV84M9309)

Clinton, DeWitt W. 1820 *A Memoir on the Antiquities of the Western Parts of the State of New York*. E. and E. Hosford, Albany.

Cohen, Leslie 2006 Dutton's Dirty Diggers. *El Palacio* 11(2):34–37.

Cole, David L. 1954 A Contribution to the Archaeology of The Dalles Region. Unpublished Master's thesis, Department of Anthropology, University of Oregon, Eugene.

Cole, Maria S. and Theodore M. Cole III 1994 Metric Variation in the Supraorbital Region in Northern Plains Indians. In *Skeletal Biology in the Great Plains: Migration, Warfare, Health, and Subsistence*, edited by Douglas W. Owsley and Richard L. Jantz, pp. 209–218. Smithsonian Institution, Washington, DC.

Cole III, Theodore M. 1994 Size and Shape of the Femur and Tibia in Northern Plains Indians. In *Skeletal Biology in the Great Plains: Migration, Warfare, Health, and Subsistence*, edited by Douglas W. Owsley and Richard L. Jantz, pp. 219–234. Smithsonian Institution, Washington, DC.

Collier, Donald, Alfred E. Hudson, and Arlo Ford 1942 *Archaeology of the Upper Columbia Region*. University of Washington Publications in Anthropology 9(1), Seattle.

Committee for the Recovery of Archaeological Remains (CRAR) 1958 *The Inter-agency Archaeological Salvage Program After Twelve Years*. University of Missouri, Columbia.

Cooper, Paul L. 1949 Recent Investigations in Fort Randall and Oahe Reservoirs, South Dakota. *American Antiquity* 14:300–310.

——— 1955 *The Archeological and Paleontological Salvage Program in the Missouri Basin 1950–1951*. Smithsonian Miscellaneous Collection. Vol. 126, No. 2, Smithsonian Institution, Washington, DC.

——— 1958 *Archeological Investigations in the Heart Butte Reservoir Area, North Dakota*. Smithsonian Institution, Bureau of American Ethnology Bulletin 169, River Basin Surveys Paper 9. Washington, DC.

Corbett, John M. (editor) 1961 Symposium on Salvage Archaeology. Paper presented at the Society for American Archaeology and the Committee for the Recovery of Archaeological Remains, Bloomington, Indiana. Manuscript on file, Midwest Archeological Center, National Park Service, Lincoln.

Cotter, John L. 1964 Northeast. In *Current Research*, edited by C. Borden. *American Antiquity* 30:237–239.

——— 1966 Northeast. In *Current Research*, edited by F.A. Riddell. *American Antiquity* 31:899–902.

——— 1993 Historical Archaeology before 1967. *Historical Archaeology* 27(1):4–9.

Crampton, C. Gregory 2009 *Ghosts of Glen Canyon: History Beneath Lake Powell*. University of Utah Press, Salt Lake City.

Crane, Horace R. and James B. Griffin 1958a University of Michigan Radiocarbon Dates II. *Science* 127:1098–1105.

——— 1958b University of Michigan Radiocarbon Dates III. *Science* 128:1117–1123.

——— 1972 University of Michigan Radiocarbon Dates. *Radiocarbon* 14(1):155–194.

Craven, Sloan L. 2003 Lithic Variation in Hafted Bifaces at the Lind Coulee Site (45GR97), Washington. Unpublished Master's thesis, Department of Anthropology, Washington State University, Pullman.

——— 2004 New Dates for the Lind Coulee Site (45GR97), Washington. *Current Research in the Pleistocene* 21:28–30.

Cressman, Luther S., with David L. Cole, Wilbur A. Davis, Thomas A. Newman, and Daniel J. Scheans 1960 Cultural Sequences at The Dalles, Oregon: A Contribution to Pacific Northwest Prehistory. *Transactions of the American Philosophical Society* 50(10):1–108. Philadelphia.

Cross, Dorothy 1956 *The Abbott Farm Site*. Archaeological Society of New Jersey, New Jersey State Museum, Trenton.

——— 1964 The 1964 Excavations in the Tocks Island Reservoir Area, Warren County, New Jersey. River Basin Surveys, National Anthropological Archives, National Museum of Natural History, Smithsonian Institution, Washington, DC.

Crozier, Daniel G. 1973 *Preliminary Archaeological Salvage Operation in a Portion of the Cowanesque Dam Flood Control Project: The Antonio Site, 36–Ti-30*. Temple University, Philadelphia.

Cumming, Robert B., Jr. 1958 *Archeological Investigations at the Tuttle Creek Dam, Kansas*. Smithsonian Institution, Bureau of American Ethnology Bulletin 169, River Basin Surveys Papers 10. Washington, DC.

Daugherty, Richard D. 1952 Archaeological Investigations in O'Sullivan Reservoir, Grant County, Washington. *American Antiquity* 17:374–383.

——— 1953 Early Man in the Columbia Intermontane Province. Unpublished Ph.D. dissertation, University of Washington, Seattle.

——— 1956a Archaeology of the Lind Coulee Site, Washington. *Proceedings of the American Philosophical Society* 100:223–278. Philadelphia.

——— 1956b *Early Man in the Columbia Intermontane Province*. University of Utah Anthropological Papers No. 24. Salt Lake City.

——— 1962 The Intermontane Western Tradition. *American Antiquity* 28:144–150.

Daugherty, Richard D. and Francis A. Riddell 1947a Appraisal of the Archeological Resources of the Anderson Ranch Reservoir, Elmore County, Idaho. Columbia Basin Project, River Basin Surveys, Smithsonian Institution. Manuscript on file, Bureau of Reclamation, Denver Office Library, Lakewood, Colorado.

——— 1947b Appraisal of the Archeological Resources of the Palisades Reservoir, Bonneville, Idaho. Columbia Basin Project, River Basin Surveys, Smithsonian Institution. Manuscript on file, Tozzer Library, Harvard University, Cambridge, Massachusetts.

——— 1948 Appraisal of the Archeological Resources of Equalizing, Long Lake, and Potholes Reservoirs in East Central Washington (with an Appendix on the Archeological and Paleontological Remains in Lind Coulee, Grant County, Washington). Columbia Basin Project, River Basin Surveys, Smithsonian Institution. Manuscript on file, Bureau of Reclamation, Denver Office Library, Lakewood, Colorado.

Davis, Arthur P. 1917 *Irrigation Works Constructed by the United States Government*. John Wiley and Sons, New York.

Davis, E. Mott 1953 Recent Data From Two Paleo-Indian Sites on Medicine Creek. *American Antiquity* 18:380–386.

——— 1954 The Culture History of the Central Great Plains Prior to the Introduction of Pottery. Unpublished Ph.D. dissertation, Harvard University, Cambridge, Massachusetts.

Davis, E. Mott 1956 *Archaeological Survey of the Big Sandy Reservoir Area, Southwestern Wyoming*. Notebook No. 2, Laboratory of Anthropology. University of Nebraska, Lincoln.

Davis, Hester A. 1972 The Crisis in American Archaeology. *Science* 176:267–272.

——— 1996 NHPA and the Practice of Archeology. *CRM* 19(6):42–44.

——— 1999 The Life and Times of Bettye J. Broyles. "I Did a Man's Work for Thirty Years." In *Grit-Tempered: Early Women Archaeologists in the Southeastern United States*, edited by Nancy Marie White, Lynne P. Sullivan, and Rochelle A. Marrinan, pp. 115–147. University Press of Florida, Gainesville.

——— 2009 Archaeologists Looked to the Future in the Past. In *Archaeology and Cultural Resource Management: Visions for the Future*, edited by Lynne Sebastian and William D. Lipe, pp. 19–40. School for Advanced Research Advanced Seminars Series, School for Advanced Research Press, Santa Fe.

Day, John S. 1981 House Votes to Kill Dickey-Lincoln Dam. *Bangor Daily News* 25 November: 13. Bangor, Maine.

Day, Kent C. and David S. Dibble 1963 *Archaeological Survey of the Flaming Gorge Reservoir Area, Wyoming- Utah*. Anthropological Papers 65. University of Utah, Salt Lake City.

de Laguna, Frederica 1961 Salvage of Ethnological Data and Its Problems. In "Symposium on Salvage Archaeology," edited by John M. Corbett. Paper presented at the Society for American Archaeology and the Committee for the Recovery of Archaeological Remains, Bloomington, Indiana, Manuscript on file, Midwest Archeological Center, National Park Service, Lincoln, Nebraska.

Deetz, James 1965 *The Dynamics of Stylistic Change in Arikara Ceramics*. Illinois Studies in Anthropology No. 4. University of Illinois Press, Urbana.

Deetz, James F. and Edwin F. Dethlefsen 1967 Death's Head, Cherub, Urn and Willow. *Natural History* 76(3):29–37.

Department of Anthropology, National Museum of Natural History, Smithsonian Institution 2002 Archaeology for the Nation: A Resource Guide to the Collections of the River Basin Surveys. A proposal for the 2002 Smithsonian Women's Committee Grants. Proposal on file, Collections and Archives Program, Department of Anthropology, Smithsonian Museum Support Center, Suitland, Maryland.

Departmental Consulting Archeologist 2009 *The Goals and Accomplishments of the Federal Archeology Program: The Secretary of the Interior's Report to Congress on the Federal Archeology Program, 1998–2003*. Archeology Program, National Park Service, Washington, DC.

——— 2010 *The Secretary of the Interior's Report to Congress on the Federal Archeology Program, 2004–2007*. Archeology Program, National Park Service, Washington, DC.

DePratter, Chester B., Charles M. Hudson, and Marvin T. Smith 1985 The Hernando de Soto Expedition: From Chiaha to Mabila. In *Alabama and the Borderlands: From Prehistory to Statehood*, edited by Reid Badger and Lawrence Clayton, pp. 108–121. University of Alabama Press, Tuscaloosa.

Dibble, David S. 1968 *Bonfire Shelter: A Stratified Bison Kill Site, Val Verde County, Texas*. Texas Memorial Museum Miscellaneous Papers No. 1. Austin.

Dincauze, Dena F. 1967 Cremation Cemeteries in Eastern Massachusetts. Unpublished Ph.D. dissertation, Harvard University, Cambridge, Massachusetts.

——— 1992 Exploring Career Styles in Archaeology. In *Rediscovering Our Past: Essays on the History of American Archaeology*, edited by Jonathan E. Reyman, pp. 131–136. World Archaeology Series, Vol. 2. Avebury, Ashgate Publishing Ltd., Aldershot, United Kingdom.

Dincauze, Dena F. and Judith W. Meyer 1977 Prehistoric Resources of East-Central New England: A Preliminary Predictive Study. Cultural Resource Management Studies,

Office of Archeology and Historic Preservation, National Park Service, Department of the Interior, Washington, DC.

Dollar, Clyde D. 1968 Some Thoughts on Theory and Method in Historical Archeology. *The Conference on Historic Site Archeology Papers* 2(2):3–30.

Dragoo, Don W. 1955 Excavations at the Johnston Site, Indiana County, Pennsylvania. *Pennsylvania Archaeologist* 25(2):85–141.

——— 1977 Prehistoric Iroquoian Occupation in the Upper Ohio Valley. In *Current Perspectives in Northeastern Anthropology*, edited by Robert F. Funk and Charles F. Hayes III, pp. 41–47. Researches and Transactions of the New York State Archeological Association, Vol. 17(1). Rochester, New York.

Dragoo, Don W. and Stanley W. Lantz 1967 *Archaeological Investigations at the Onoville Bridge Site (30CA5) and at Other Locations in the Allegheny Reservoir in Pennsylvania and New York*. U.S. National Park Service Report No. 14–10–0529–2878 and No. 950–78. Carnegie Museum of Natural History, Pittsburgh, Pennsylvania.

——— 1971 *Archeological Salvage of Selected Sites in the Allegheny Reservoir in Pennsylvania and New York, 1969–1971*. U.S. National Park Service Report No. 14–10–5–950–28 and 14–10–5–950–48. Carnegie Museum of Natural History, Pittsburgh, Pennsylvania.

——— 1973 *Archaeological Salvage of Selected Sites in the Allegheny Reservoir in New York, 1971–1972*. U.S. National Park Service Project No. CX-4000–3–0027. Carnegie Museum of Natural History, Pittsburgh, Pennsylvania.

——— 1975 *Archaeological Salvage of Selected Sites in the Allegheny Reservoir in New York, 1973–1974*. National Park Service Report. Carnegie Museum of Natural History, Pittsburgh, Pennsylvania.

Drucker, Philip 1947a Preliminary Statement of the Archeological Remains in the McNary Reservoir, Columbia River, Oregon-Washington. Columbia Basin Project, River Basin Surveys, Smithsonian Institution, Eugene. Manuscript on file, Bancroft Library, University of California Survey Collection, University of California, Berkeley.

——— 1947b Appraisal of the Archaeological Resources of Isabella Reservoir, Kern County, California. Smithsonian Institution, River Basin Surveys, West Coast Projects, Appraisals. Manuscript on file, Bancroft Library, University of California Survey Collection, University of California, Berkeley.

——— 1948a Appraisal of the Archaeological Resources of Kelsey Creek Reservoir, Lake County, California. Smithsonian Institution, River Basin Surveys, Pacific Coast Area, Appraisals. Manuscript on file, Bancroft Library, University of California Survey Collection, University of California, Berkeley. https://core.tdar.org/document/391106

——— 1948b Appraisal of the Archaeological Resources of Monticello Reservoir Area, Napa County, California. Smithsonian Institution, River Basin Surveys, Pacific Coast Area, Appraisals. Manuscript on file, Bancroft Library, University of California Survey Collection, University of California, Berkeley. https://core.tdar.org/document/391107

——— 1948c Appraisal of the Archaeological Resources of Pine Flat Reservoir, Fresno County, California. Smithsonian Institution, River Basin Surveys, Columbia Basin Projects, Appraisals. Manuscript on file, Bancroft Library, University of California Survey Collection, University of California, Berkeley.

——— 1948d Appraisal of the Archaeological Resources of Success Reservoir, Tulare County, California. Smithsonian Institution, River Basin Surveys, Pacific Coast Area, Appraisals. Manuscript on file, Bancroft Library, University of California Survey Collection, University of California, Berkeley.

——— 1948e Appraisal of the Archaeological Resources of Wilson Valley Reservoir, Lake County, California. Smithsonian Institution, River Basin Surveys, Pacific Coast Area, Appraisals. Manuscript on file, Bancroft Library, University of California Survey Collection, University of California, Berkeley.

Drucker, Philip 1948f Archeological Appraisal of Indian Valley Reservoir, Lake Co., Sly Park Reservoir, El Dorado County, and Dry Creek Reservoir, Sonoma County, California. Smithsonian Institution, River Basin Surveys, Pacific Coast Area, Appraisals. Manuscript on file, Bancroft Library, University of California Survey Collection, University of California, Berkeley.

——— 1948g Preliminary Appraisal of Archaeological Resources of Coyote Valley Reservoir, Mendocino County, California. Smithsonian Institution, River Basin Surveys, Pacific Coast Area, Appraisals. Manuscript on file, Bancroft Library, University of California Survey Collection, University of California, Berkeley.

——— 1948h Preliminary Appraisal of the Archaeological Resources of Folsom Reservoir, Placer, El Dorado, and Sacramento Counties, California. Smithsonian Institution, River Basin Surveys, Pacific Coast Area, Appraisals. Manuscript on file, Bancroft Library, University of California Survey Collection, University of California, Berkeley.

——— 1948i Appraisal of the Archaeological Resources of the McNary Reservoir, Oregon and Washington. Columbia Basin Project, River Basin Surveys, Smithsonian Institution. Manuscript on file, Oregon State University Library, Corvallis, Oregon.

——— ca. 1948 Appraisal of the Archaeological Resources of Terminus Reservoir, Tulare County, California. Smithsonian Institution, River Basin Surveys. Manuscript on file, Bancroft Library, University of California Survey Collection, University of California, Berkeley.

——— 1951 *The Northern and Central Nootkan Tribes*. Smithsonian Institution, Bureau of American Ethnology Bulletin 144. Washington, DC.

Duffield, Lathel F. 1963 *The Strawn Creek Site: A Mixed Archaic and Neo-American Site at Navarro Mills Reservoir, Navarro County, Texas*. The Texas Archaeological Salvage Project, The University of Texas, Austin.

Duffield, Lathel F. and Edward B. Jelks 1961 *The Pearson Site: A Historic Indian Site in Ferrell's Bridge Reservoir, Rains County, Texas*. Archaeology Series No. 4. Department of Anthropology, University of Texas, Austin.

Dumond, Don E. and Rick Minor 1983 *Archaeology in the John Day Reservoir: The Wildcat Canyon Site, 35–GM-9*. University of Oregon Anthropological Papers No. 30. Eugene.

Earle, Timothy 2008 Cultural Anthropology and Archaeology: Theoretical Dialogues. *Handbook of Archaeological Theories*, edited by R. Alexander Bentley, Herbert D.G. Maschner, and Christopher Chippindale, pp. 187–202. AltaMira Press, New York.

Eddy, Frank W. 1966 *Prehistory in the Navajo Reservoir District, Northwestern New Mexico*. Museum of New Mexico Papers in Anthropology 15, Pts. 1 and 2. Santa Fe.

——— 1972 Cultural Ecology and the Prehistory of the Navajo Reservoir District. *Southwestern Lore* 38:1–75.

Edwards, Guy D. and Merrill J. Mattes 1948 Project Report on Historic Sites in the Fort Randall Reservoir Area, Missouri River: Gregory, Charles Mix, Lyman, Brule, and Buffalo Counties, South Dakota. Department of the Interior, National Park Service, Region Two. Submitted to the Corps of Engineers, Omaha District. Manuscript on file, National Park Service, Midwest Archaeological Center, Lincoln, Nebraska.

Elsasser, Albert B. ca. late 1940s Salvage Archaeology at Sites NAP-74 and NAP-95 in the Monticello Reservoir Area, California, Part 2. Manuscript on file, Sonoma State Information Center, Sonoma State College, Sonoma, California.

Enbysk, Betty Joyce 1956 Vertebrate and Mollusca from Lind Coulee, Washington. In "Archaeology of the Lind Coulee site, Washington" by Richard D. Daugherty, pp. 267–278. *Proceedings of the American Philosophical Society* 100(3):223–278. Philadelphia.

Epstein, Jeremiah F. 1959 Burins from Texas. Manuscript on file, Texas Archeological Research Laboratory, The University of Texas, Austin.

Evans, Clifford 1973 Smithsonian Institution. In *Reports Presented at the Annual Meeting of The Committee for the Recovery of Archaeological Remains, March 22–23, 1973*: pp. 132–133. U.S. Department of the Interior, National Park Service, Washington, DC.

Fagan, Brian M. 2005 *A Brief History of Archaeology: Classical Times to the Twenty-First Century.* Pearson Prentice-Hall, Upper Saddle River, New Jersey.

Fenenga, Franklin 1947a Preliminary Appraisal of the Archaeological Resources of the Coyote Valley Reservoir, Mendocino County, California. Report submitted to the National Park Service, Western Region, San Francisco. Manuscript 31 of the University of California Archaeological Research Facility. On file, Bancroft Library, University of California, Berkeley.

——— 1947b Preliminary Appraisal of the Archaeological Resources of the Pine Flat Reservoir, Fresno County, California. Manuscript 3 of the University of California Archaeological Research Facility. On file, Bancroft Library, University of California, Berkeley.

——— 1947c Preliminary Appraisal of the Archaeological Resources of Success Reservoir, Tulare County, California. Smithsonian Institution, River Basin Surveys, Pacific Coast Area, Appraisals. Manuscript 12 of the University of California Archaeological Research Facility. Manuscript on file, Bancroft Library, University of California, Berkeley.

——— 1947d Preliminary Survey of Archeological Resources in the Isabella Reservoir, Kern County, California. Smithsonian Institution, River Basin Surveys, West Coast Projects, Appraisals. Manuscript on file, Bancroft Library, University of California, Berkeley.

——— ca. 1950a An Account of the Archaeological Field Work Conducted at Site ELD-1 in the Folsom Reservoir, Eldorado [*sic*] County, California. Manuscript on file, National Park Service, Western Regional Office, San Francisco.

——— ca. 1950b A Preliminary Account of the Archaeological Field Work Conducted in the Terminus Reservoir, in Tulare County, California in June and July 1950. Manuscript on file, National Park Service, Western Regional Office, San Francisco.

——— 1952 The Archaeology of the Slick Rock Village, Tulare County, California. *American Antiquity* 17:339–347.

Ferguson, Bobbie 2001a An Inquiry into the Ownership of Materials Recovered by the Investigations of the River Basin Surveys. Report on file, Office of Policy and Administration, U.S. Department of the Interior, Bureau of Reclamation, Denver.

——— 2001b An Addendum to: An Inquiry into the Ownership of Materials Recovered by the Investigations of the River Basin Surveys. Report on file, Office of Policy and Administration, U.S. Department of the Interior, Bureau of Reclamation, Denver.

Fiorillo, Anthony R. and Edward Daschle 1990 E. D. Cope's Expedition to the Dakota's Revisited. *Earth Sciences History* 9(1):57–61.

Fiske, Moses 1820 Conjectures Respecting the Ancient Inhabitants of North America. *Archaeologia Americana: Transactions and Collections of the American Antiquarian Society* 11:300–307.

Fontana, Bernard L. 1965 On the Meaning of Historic Sites Archaeology. *American Antiquity* 31:61–65.

——— 1968 A Reply to "Some Thoughts on Theory and Method in Historical Archeology" by Clyde Dollar. *The Conference on Historic Site Archeology Papers* 2(2):75–78.

Ford, James A. 1936 Analysis of Indian Village Site Collections from Louisiana and Mississippi. *Louisiana Geological Survey, Anthropological Study* (2):1–285. Baton Rouge.

Ford, James A. and Gordon R. Willey 1941 An Interpretation of the Prehistory of the Eastern United States. *American Anthropologist* 43:325–363.

Foster, Gene 1952 A Brief Archaeological Summary of Glen Canyon. *Plateau* 25(2):21–26.

——— 1954 Petrographic Art in Glen Canyon. *Plateau* 27(1):6–18.

Fowler, Catherine S. 1972 Comparative Numic Ethnobiology. Unpublished Ph.D. dissertation, University of Pittsburgh, Pennsylvania.

——— 2010 Bertha P. Dutton and the Girl Scouts in the Southwest, 1947–1957. In *Threads, Tints and Edification: Papers in Honor of Glenna Dean*, edited by Emily J. Brown, Karen Armstrong, David M. Brugge, and Carol Condie, pp. 61–71. Papers of the Archaeological Society of New Mexico 36. Albuquerque.

Fowler, Don D. 1982 Cultural Resource Management. In *Advances in Archaeological Method and Theory*, edited by M. B. Schiffer, pp. 1–50. Academic Press, New York.

——— 1986 Conserving American Archaeological Resources. In *American Archaeology Past and Future: A Celebration of the Society for American Archaeology, 1935–1985*, edited by David J. Meltzer, Don D. Fowler, and Jeremy A. Sabloff, pp. 135–162. Smithsonian Press, Washington, DC.

——— 2011 *The Glen Canyon Country, a Personal Memoir*. University of Utah Press, Salt Lake City.

Fowler, Melvin L. and Howard Winters 1956 *Modoc Rock Shelter: Preliminary Report.* Illinois State Museum Report of Investigations 4. Springfield, Illinois.

Francis, Julie 1994 The Middle Fork of the Powder River: 1947–1985. In "40 Something: The Smithsonian Institution River Basin Surveys." edited by Kimball Banks. *North Dakota Archaeology: Journal of the North Dakota Archaeological Association* 5:177–189.

Fraser, Dorothy E. 1941 The Cultural Significance of an Archaeological Site on the Middle Missouri River. Unpublished Master's thesis, Columbia University, New York.

Fredrickson, David A. 1949 Appraisal of the Archeological Resources of New Melones Reservoir, Calaveras and Tuolumne Counties, California. Pacific Coast Division, River Basin Surveys, Smithsonian Institution, Berkeley, California, and Washington DC. On file, National Park Service, Western Regional Office, San Francisco.

Fryman, Frank B., Jr. and Mildred L. Fryman 1967 An Archeological and Historical Survey of the Proposed Wolf Creek Dam Cargo Life Area, Russell County, Kentucky. Manuscript on file, National Park Service, Southeast Archeological Center, Tallahassee.

Fryman, Frank B., Jr., Mildred L. Fryman, and Michael J. Rodeffer 1967a A Preliminary Survey of the Archeological and Historical Resources of the Proposed Parker Branch Reservoir, Laurel, Jackson and Rockcastle Counties, Kentucky. Manuscript on file, National Park Service, Southeast Archeological Center, Tallahassee.

——— 1967b A Preliminary Survey of the Archeological and Historical Resources of the Proposed Ulvah Reservoir, Letcher County, Kentucky. Manuscript on file, National Park Service, Southeast Archeological Center, Tallahassee.

Fryxell, Roald 1963 *Late Glacial and Post Glacial Geological and Archaeological Chronology of the Columbia Plateau, Washington.* Washington State University, Laboratory of Anthropology, Reports of Investigations No. 23. Pullman.

Fryxell, Roald and Richard D. Daugherty 1962 *Interim Report: Archaeological Salvage in the Lower Monumental Reservoir, Washington.* Washington State University, Laboratory of Anthropology, Reports of Investigations No. 21. Pullman.

Fryxell, Roald and Bennie C. Keel 1969 Emergency Salvage Excavations for the Recovery of Early Human Remains and Related Scientific Materials from the Marmes Rockshelter Archaeological Site, Southeastern Washington, May 3–December 15, 1968. Unpublished final report to the U.S. Army Corps of Engineers, Walla Walla District. On file, Washington State University, Laboratory of Anthropology, Pullman.

Funkhouser, William D. 1938 A Study of the Physical Anthropology and Pathology of the Osteological Material from the Norris Basin. In *An Archaeological Survey of the Norris Basin in Eastern Tennessee*, by William S. Webb. Smithsonian Institution, Bureau of American Ethnology, *Bulletin* 118:225–251. Washington, DC.

Gaines, Xerpha M. 1957 Plants in Glen Canyon. *Plateau* 30(2):31–34.

——— ca. 1960 Some Plants of Glen Canyon and Their Probable Uses by Indians. Manuscript on file, Museum of Northern Arizona Library Separate file #26116. Flagstaff.

Gallatin, Albert 1836 A Synopsis of the Indian Tribes within the United States East of the Rocky Mountains and in the British and Russian Possessions in North America. *Archaeologia Americana: Transactions and Collections of the American Antiquarian Society* 2:1–422. Worcester, Massachusetts.

Galligan, Elizabeth Ann 2012 *Secrets of the Plumed Serpent*. ABQ Press, Albuquerque, New Mexico.

Gazin, Charles L. 1956 *Paleocene Mammalian Faunas of the Bison Basin in South-Central Wyoming*. Smithsonian Miscellaneous Collections, Vol. 131, No. 6, pp. 1–57. Smithsonian Institution, Washington, DC.

Gebhard, David S. 1960 Prehistoric Paintings of the Diablo Region of West Texas. *Publication in Art and Science*. Roswell Museum and Art Center, Roswell, New Mexico.

George, Richard L. 1978 The McJunkin Site, A Preliminary Report. *Pennsylvania Archaeologist* 48(4):33–47.

——— 1997 McFate Artifacts in a Monongahela Context: McJunkin, Johnston, and Squirrel Hill. *Pennsylvania Archaeologist* 67(1):35–44.

Gill, George W. 2008 Introduction: An Overview of the Region. In *Skeletal Biology and Bioarchaeology of the Northwestern Plains*, edited by George W. Gill and Rick L. Weathermon, pp. 1–13. The University of Utah Press, Salt Lake City.

Gill, George W. and Rick L. Weathermon (editors) 2008 *Skeletal Biology and Bioarchaeology of the Northwestern Plains*. The University of Utah Press, Salt Lake City.

Glass, James A. 1990 *The Beginnings of a New National Historic Preservation Program, 1957 to 1969*. American Association for State and Local History, Nashville, Tennessee.

Glenn, James R. 1994 The River Basin Surveys Program. In "40 Something: The Smithsonian Institution River Basin Surveys," edited by Kimball Banks. *North Dakota Archaeology: Journal of the North Dakota Archaeological Association* 5:5–14.

Grabert, Garland F. 1968 *North-Central Washington Prehistory*. University of Washington. Department of Anthropology Reports in Archaeology No. 1. Seattle.

——— 1970 Prehistoric Cultural Stability in the Okanogan Valley of Washington and British Columbia. Unpublished Ph.D. dissertation, Department of Anthropology, University of Washington, Seattle.

Gradwohl, David Mayer 1995 Mildred Mott Wedel 1908–1996. 40(154):399–403.

——— 1996 Waldo R. Wedel 1908–1996. *Plains Anthropologist* 41(158):317–332.

——— 1997 Accidental Plains Archeologist: Neither Courage Nor Naiveté? *The Kansas Anthropologist* 18(2):19–43.

——— 2003 From Turin to 2002: Reflections on My Career in Iowa Archaeology. In "Always on the Edge (in the Prairie-Plains): Essays in Honor of David Mayer Gradwohl," edited by Larry J. Zimmerman. *Journal of the Iowa Archeological Society* 50:11–50.

Grady, Mark and William Lipe 1977 The Role of Preservation in Conservation Archaeology. *Proceedings of the Society for Conservation Archeology, 1976*, pp. 1–11.

Grange, Roger T., Jr. 1990 J. C. Harrington Medal in Historical Archaeology, Carlyle Shree Smith 1989. *Historical Archaeology* 24(2):1–3.

——— 2006 When I Was a Lad, I served a Term . . . Minor Adventures in Plains Archaeology in the 1950s and 1960s. In *Plains Archaeology's Past: A Collection of Personal Narratives*, edited by M. F. Hawley and V. A. Wulfkuhle. *Plains Anthropologist* 51(200):683–700.

Grater, Barbara Alice 1966 The Archaeology of the Votaw Site: Lower Snake River, Washington. Unpublished Master's thesis, San Francisco State College, San Francisco.

Greengo, Robert E. 1982 Studies in Plateau Prehistory, Priest Rapids and Wanapum Dam Reservoir Areas, Columbia River, Washington. University of Washington, Department of Anthropology. Manuscript on file, U.S. Department of Interior, National Park Service, Western Regional Office, San Francisco.

——— 1986 *Prehistory of the Priest Rapids–Wanapum Region, Columbia River, Washington.* BAR International Series 290. British Archaeological Reports, Oxford.

Grieder, Terence 1965 Report on a Study of the Pictographs in Satan Canyon, Val Verde County, Texas. Manuscript on file, The Texas Archeological Research Laboratory, The University of Texas, Austin.

Griffin, James B. 1936 The Cultural Significance of the Ceramic Remains from the Norris Basin. Unpublished Ph.D. dissertation, Department of Anthropology, University of Michigan, Ann Arbor.

——— 1938 The Ceramic Remains from Norris Basin, Tennessee. In *An Archaeological Survey of the Norris Basin in Eastern Tennessee*, by W. S. Webb. Smithsonian Institution, Bureau of American Ethnology, *Bulletin* 118:253–358. Washington, DC.

Gruhn, Ruth 1972 Letter to *Science*. *Science* 176(4033):353–354.

Guilday, John 1955 Animal Remains from an Indian Village Site, Indiana County, Pennsylvania. *Pennsylvania Archaeologist* 25(2):142–147.

Gunnerson, James H. 2006 From the Plains and Back Again. In *Plains Archaeology's Past: A Collection of Personal Narratives*, edited by M. F. Hawley and V. A. Wulfkuhle. *Plains Anthropologist* 51(200):615–632.

Gunnerson, James H. and Delores A. Gunnerson 1952 Appraisal of the Archeological and Paleontological Resources of the Lower Platte Basin, Nebraska: Supplement. Smithsonian Institution River Basin Surveys Missouri Basin Projects Appraisal. Manuscript on file, National Park Service, Midwest Archeological Center, Lincoln, Nebraska.

Gunnerson, James H., William Lipe, Robert H. Lister, and Dee Ann Suhm 1959 Findings. In *The Glen Canyon Archeological Survey, Part I*, by Don D. Fowler, James H. Gunnerson, Jesse D. Jennings, Robert H. Lister, Dee Ann Suhm, and Ted Weller, pp. 14–26. University of Utah Papers, 39. Salt Lake City.

Hall, Robert L. 2006 A Plains Experience and Beyond. In *Plains Archaeology's Past: A Collection of Personal Narratives*, edited, by M. F. and V. A. Wulfkuhle. *Plains Anthropologist* 51(200):633–648.

Haring, Inez M. 1961 Mosses of the Glen Canyon Area. *Plateau* 33(4):120–122.

Harrington, J. (Jean) C. 1947 *Buggs Island Reservoir Project*. Archeological Society of Virginia Quarterly Bulletin 1(3). Richmond, Virginia.

Hartman, Joseph H. and Allen J. Kihm 1991 Stratigraphic Distribution of *Titanoides* (Mammalian: Pantodonta) in the Fort Union Group (Paleocene) of North Dakota. In *Sixth International Williston Basin Symposium*, edited by J. E. Christopher and F. M. Haidl, pp. 207–215. Special Publication Number 11 of the Saskatchewan Geological Society.

Harvey, Amy Evelyn 1960 Temporal-Cultural Analysis of Archaeological Remains in a Group of Rock Shelters in Southwest Missouri. In *Archaeological Investigations in the Table Rock Reservoir Area*, by Carl. H. Chapman, Amy Evelyn Harvey, Dale R. Henning, Richard A. Marshall, Rolland E. Pangborn, and John R. Vincent, pp. 589–794. Report to National Park Service on Contract 14–10–333–96 by the University of Missouri. On file, National Park Service, Midwest Archeological Center, Lincoln, Nebraska.

Harvey, Mark 1994 *A Symbol of Wilderness: Echo Park and the American Conservation Movement*. University of New Mexico Press, Albuquerque.

Hata, Nadine Ishitani 1992 *The Historic Preservation Movement in California, 1940–1976*. California Department of Parks and Recreation, Office of Historic Preservation. Sacramento.

Hattori, Eugene M. 2005 *Mary K. Rusco*. Manuscript on file, Nevada State Museum, Carson City.

Hawley, Florence M. 1938 Tree Ring Dating for Southeastern Mounds. In *An Archaeological Survey of the Norris Basin in Eastern Tennessee*, by W. S. Webb. Smithsonian Institution, Bureau of American Ethnology, Bulletin 118:359–362. Washington, DC.

Hawley, Marlin F. 2006 Introduction. In *Plains Archaeology's Past: A Collection of Personal Narratives*, edited by M. F. Hawley and V. A. Wulfkuhle. *Plains Anthropologist* 51(200):487–520.

Heizer, Robert F. 1964 The Western Coast of North America. In *Prehistoric Man in the New World*, edited by Jesse D. Jennings and Edward Norbeck, pp. 117–148. Published for William Marsh Rice University by the University of Chicago Press, Chicago and London.

——— 1972 A Brief History of the UCAS and ARF, 1948–1972. *Reports of the University of California Archaeological Survey* 75:37–41. University of California, Berkeley.

Heizer, Robert F. (vol. editor). 1978 *California*. Handbook of North American Indians, Vol. 8, William C. Sturtevant, general editor, Smithsonian Institution, Washington, DC.

Henning, Amy Evelyn 1964 Starr Mound. In *Archaeological Excavations, Joanna Reservoir, Missouri*, edited by Dale E. Henning, Amy E. Henning, and Kenneth W. Cole. Report prepared by the University of Missouri, Columbia. On file, National Park Service, Midwest Archeological Center, Lincoln, Nebraska.

Henning, Dale R. 2006 A Midwesterner's Ventures onto the Plains. In *Plains Archaeology's Past: A Collection of Personal Narratives*, edited by M. F. Hawley and V. A. Wulfkuhle. *Plains Anthropologist* 51(200):785–796.

Hester, James J. 1963 A Training Program in Salvage Archaeology. *American Antiquity* 28:392–394.

Hewes, Gordon W. 1948 Burial Mounds in the Baldhill Area, North Dakota. *American Antiquity* 14:322–328.

Hicks, Brent A., editor 2004 *Marmes Rockshelter: A Final Report on 11,000 Years of Cultural Use*. Washington State University Press, Pullman, Washington.

Hoganson, John W. and Edward C. Murphy 2003 *Geology of the Lewis and Clark Trail in North Dakota*. Mountain Press Publishing Company. Missoula, Montana.

Holder, Preston and Joyce Wike 1949 The Frontier Culture Complex, a Preliminary Report on a Prehistoric Hunter's Camp in Southwestern Nebraska. *American Antiquity* 24:260–266.

——— 1950 The Allen Site (FT-50); Archeological Evidence of an Early Hunter's Camp on Medicine Creek, Frontier County, Nebraska. In *Proceedings of the Sixth Plains Archaeological Conference, 1948*, edited by Jesse Jennings, pp. 105–107. Anthropological Papers No. 11. Department of Anthropology, University of Utah, Salt Lake City.

Holland, C. G. 1969 *The Revised Blue Ridge Project: An Archeological Survey and Summary*. Prepared by the Smithsonian Institution. On file at the North Carolina Office of State Archaeology, Raleigh.

Holschlag, Stephanie Lynn 1975 Pot Creek Pueblo and the Question of Prehistoric Northern Tiwa Household Configuration. Unpublished Ph.D. dissertation, Washington State University, Pullman.

Hoover, William J. 2006 *Kinzua: From Cornplanter to the Corps*. iUniverse, Inc. New York.

Hosmer, Charles R., Jr. 1981 *Preservation Comes of Age: From Williamsburg to the National Trust, 1926–1949*. University of Virginia Press, Charlottesville.

Hoyme, Lucile E. and William M. Bass 1962 Human Skeletal Remains from the Tollifero (Ha6) and Clarksville (Mc14) Sites, John H. Kerr Reservoir Basin, Virginia. In *Inter-Agency Archeological Salvage Program: River Basin Surveys Papers* 25, edited by Frank H. H. Roberts, Jr., pp. 329–400. Smithsonian Institution, Bureau of American Ethnology Bulletin 182. Washington DC.

Hurt, Wesley R. 2006 Recollections of a Missouri Valley Salvage Archaeologist. In *Plains Archaeology's Past: A Collection of Personal Narratives*, edited by M. F. Hawley and V. A. Wulfkuhle. *Plains Anthropologist* 51(200):587–596.

Husted, Wilfred M. 2006 Archaeology in the Heartland. In *Plains Archaeology's Past: A Collection of Personal Narratives*, edited by M. F. Hawley and V. A. Wulfkuhle. *Plains Anthropologist* 51(200):745–758.

Interagency Archeological Services 1976a *A Status Report to the Archeological Community*. National Park Service, Washington, DC.

——— 1976b The National Register and Archeology: Suggestions to Archeologists Regarding Determinations of Eligibility. Paper presented at the 41st Annual Meeting of the Society for American Archaeology, St. Louis.

Irwin, Ann M. and Ula Moody 1976 *The Lind Coulee Site (45GR97) 1973 Field Season*. Washington Archaeological Research Center Project Report 36. Pullman.

——— 1977 *The Lind Coulee Site (45GR97) 1974 Field Season*. Washington Archaeological Research Center Project Report 53. Pullman.

Jackson, Donald D. 1962 *Letters of the Lewis and Clark Expedition with Related Documents, 1783–1854*. University of Illinois Press, Urbana.

Jantz, Richard L. 1977 Craniometric Relationships of Plains Populations: Historical and Evolutionary Implications in Trends. In *Middle Missouri Prehistory: A Festschrift Honoring the Contributions of Donald J. Lehmer*, edited by W. Raymond Wood, pp. 162–176. *Plains Anthropologist* Vol. 22, No. 78, Part 2, Memoir 13.

Jantz, Richard L. and Douglas W. Owsley 1982 Osteological Analysis of the Burial from the McClure Site. *Plains Anthropologist* 27(98):54–58.

Jelks, Edward B. 1953 Excavations at the Blum Rockshelter. *Bulletin of the Texas Archeological Society* 24:189–207.

——— 1961 *Excavations at Texarkana Reservoir, Sulphur River, Texas*. Smithsonian Institution, Bureau of American Ethnology Bulletin 179, River Basin Surveys Papers 21. Washington, DC.

——— 1962 *The Kyle Site: A Stratified Central Texas Aspect Site in Hill County, Texas*. Archaeology Series No. 5. Department of Anthropology, University of Texas, Austin.

——— 1965 The Archaeology of the McGee Bend Reservoir, Texas. Unpublished Ph.D. dissertation, The University of Texas, Austin.

——— 1970 Documentary Evidence of Indian Occupation at the Stansbury Site (41–39B1–1). *Bulletin of the Texas Archeological Society* 41:277–283.

——— 1993 The Founding Meeting of the Society for Historical Archaeology, January 6, 1967. *Historical Archaeology* 27(1):10–11.

——— 1995 Letter to the James Smithson Society in Support of a proposed Pilot Study of the River Basin Surveys Program, March 7, 1995. Letter on file, Collections and Archives Program, Department of Anthropology, Smithsonian Museum Support Center, Suitland, Maryland.

——— 2006 Reminiscences of Archaeology in Texas: 1947–1968. In *Plains Archaeology's Past: A Collection of Personal Narratives*, edited by M. F. Hawley and V. A. Wulfkuhle. *Plains Anthropologist* 51(200):597–614.

——— 2012 Discussant. In "Critical Thinking in Archaeology: Papers in Memory of *Dee Ann Story*," symposium organized by Nancy A. Kenmotsu, 77th Annual Meeting of the Society for American Archaeology, Memphis.

Jenkins, Dennis L., Loren G. Davis, Thomas W. Stafford, Jr., Paul F. Campos, Bryan Hockett, George T. Jones, Linda Scott Cummings, Chad Yost, Thomas J. Connolly, Robert M. Yohe II, Summer C. Gibbons, Maanasa Raghavan, Morten Rasmussen, Johanna L. A. Paijmans, Michael Hofreiter, Brian M. Kemp, Jodi Lynn Barta, Cara Monroe, M. Thomas P. Gilbert, and Eske Willerslev 2012 Clovis Age Western Stemmed Projectile Points and Human Coprolites at the Paisley Caves. *Science* 337:223–228.

Jenkins, Ned J. and Richard A. Krause 1986 *The Tombigbee Watershed and Southeastern Prehistory.* University of Alabama Press, Tuscaloosa.

——— 2009 The Woodland-Mississippian Interface in Alabama, Ca. A.D. 1075–1200: An Adaptive Radiation? *Southeastern Archaeology* 28(2):202–219.

Jenkins, Sarah L. 2011 An Analysis of the Faunal Remains from Windust Cave C (45FR46), Washington. Unpublished Master's thesis, Department of Anthropology, Washington State University, Pullman.

Jennings, Jesse D. 1963 Administration of Contract Emergency Archaeological Programs. *American Antiquity* 28:282–285.

——— 1966 *Glen Canyon: A Summary.* University of Utah Anthropological Papers No. 81, Salt Lake City.

——— 1985 River Basin Surveys: Origins, Operations, and Results, 1945–1969. *American Antiquity* 50:281–296.

——— 1986 American Archaeology, 1930–1985. In *American Archaeology Past and Future; A Celebration of the Society for American Archaeology 1935–1985*, edited by David J. Meltzer, Don D. Fowler, and Jeremy A. Sabloff, pp. 53–62. Published for the Society for American Archaeology by the Smithsonian Institution Press, Washington DC.

——— 1989 *Prehistory of North America,* 3rd edition. Mayfield, Mountain View, California.

——— 1994 *Accidental Archaeologist: Memoirs of Jesse D. Jennings.* University of Utah Press, Salt Lake City.

Jennings, Jesse D. and Floyd W. Sharrock 1965 The Glen Canyon: A Multidisciplinary Project. *Utah Historical Quarterly* 33(1):34–50.

Johnson, Ann Mary 1977 The John Ketchen and Durkin Sites, 39ST223 and 29ST238: Extended Middle Missouri Components in the Northern Big Bend Region, South Dakota. Ph.D. dissertation, University of Missouri, Columbia.

——— 1979 Extended Middle Missouri Components in the Big Bend Region, South Dakota. *Special Publication of the South Dakota Archaeological Society 1.* Vermillion.

Johnson, Alfred E. 1968 Archeological Investigations in the Clinton Reservoir Area, Eastern Kansas. University of Kansas Museum of Anthropology, Interagency Archeological Salvage Program Report. Manuscript on file, National Park Service, Midwest Archeological Center, Lincoln, Nebraska.

Johnson, Ann Stofer 1958 Similarities in Hohokam and Chalchihuites Artifacts. *American Antiquity* 24:126–130.

Johnson, Frederick 1947 The Work of the Committee for the Recovery of Archaeological Remains: Aims, History, and Activities to Date. *American Antiquity* 12:212–215.

——— 1951a The Inter-Agency Archaeological Salvage Program in the United States. *Archaeology* 4:25–40.

——— 1951b Archaeological Researches in the Bluestone Reservoir, Summers County, West Virginia. *American Journal of Archaeology* 55(3):255–268.

Johnson, LeRoy, Jr. 1961 The Devil's Mouth Site: A River Terrace Midden at Diablo Reservoir, Texas. *Bulletin of the Texas Archeological Society* 30(1959):253–285.

——— 1963 Pollen Analysis of Two Archeological Sites at Amistad Reservoir, Texas. *Texas Journal of Science* 15(2):203–217.

Johnson, William C. 1984 Archaeological Researches in the Bluestone Reservoir, Summers County, West Virginia. Unpublished Ph.D. dissertation, Department of Anthropology, University of Pittsburgh, Pennsylvania.

——— 1999 Tracing the Glaciated Allegheny Plateau Tradition: A Suggested Culture History Sequence for the Late Woodland Period (ca. A. D. 1000–1600) in the Glaciated Allegheny Plateau of Northwestern Pennsylvania (Five Years Later). Paper presented at the 69th Annual Meeting of the Society for Pennsylvania Archaeology, Brookville, Pennsylvania.

Johnson, William C., James M. Adovasio, Jack Donahue, Phillip T. Fitzgibbons, Thomas C. East, John P. Marwitt, Joseph L. Yedlowski, and Jan D. Applegarth 1983 *Archaeological Researches in the Bluestone Reservoir, Summers County, West Virginia.* Report Prepared by the University of Pittsburgh. Manuscript on file, West Virginia State Historic Preservation Office. Charleston.

Johnson, William C. and Andrew J. Myers 2004 Population Continuity and Dispersal: Cordage Twist Analysis and the Late Woodland in the Glaciated Allegheny Plateau of Northwestern Pennsylvania. In *Perishable Material Culture in the Northeast*, edited by Penelope Ballard Drooker, pp. 87–128. New York State Museum Bulletin 500. Albany.

Johnston, Richard B. 1982 Archaeology of the McClure Site (39HU7) and the Protohistoric Period in the Big Bend Region of South Dakota. *Plains Anthropologist* Memoir 18.

Jones, Joseph 1876 *Explorations of the Aboriginal Remains of Tennessee.* Smithsonian Contributions to Knowledge No. 259. Smithsonian Institution. Washington DC.

Kahler, Herbert E. 1947 The Role of the National Park Service in River Basin Archaeology with Particular Reference to Inter-Bureau Agreements. *American Antiquity* 12:215–216.

Kehoe, Alice B. 1964 The Ghost Dance Religion in Saskatchewan: A Functional Analysis. Unpublished Ph.D. dissertation, Harvard University, Cambridge.

——— 1998 *The Land of Prehistory: A Critical History of American Archaeology.* Routledge, New York.

——— 2010 Retrospective: My Life in Prairie Archaeology. *Canadian Journal of Archaeology* 34:131–147.

Kelly, Arthur R. 1940 Archaeology in the National Park Service. *American Antiquity* 5:274–282.

Kelley, J. Charles 1947 Lehmann Rock Shelter, a Stratified Site of the Toyah, Round Rock, and Uvalde Foci. *Bulletin of the Texas Archeological and Paleontological Society* 18:115–128.

Kelley, Jane H. 1966a Archaeological Investigations in the Tuttle Creek Reservoir Area. Manuscript on file, National Park Service, Midwest Archeological Center, Lincoln, Nebraska.

——— 1966b The Archaeology of the Si. Unpublished Ph.D. dissertation, Harvard University, Cambridge, Massachusetts.

——— 1992 Being and Becoming. In *Rediscovering Our Past: Essays on the History of American Archaeology*, edited by Jonathan E. Reyman, pp. 81–90. Ashgate Publishing Company, Burlington, Vermont.

Kennedy, J. A. 1968 Letter to Dr. Warren W. Caldwell, Director, River Basin Surveys, Lincoln, Nebraska, from J. A. Kennedy, Smithsonian Institution Equal Employment Opportunity Officer. River Basin Surveys, National Anthropological Archives, National Museum of Natural History, Smithsonian Institution, Washington, DC.

Kiehl, Mary 1953 The Glen Elder and White Rock Sites in North-Central Kansas. *Proceedings* 63rd *Annual Meeting of the Nebraska Academy of Sciences and Affiliated Societies*, p. 4. Lincoln, Nebraska.

King, Mary Elizabeth 1965a Prehistoric Textiles from the Gila Bend Area. In *Salvage Archaeology in Painted Rocks Reservoir, Western Arizona*, edited by William W. Wasley and Alfred E. Johnson, Appendix C. University of Arizona Anthropological Papers 9. University of Arizona Press, Tucson.

——— 1965b Textiles and Basketry of the Paracas Period, Ica Valley, Peru. Unpublished Ph.D. dissertation, University of Arizona, Tucson.

King, Thomas F. 1971 A Conflict of Values in American Archaeology. *American Antiquity* 36:255–262.

——— 1977 Issues in Contract Archaeology. *Archaeology* 30(5):352–353.

——— 1978a *The Archeological Survey: Methods and Uses.* Cultural Resource Management Studies, Office of Archeology and Historic Preservation, National Park Service, Department of the Interior, Washington, DC.

King, Thomas, F. 1978b Archeology and Historic Preservation: A Case for Convergence. In *Social Archeology: Beyond Substance and Dating*, edited by Charles L. Redman, Mary Jane Berman, Edward V. Curtin, William T. Langhorne, Jr., Nina M. Versaggi, and Jeffery C. Wanser, pp. 431–437. Academic Press, New York.

——— 2004 *Cultural Resource Laws and Practice: An Introductory Guide*, 2nd ed. AltaMira Press, Walnut Creek, California.

King, Thomas F. and Margaret M. Lyneis 1978 Preservation: A Developing Focus of American Archaeology. *American Anthropologist* 80:873–893.

King, Thomas F., Patricia Parker Hickman, and Gary Berg 1977 *Anthropology in Historic Preservation: Caring for Culture's Clutter*. Academic Press, New York.

Kingsborough, Edward 1830 *Antiquities of Mexico*. Augustine Aglio, London.

Kinsey, W. Fred (editor) 1972 Archeology in the Upper Delaware Valley: A Study of the Cultural Chronology of the Tocks Island Reservoir. *Anthropological Series* 2. Pennsylvania Historical and Museum Commission, Harrisburg.

Kintigh, Keith W. 2006 The Promise and Challenge of Archaeological Data Integration. *American Antiquity* 71:567–578.

Kirk, Ruth 1970 *The Oldest Man in America: An Adventure in Archaeology*. Harcourt Brace Jovanovich, Inc., New York.

Kivett, Marvin F. 1953 *The Woodruff Ossuary, a Prehistoric Burial Site in Phillips County, Kansas*. Smithsonian Institution, Bureau of American Ethnology Bulletin 154, River Basin Surveys Paper 3. Washington, DC.

——— 1958 The Oacoma Sites, 39LM26 and 39LM27, Lyman County, South Dakota. Manuscript on file, National Park Service, Midwest Archeological Center, Lincoln, Nebraska.

Kivett, Marvin F. and Richard E. Jensen 1976 *Archaeological Investigations at the Crow Creek Site (39BF11), Fort Randall Reservoir Area, South Dakota*. Publications in Anthropology No. 7, Nebraska State Historical Society, Lincoln.

Klepinger, Linda and Dale R. Henning 1976 The Hatten Site: A Two Component Burial Site in Northeast Missouri. *The Missouri Archaeologist* 37:92–170.

Knudson, Ruthann 1986 Contemporary Cultural Resource Management. In *American Archaeology: Past and Future*, edited by D. J. Meltzer, D. D. Fowler, and J. A. Sabloff, pp. 395–413. Smithsonian Institution Press, Washington and London.

——— 2002 Medicine Creek is a Paleoindian Cultural Ecotone: The Red Smoke Assemblage. In *Seventy Years of Archaeology in the Medicine Creek Valley of Nebraska*, edited by Donna Roper, pp. 84–141+bibliographic references. The University of Alabama Press, Tuscaloosa, Alabama.

——— 2004 Wormington, Hannah Marie, September 5, 1914–May 31, 1994. Archaeologist. In *Notable American Women, A Biographical Dictionary Completing the Twentieth Century*, edited by Susan Ware, pp. 701–703. The Belknap Press of Harvard University Press, Cambridge.

Knudson, Ruthann, Jeanne Moe, and Alfred Bowers 1983 *Investigation of the Anton Rygh Site (39CA4), Campbell County, South Dakota, 1957–1959*. University of Idaho Anthropological Research Manuscript Series, No. 75. Moscow, Idaho.

Knudson, Ruthann and Don Wyckoff 2013 Medicine Creek Paleoindians and E. Mott Davis, 1947–1954. Manuscript on file, Knudson Associates, Great Falls, Montana.

Kohler, Herbert E. 1953 *Report on Inter-agency Salvage Program*. Department of the Interior, National Park Service. Washington, DC.

Kraft, Herbert C. 1972 The Miller Field Site, Warren County New Jersey. In *Archeology in the Upper Delaware Valley: A Study of the Cultural Chronology of the Tocks Island Reservoir*. Anthropological Series 2, edited by W. Fred Kinsey, pp. 1–54. Pennsylvania Historical and Museum Commission, Harrisburg.

Kraft, Herbert C. 1976 *The Archaeology of the Pahaquarra Site: A Preliminary Report*. Report or the United States National Park Service. Archaeological Research Center, Seton Hall University, South Orange, New Jersey.

Krause, Richard A. 1972 *The Leavenworth Site: Archaeology of an Historic Arikara Community*. University of Kansas Publications in Anthropology, No. 3. Lawrence.

——— 1985 Trend and Trajectory in American Archaeology: Some Questions About the Mississippian Period in Southeastern Prehistory. In *Alabama and the Borderlands: From Prehistory to Statehood*, edited by Reid Badger and Lawrence Clayton, pp. 17–39, The University of Alabama Press, Tuscaloosa, Alabama.

——— 1994 Paper Sacks, Paste-Board Boxes and Intellectual Bins: The River Basin Salvage Program and Archaeological Classification. In "40 Something: The Smithsonian Institution River Basin Surveys," edited by Kimball Banks. *North Dakota Archaeology: Journal of the North Dakota Archaeological Association* 5:27–38.

——— 2012 The Import of Earthen and Stone Tumuli in North American Prehistory. *Journal of Alabama Archaeology* 56(1):7–32.

Krieger, Alex D. 1946 *Culture Complexes and Chronology in Northern Texas*. University of Texas Publication No. 4640. Austin.

Krieger, Herbert W. 1927 *Archeological Investigations in the Columbia River Valley*. Smithsonian Miscellaneous Collections 78:187–200. Washington, DC.

——— 1928 A Prehistoric Pit House Village Site on the Columbia River at Wahluke, Grant County, Washington. *Proceedings of the U.S. National Museum* 73(11):1–29. Washington, DC.

——— 1935 Salvaging Early Cultural Remains in the Valley of the Lower Columbia River. *Annual Pamphlet Explorations and Field-Work of the Smithsonian Institution in 1934*: 53–56. Washington, DC.

Lawson, Michael L. 2012 *Dammed Indians Revisited: The Continuing History of the Pick-Sloan Plan and the Missouri River Sioux*. South Dakota State Historical Society Press. Pierre, South Dakota.

Lee, Ronald F. 2000 *The Antiquities Act of 1906*. National Park Service, Department of the Interior, Washington, DC. Originally published 1970, electronic edition 2000.

Lee, Ronald F., John O. Brew, and E. A. Connelly 1966 Report of the Special Committee on Historic Preservation. National Park Service, Department of the Interior, Washington, DC.

Lees, William B. 1985 Dakota Acculturation during the Early Reservation Period: Evidence from the Deerfly Site (39ML39), South Dakota. *Plains Anthropologist* 30(108):103–122.

——— 2001 The Impact of the River Basin Surveys Programs in Historical Archaeology. Paper presented at the 66th Annual Meeting of the Society for Historical Archaeology, New Orleans.

——— 2008 Historical Archaeology as a Discipline. In *Encyclopedia of Archaeology*, edited by Deborah M. Pearsall. Vol. 2, pp. 1438–1443, Academic Press, New York.

Lehman, Url 1973 *The Bone Hunters*. Columbia University Press. New York.

Lehmer, Donald J. 1954 *Archeological Investigations in the Oahe Dam Area, South Dakota, 1950–51*. Smithsonian Institution, Bureau of American Ethnology Bulletin 158. Washington, DC.

——— 1965 Salvage Archaeology in the Middle Missouri, A Summary and Evaluation. Report prepared for the National Park Service. On file, National Park Service Library, Denver Service Center, Denver, Colorado.

——— 1966 *Fire Heart Creek Site*. Smithsonian Institution, Publications in Salvage Archaeology, 1. Washington, DC.

——— 1971 *Introduction to Middle Missouri Archeology*. Anthropological Papers 1, National Park Service. U.S. Department of the Interior, Washington, DC.

Leidy, Joseph 1869 Extinct Mammalian Fauna of Dakota and Nebraska, with an Introduction on the Geology of the Tertiary Formations of Dakota and Nebraska by F. V. Hayden. *Journal of the Academy of Natural Sciences of Philadelphia* 7(2):1–472.

Leone, Mark P. 1973 Archaeology as the Science of Technology: Mormon Town Plans and Fences. In *Research and Theory in Current Archaeology*, edited by Charles Redman, pp. 125–150. John Wiley and Sons, New York.

Leonhardy, Frank C. 1968 *An Opinion on Archaeological Interpretation in the Plateau*. Eastern New Mexico University, Contributions in Anthropology 1:27–31. Portales, New Mexico.

——— 1970 Artifact Assemblages and Archaeological Units at Granite Point Locality 1 (45WT41), Southeastern Washington. Unpublished Ph.D. dissertation, Department of Anthropology, Washington State University, Pullman.

Leonhardy, Frank C. and David G. Rice 1970 A Proposed Cultural Typology for the Lower Snake River Region, Southeastern Washington. *Northwest Anthropological Research Notes* 4:1–29.

Lewis, Thomas M. N. and Madeline D. Kneberg 1941 *The Prehistory of the Chickamauga Basin in Tennessee: A Preview.* Tennessee Anthropology Papers No. 1, Division of Anthropology. University of Tennessee, Knoxville.

——— 1946 *Hiwassee Island: An Archaeological Account of Four Tennessee Indian Peoples*. The University of Tennessee Press, Knoxville.

Lewis, Thomas M.N. and Madeline K. Lewis 1961 *Eva: An Archaic Site*. The University of Tennessee Press, Knoxville.

——— 1995 *The Prehistory of the Chickamauga Basin in Tennessee.* 2 vols. Compiled and edited by Lynne P. Sullivan. University of Tennessee Press, Knoxville.

Lightfoot, Kent G. 1995 Culture Contact Studies: Redefining the Relationship Between Prehistoric and Historical Archaeology. *American Antiquity* 60:199–217.

Lindsay, Alexander J. Jr., J. Richard Ambler, Mary Anne Stein, and Philip M. Hobler 1969 *Survey and Excavations North and East of Navajo Mountain, Utah, 1959–1962.* Museum of Northern Arizona Bulletin No. 45, Glen Canyon Series No. 8, Flagstaff, Arizona.

Lindsay, Alexander J. Jr., and William D. Lipe 1974 Introduction. In *Proceedings of the 1974 Cultural Resource Management Conference, Federal Center, Denver, Colorado*, edited by William D. Lipe and Alexander J. Lindsay, Jr., pp. vii–xiii. Museum of Northern Arizona, Technical Series 14, Flagstaff.

Lipe, William D. 1974a A Conservation Model for American Archaeology. *The Kiva* 39(3–4):213–245.

——— 1974b Introduction of J. O. Brew. In *Proceedings of the 1974 Cultural Resource Management Conference, Federal Center, Denver, Colorado*, edited by William D. Lipe and Alexander J. Lindsay, pp. 137–139. Museum of Northern Arizona Technical Series 14, Flagstaff.

——— 1978 Contracts, Bureaucrats and Research: Some Emerging Problems of Conservation Archeology in the United States. In *Archaeological Essays in Honor of Irving B. Rouse*, edited by Robert C. Dunnell and Edwin S. Hall, Jr., pp. 121–147. Mouton Publishers, The Hague, Nederlands.

——— 1996 In Defense of Digging: Archeological Preservation as a Means, not an End. *CRM* 19(7):23–27.

——— 1999 View from the Lake: Legacies of the Dolores Archaeological Program. In *The Dolores Legacy: A User's Guide to the Dolores Archaeological Program Data.* Electronic document, complied by Richard H. Wilshusen, Anasazi Heritage Center, www.blm.gov/pgdata/etc/medialib/blm/co/field_offices/ahc/documents.Par.24258.File.dat/doloresguide.pdf, accessed July 1, 2012.

——— 2012 Why Did We Do It That Way? The University of Utah Glen Canyon Project in Retrospect. In *Papers of the Archaeological Society of New Mexico* 34:87–104. Albuquerque, New Mexico.

Lipe, William D. and Alexander J. Lindsay, Jr. 1974 *Proceedings of the 1974 Cultural Resource Management Conference (Federal Center, Denver)*. Museum of Northern Arizona Technical Series 14. Flagstaff.

Lister, Florence C. 1964 *Kaiparowits Plateau and Glen Canyon Prehistory: An Interpretation Based on Ceramics*. University of Utah Anthropological Papers, 71, Glen Canyon Series 23. Salt Lake City.

Little, Michael 2010 Franz Boas' Place in American Physical Anthropology and Its Institutions. In *Histories of American Physical Anthropology in the Twentieth Century*, edited by Michael A. Little and Kenneth A. R. Kennedy, pp. 55–85. Lexington Books, Lanham, Maryland.

Little, Michael A. and Kenneth A. R. Kennedy 2010 *Histories of American Physical Anthropology in the Twentieth Century*. Lexington Books, Lanham, Maryland.

Longacre, William A. 1981 CRM Publication: A Review Essay. *Journal of Field Archaeology* 8:487–491.

Longacre, William A. and Gwinn Vivian 1972 Letter to *Science*, Salvage Archeology. *Science* 178(4063):811–812.

Lohse, Ernest S. and Roderick Sprague 1998 History of Research. In *Plateau*, edited by Deward E. Walker, Jr., pp. 8–28. Handbook of North American Indians, Vol. 12, William C. Sturtevant, general editor, Smithsonian Institution, Washington, DC.

Lyman, R. Lee 1976. *Exploratory Archaeological Research Along Rufus Woods Lake, Upper Columbia River Region, North-Central Washington, 1975*. Washington Archaeological Research Center Project Reports No. 29. Pullman.

——— 2000 Building Cultural Chronology in Eastern Washington: The Influence of Geochronology, Index Fossils, and Radiocarbon Dating. *Geoarchaeology* 15:609–648.

Lyman, R. Lee and Michael J. O'Brien 2003 Cultural Traits: Units of Analysis in Early Twentieth-Century Anthropology. *Journal of Anthropological Research* 59:225–250.

Lyneis, Margaret M. 1968 Cultural Ecology of Lakeside Adaptation in the Western Great Basin. Unpublished Ph.D. dissertation, University of California, Los Angeles.

——— 1992 *The Main Ridge Community at Lost City*. University of Utah Anthropological Papers 117. Salt Lake City.

Lyon, Edwin A. 1996 *A New Deal for Southeastern Archaeology*. University of Alabama Press, Tuscaloosa.

Lyons, Thomas R. and Thomas E. Avery 1977 *Remote Sensing: A Handbook for Archeologists and Cultural Resource Managers*. National Park Service, Washington, DC.

Mallory, Oscar L. 1966 An Archaeological Appraisal of the Pipestem Reservoir, Stutsman County, North Dakota, September 1966. Smithsonian Institution River Basin Surveys Missouri Basin Project Appraisal. Report on file, National Park Service Midwest Archeological Center, Lincoln, Nebraska.

Malouf, Carling 1952 Members of the Montana State University Party of 1952. Memorandum on file, River Basin Surveys Papers, National Anthropological Archives, National Museum of Natural History, Smithsonian Institution, Washington, DC.

——— 1963 *Crow-Flies-High (32MZ1), a Historic Hidatsa Village in the Garrison Reservoir Area, North Dakota*. Smithsonian Institution, Bureau of American Ethnology Bulletin 185, River Basin Survey Papers 29. Washington, DC.

Marchiando, Patricia 1972 The Bell-Browning Site (28Sx19). In *Archaeology in the Upper Delaware Valley: A Study of the Cultural Chronology of the Tocks Island Reservoir*, Anthropological Series 2, edited by W. F. Kinsey, pp. 131–158. Pennsylvania Historical and Museum Commission, Harrisburg.

Marshall, James O. 1966 Appraisal of the Archeological Resources of the Big Hill Reservoir, Labette County, Kansas, May 20, 1966. Manuscript on file, Kansas State Historical Society, Topeka.

Marshall, James O. 1969a The Glen Elder Focus: The Cultural Affiliations of Archeological Material from the Glen Elder Site, 14ML1. Report submitted to the National Park Service, Midwest Region, U.S. Department of the Interior, Omaha. Manuscript on file, University of Nebraska Library, Lincoln, Nebraska.

——— 1969b Appraisal of the Archeological Resources of the Grove Reservoir, Shawnee County, Kansas, May 15, 1969. Manuscript on file, Kansas State Historical Society, Topeka.

——— 1972 The Archeology of the Elk City Reservoir: A Local Archeological Sequence in Southeast Kansas. Manuscript on file, Kansas State Historical Society, Topeka.

Marshall, James O. and Thomas A. Witty, Jr. 1967 The Bogan Site, 14GE1, an Historic Pawnee Village—an Appraisal of an Archeological Site in the Milford Reservoir, Geary County, Kansas, December 1967. Manuscript on file, Kansas State Historical Society, Topeka.

——— 1968 *Archeological Survey of the Lower Salt and Plum Creek Valley, Leavenworth County, Kansas.* Kansas State Historical Society Anthropological Series 4. Topeka.

Martin, Paul S., George I. Quimby, and Donald Collier 1947 *Indians before Columbus: 20,000 Years of North American History Revealed by Archaeology.* University of Chicago Press, Chicago.

Maslowski, Robert F. 1996 Cordage Twist and Ethnicity. In *A Most Indispensable Art: Native Fiber Industries from Eastern North America*, edited by James Petersen, pp. 88–99. University of Tennessee Press, Knoxville.

Mason, Carol I. 1992 Women in American Archaeology in the 1950s. In *Rediscovering Our Past: Essays on the History of American Archaeology*, edited by Jonathan E. Reyman, pp. 91–101. Ashgate Publishing Company, Burlington, Vermont.

Matson, R. G. and Timothy Al Kohler (editors) 2006 *Tracking Ancient Footsteps. William D. Lipe's Contributions to Southwestern Prehistory and Public Archaeology.* Washington State University Press, Pullman.

Mattes, Merrill J. 1949 Historic Sites in the Fort Randall Reservoir Area. *South Dakota Historical Collections* 24:470–577.

——— 1960 Historic Sites Archeology on the Upper Missouri. In *Bureau of American Ethnology Bulletin 176, Interagency Archeological Salvage Program, River Basin Survey Papers* 15, pp. 5–24. Smithsonian Institution, Washington, DC.

Mayer-Oakes, William J. 1955 *Prehistory of the Upper Ohio Valley: An Introductory Archaeological Study.* Annals of the Carnegie Museum of Natural History, 34. Pittsburg.

——— 1974 Announcement of the Formation of the American Society for Conservation Archaeology (ASCA). In *Proceedings of the 1974 Cultural Resource Management Conference, Federal Center, Denver, Colorado*, edited by William D. Lipe and Alexander J. Lindsay, Jr, pp. 141–143. Museum of Northern Arizona, Technical Series 14. Flagstaff.

McClurkan, Burney B., Edward B. Jelks, and Harald P. Jensen 1980 Jonas Short and Coral Snake Mounds: A Comparison. In *Caddoan and Poverty Point Archaeology: Essays in Honor of Clarence Hungerford Webb*, edited by Jon L. Gibson, pp. 173–206. Bulletin of the Louisiana Archaeological Society No. 6. New Orleans.

McGimsey, Charles R., III 1972 *Public Archaeology*. Seminar Press, New York.

——— 1981 Archeology: A Profession in Transition. *Early Man* Autumn 1981:28–32.

——— 1985 "This, too, Will Pass": Moss-Bennett in Perspective. *American Antiquity* 50:326–331.

——— 1991 Foreword, Protecting the Past: Cultural Resource Management—A Personal Perspective. In *Protecting the Past*, edited by George S. Smith and John E. Ehrenhard, pp. xvii–xxii. CRC Press, Boca Raton, Florida.

——— 2004 *CRM on CRM: Charles R. McGimsey III on Cultural Resource Management.* Arkansas Archeological Survey Research Series 61. Fayetteville.

McGimsey, Charles R., III 2006a Point: The Life and Hard Times of the Archaeological and Historic Preservation Act in Washington, DC. An Assessment 30 Years After. *The SAA Archaeological Record* 6(5):6.

——— 2006b Rebuttal. *The SAA Archaeological Record* 6(5):10.

McGimsey, Charles R., III, Carl Chapman, and Hester A. Davis 1970 *Stewards of the Past.* Mississippi Alluvial Valley Archaeological Program, University of Missouri, Columbia.

——— 1991 The Utz Site: An Oneota Village in Central Missouri. *Missouri Archaeologist* 52.

McGimsey, Charles R., III and Hester A. Davis, editors 1977 *The Management of Archaeological Resources: The Airlie House Report.* Society for American Archaeology, Washington, DC.

McManamon, Francis, P. 1996 The Antiquities Act: Setting Basic Preservation Priorities. *CRM* 19(7):18–23.

——— 2000 Cultural Resource Management (CRM). In *Archaeological Method and Theory: An Encyclopedia*, edited by Linda Ellis, pp. 138–142. Garland Publishing Co., New York and London.

——— 2001 Cultural Resources and Protection under United States Law. *Connecticut Journal of International Law* 16(2):247–282.

——— 2006a The Foundation for American Public Archaeology. In *The Antiquities Act: A Century of American Archaeology, Historic Preservation, and Nature Conservation*, edited by David Harmon, Francis P. McManamon, and Dwight T. Pitcaithley, pp. 153–175. The University of Arizona Press, Tucson.

——— 2006b Counterpoint: National Leadership and Coordination for Federal Archaeology. *The SAA Archaeological Record* 6(5):7–9.

McManamon, Francis P. and Keith W. Kintigh 2010 Digital Antiquity: Transforming Archaeological Data into Knowledge. *The SAA Archaeological Record* 10(2):37–40.

McManamon, Francis P. and Fred Wendorf D. 2000 "Dam Good Archeology"—We're Glad It Got Done! The Historical Importance of Reservoir Archeology. *CRM* 23(1):41–46.

McNutt, Charles H. 1995 Robert L. Stephenson: The Plains Years. *Plains Anthropologist* 40(151):77–80.

Means, Bernard K. 1998 Archaeological Past and Present: Field Methodology from 1930s: Relief Excavations in Somerset County, Pennsylvania and Its Relevance to Modern Archaeological Interpretations. *Journal of Middle Atlantic Archaeology* 14:39–63.

——— 2010 Archaeology during the Great Depression: New Deal Excavations in Somerset County, Pennsylvania. *Antiquity* 84(325) Project Gallery: www.antiquity.ac.uk/projgall/means325/

——— 2011a On the Development of a GIS for New Deal Archaeology. *Bulletin of the History of Archaeology* 21(1):59–61.

——— 2011b The Future Meets the Past, Digital Mapping of the New Deal Archaeology Projects Across the Lower 48 States. *The SAA Archaeological Record* 11(3):29–33.

Means, Bernard K. (editor) 2013 *Shovel Ready: Archaeology and Roosevelt's New Deal for America.* University of Alabama Press, Tuscaloosa, Alabama.

Meek, Fielding B. 1876 *Report of the Invertebrate Cretaceous and Tertiary Fossils of the Upper Missouri Country: F. V. Hayden Survey.* U.S. Geological Survey of the Territories, Monograph 9. Washington, DC.

Meighan, Clement W. 1958 Progress Report on Excavations at TUL-1 Archaeological Site, Terminus Reservoir Area, Tulare County, California. Manuscript on file, National Park Service, Pacific West Region Library, San Francisco.

Meighan, Clement W., Brian D. Dillon, Douglas V. Armstrong, Leslie Conton, Richard H. Dillon, Paul Farnsworth, Elliott A. Gehr, and David S. Whitley 1988a *Lake Kaweah Intensive Cultural Resources Survey.* Institute of Archaeology, University of California, Los Angeles. Submitted to the Sacramento District, U.S. Army Corps of Engineers, Sacramento. Manuscript on file, Autry Library and Braun Research Center, Los Angeles.

Meighan, Clement W., Brian D. Dillon, Douglas V. Armstrong, Leslie Conton, Richard H. Dillon, Paul Farnsworth, Elliott A. Gehr, and David S. Whitley 1988b *Success Lake Intensive Cultural Resources Survey.* Institute of Archaeology, University of California, Los Angeles. Submitted to the Sacramento District, U.S. Army Corps of Engineers, Sacramento.

Metcalf, George and Theodore L. White 1953 Appraisal of the Archeological and Paleontological Resources of the Garrison Reservoir, North Dakota. Supplement. Smithsonian Institution, Bureau of American Ethnology, Washington, DC. Manuscript on file, State Historical Society of North Dakota State Archives, Bismarck.

Meyer, Roy W. 1977 *The Village Indians of the Upper Missouri: The Madan, Hidatsas and Arikaras.* University of Nebraska Press, Lincoln.

Miller, Carl F. 1947 Preliminary Report of the Buggs Island Reservoir in Mecklenburg, Halifax, Charlotte Counties, Virginia, and Warren, Vance, and Granville Counties North Carolina. Manuscript on file, River Basin Surveys Collections, Smithsonian Institution, Washington, DC.

——— 1949a The Lake Spring Site, Columbia County, Georgia. *American Antiquity* 15: 38–51.

——— 1949b Early Cultural Manifestations Exposed by the Archeological Survey of the Buggs Island Reservoir in Southern Virginia and Northern North Carolina [Warren, Vance, and Granville Counties]. *Washington Academy of Sciences Journal* 38(12).

——— 1949c Appraisal of the Archeological Resources, Buggs Island Reservoir in Mecklenburg, Halifax, Charlotte Counties, Virginia; Warren, Vance, and Granville Counties, North Carolina. *Archaeological Society of Virginia* 4(1).

——— 1950 Early Cultural Horizons in the Southeastern United States. *American Antiquity* 15:273–288.

——— 1956 Supplementary Appraisal of the Archeological Resources of Four Reservoirs in the Missouri River Basin: Merritt Reservoir, Cherry County, Nebraska; Pactola Reservoir, Pennington County, South Dakota; Tiber Reservoir, Toole County, Montana; Glendo Reservoir, Platte County, Wyoming. Smithsonian Institution River Basin Surveys Missouri Basin Project Appraisal. Manuscript on file, National Park Service, Midwest Archeological Center, Lincoln, Nebraska.

——— 1957 Radiocarbon Dates from an Early Archaic Deposit in Russell Cave, Alabama. *American Antiquity* 23:84.

——— 1958 Russell Cave: New Light on Stone Age Life. *National Geographic* 113: 426–443.

——— 1960a The Use of Chenopodium Seeds as a Source of Food by the Early Peoples in Russell Cave, Alabama. *Southern Indian Studies* 12:31–32.

——— 1960b *The Excavation and Investigation of Fort Lookout Trading Post II (39LM57) in the Fort Randall Reservoir, South Dakota.* Smithsonian Institution, Bureau of American Ethnology Bulletin 176, River Basin Surveys Papers 17. Washington, DC.

——— 1962 *Archeology of the John H. Kerr Reservoir Basin, Roanoke River Virginia—North Carolina.* Smithsonian Institution, Bureau of American Ethnology Bulletin 182, Interagency Archeological Salvage Program, River Basin Survey Papers 25. Washington, DC.

——— 1964 *Archeological Investigations at the Hosterman Site (39PO7), Oahe Reservoir Area, Potter County, South Dakota, 1956.* Smithsonian Institution, Bureau of American Ethnology Bulletin 189, River Basin Surveys Paper 35. Washington, DC.

Miller, George L. 1998 J. C. Harrington, 1901–1998. *Historical Archaeology* 32(4):1–7.

Mills, John E. and Carolyn Osborne 1952 Material Culture from an Upper Coulee Rockshelter. *American Antiquity* 17:352–359.

Minnesota Historical Society 2012a G. Hubert Smith: An Inventory of His Papers at the Minnesota Historical Society Manuscripts Collection. Electronic document. http://www.mnhs.org/library/findaids/PO816.xml, accessed April 12, 2012.

Minnesota Historical Society 2012b Alan R. Woodworth: An Inventory of His Papers at the Minnesota Historical Society Manuscripts Collection. Electronic document. http://www.mnhs.org/library/findaids/00339.xml, accessed April 12, 2012.

Mitchell, Seth T. 2011 Understanding the Occupational History of the Monongahela Johnston Village Site through Total Artifact Design. Unpublished Master's thesis, Department of Anthropology, Indiana University of Pennsylvania, Indiana, Pennsylvania.

Mohr, Albert 1948 The Excavation of Site CA-Sta-6. Manuscript 44 of the University of California Archaeological Research Facility. Manuscript on file, Bancroft Library, University of California, Berkeley.

——— 1949 Appraisal of the Archaeological Resources of Black Butte Reservoir, Glenn and Tehama Counties, California. Smithsonian Institution, River Basin Surveys, Pacific Coast Area, Appraisals. Manuscript on file, Bancroft Library, University of California, Berkeley.

——— 1951 Archaeological Appraisal of Burns Reservoir, Merced County, and Bear and Owens Reservoirs, Mariposa County, California. Smithsonian Institution, River Basin Surveys, West Coast Projects, Appraisals. Manuscript on file, Bancroft Library, University of California, Berkeley.

——— ca. 1950 Appraisal of the Archaeological Resources of Cachuma Reservoir, Santa Barbara County, California. Manuscript on file, National Park Service, Western Regional Office, San Francisco.

Monseth, Ann, Henry Irwin, Roald Fryxell, and Carl E. Gustafson 1973 The Lind Coulee Archaeological Site, Re-excavation of an Early Man Campsite in Central Washington. Washington State University report to the National Park Service, Western Regional Office, San Francisco. Manuscript on file, Penrose Library, Whitman College, Walla Walla, Washington.

Moore, Clarence B. 1894 Certain Sand Mounds of the St. John's River, Florida. Part I. *Journal of the Academy of Natural Sciences of Philadelphia*, 2d ser: 5–128.

——— 1907 Moundville, Revisited. *Journal of the Academy of Natural Sciences of Philadelphia*, 2d ser: 337–405.

Moratto, Michael J. 1984 *California Archaeology.* Academic Press, Orlando and New York.

——— 2002 Culture History of the New Melones Reservoir Area, Calaveras and Tuolumne Counties, California. In *Essays in California Archaeology: A Memorial to Franklin Fenenga*, edited by William J. Wallace and Francis Riddell, pp. 25–54. Contributions of the University of California Archaeological Research Facility 60. University of California, Berkeley.

Mott, Mildred 1938 The Relation of Historic Indian Tribes to Archaeological Manifestations in Iowa. *Iowa Journal of History and Politics* 36(3):227–315.

Moulton, Gary E. (editor) 1987 *The Journals of the Lewis and Clark Expedition,* Vol. 3. University of Nebraska Press, Lincoln.

——— 1993 *The Journals of the Lewis and Clark Expedition,* Vol. 8. University of Nebraska Press, Lincoln.

Mulloy, William T. 1958 *A Preliminary Historical Outline for the Northwestern Plains.* University of Wyoming Publications, Vol. 22, Nos. 2–3. Laramie.

Murtagh, William 1988 *Keeping Time: The History and Theory of Preservation in America.* The Main Street Press, Pittstown, New Jersey.

Nabokov, Peter (editor) 1992 Gone Forever, (Buffalo Bird Woman, Hidatsa). In *Native American Testimony—A Chronicle of Indian-White Relations From Prophecy to the Present 1942–1992*. Chapter 9, No. 7, pp. 182–183. Viking Penguin, New York.

Nammour, Valerie Wheeler 1974 Drums and Guns: A Cross-Cultural Study of the Nature of War. Unpublished Ph.D. dissertation, University of Oregon, Eugene.

National Park Service 1977a Guidelines for the Location and Identification of Historic Properties Containing Scientific, Prehistoric, Historic, or Archeological Data. *Federal Register* 42(19), Friday, 28 January 1977.

National Park Service 1977b 36 CFR 66: Recovery of Scientific, Prehistoric, Historic, and Archeological Data: Methods, Standards, and Reporting Requirements. *Federal Register* 42(19), Friday, 28 January 1977.

——— 1977c Determinations of Eligibility for Inclusion in the National Register of Historic Places. *Federal Register* 42(183), Wednesday, 21 September 1977.

——— 1983 Secretary of the Interior's Standards and Guidelines for Archeology and Historic Preservation. National Park Service, Washington, DC. Electronic document, http://www.cr.nps.gov/local-law/arch_stnds_0.htm, accessed October 4, 2012.

——— 2012 Delaware Water Gap, National Recreation Area. Stories: Tocks Island Dam Controversy. Electronic document, www.nps.gov/dewa/historyculture/stories-tocks.htm, accessed May 1, 2012.

Nelson, Charles M. 1969 *The Sunset Creek Site (45–KT-28) and Its Place in Plateau Prehistory.* Washington State University Laboratory of Anthropology Report of Investigations 47. Pullman.

Neuman, Robert W. 1975 *The Sonota Complex and Associated Sites on the Northern Great Plains*. Nebraska State Historical Society, Publications in Anthropology, No. 6. Lincoln.

——— 2003 Strangers in Paradise. *Journal of the Iowa Archeological Society* 50:51–62.

Neumann, Thomas W. and Robert M. Sanford 2001 *Practicing Archaeology: A Training Manual for Cultural Resources Archaeology*. AltaMira Press, Walnut Creek, California.

Noël Hume, Ivor and Henry M. Miller 2011 Ivor Noël Hume: Historical Archaeologist. *The Public Historian* 33(1):9–32.

Nusbaum, Jesse L. 1956 Introduction. In *Pipeline Archaeology*, edited by Fred Wendorf, Nancy Fox, and Orian L. Lewis. Laboratory of Anthropology, Santa Fe.

Ohio Department of Natural Resources 2012a Deer Creek Lake. Electronic document, http://www.dnr.state.oh.us/Home/FishingSubhomePage/LakeMapLandingPage/DeerCreekLakeFishingMap/tabid/19509/Default.aspx, accessed September 2012.

——— 2012b Paint Creek Lake Wildlife Area. Electronic document, http://www.dnr.state.oh.us/Home/wild_resourcessubhomepage/WildlifeAreaMaps/SouthwestOhioWildlifeAreaMaps/PaintCreekLakeWildlifeArea/tabid/19788/Default.aspx, accessed September 10, 2012

Ohlson, Kristin 2011–2012 The Dean of Texas Archaeology. *American Archaeology* 15:39–43.

Olberding, Susan Deaver 2011 *Sunset on Glen Canyon: The Words and Images of Two Remarkable Women: Katharine Bartlett and Gene Field Foster.* Fort Valley Publishing. Flagstaff, Arizona.

Ortman, Scott G. 2012 *Winds from the North: Tewa Origins and Historical Anthropology.* University of Utah Press, Salt Lake City.

Osborne, Carolyn 1969 Perishables from Eight Rockshelters on Quilomene Bar. Appendix C. In *The Sunset Creek Site (45–KT-28) and Its Place in Plateau Prehistory*, by Charles M. Nelson, pp. 243–248. Washington State University Laboratory of Anthropology Report of Investigations 47. Pullman.

Osborne, Douglas 1948a Preliminary Appraisal of the Archeological Resources of the Mariposa Reservoir, Mariposa County, California. Smithsonian Institution, River Basin Surveys, Pacific Coast Area, Appraisals. Manuscript on file, Morgan Library, Colorado State University, Fort Collins, Colorado.

——— 1948b An Appraisal of the Archeological Resources of the Ice Harbor, Lower Monumental, Little Goose and Lower Granite Reservoirs, Snake River, Washington. Columbia Basin Project, River Basin Surveys. Manuscript on file, Washington State Department of Archaeology and Historic Preservation, Olympia.

Osborne, Douglas 1949 Appraisal of the Archeological Resources of Chief Joseph Reservoir on the Upper Columbia River. Columbia Basin Project, River Basin Surveys. Manuscript on file, National Park Service, Pacific West Regional Library, San Francisco.

Osborne, Douglas 1951 *Excavations in the McNary Reservoir Basin near Umatilla, Oregon*. Unpublished Ph.D. dissertation, Department of Anthropology, University of California, Berkeley.

——— 1953 Archaeological Occurrence of Pronghorn Antelope, Bison and Horse in the Columbia Plateau. *Scientific Monthly* 77:260–269.

——— 1956 *Evidence of the Early Lithic in the Pacific Northwest*. Research Studies of the State College of Washington 24:38–44.

——— 1957 *Excavations in the McNary Reservoir Basin Near Umatilla, Oregon*. Smithsonian Institution, Bureau of American Ethnology Bulletin 166, River Basin Surveys Paper 8. Washington, DC.

——— 1967 *Archaeological Tests in the Lower Grand Coulee, Washington*. Occasional Papers of the Idaho State University Museum No. 20. Pocatello.

Osborne, H. Douglas Alan L. Bryan, and Robert Crabtree 1961 The Sheep Island Site and the Mid-Columbia Valley. In *Smithsonian Institution, Bureau of American Ethnology Bulletin* 179, *River Basin Surveys Papers* 24, pp. 267–306. Washington, DC.

Osborne, H. Douglas and Robert Crabtree 1961 Two Sites in the Upper McNary Reservoir. *Tebiwa* 4(2):19–36.

Osborne, H. Douglas Robert Crabtree, and Alan Bryan 1952 Archeological Investigations in the Chief Joseph Reservoir. *American Antiquity* 27:360–373.

Owsley, Douglas W. 1975 A Demographic Analysis of Skeletons from the Larson Site (39WW2) Walworth County, South Dakota. Unpublished Master's thesis, Anthropology Department, University of Tennessee, Knoxville.

Owsley, Douglas W. and William M. Bass 1979 A Demographic Analysis of Skeletons from the Larson Site (39WW2) Walworth County, South Dakota. *Journal of Physical Anthropology* 51(2):145–154.

Owsley, Douglas W. and Richard L. Jantz 1994 An Integrative Approach to Great Plains Skeletal Biology. In *Skeletal Biology in the Great Plains: Migration, Warfare, Health, and Subsistence*, edited by Douglas W. Owsley and Richard L. Jantz, pp. 3–8. Smithsonian Institution, Washington, DC.

Owsley, Douglas W. H. E. Berryman, and William M. Bass 1977 Demographic and Osteological Evidence for Warfare at the Larson Site, South Dakota. In *Middle Missouri Prehistory: A Festschrift Honoring the Contributions of Donald J. Lehmer*, edited by W. Raymond Wood *Plains Anthropologist* Vol. 22, No. 78, Part 2, Memoir 13, pp. 119–131.

Parent, Annette R. 1962 Hurry-up Archaeology. *Pacific Discovery* 15(6):2–9.

Patrick, Doreena, James E. Martin, David C. Parris, and David E. Grandstaff 2007 Rare Earth Element Determination of the Stratigraphic Position of the Holotype of *Mosasaurus Missouriensis* (Harlan), the First Fossil Reptile from the American West. In *The Geology and Paleontology of the Late Cretaceous Marine Deposits of the Dakotas,* edited by James E. Martin and David C. Parris, pp. 71–83. Geological Society of America Special Paper 427. Boulder, Colorado.

Peak & Associates 1981 Merced County Streams Project, Burns Reservoir, California, Intensive Cultural Resources Survey. Prepared by Peak & Associates, Inc., Sacramento, California. Submitted to the U.S. Army Corps of Engineers, Sacramento District, Sacramento, California. Manuscript on file, Defense Tech. Information Center Elec. Rep. Fort Belvoir, Virginia.

Perry, Elizabeth M., Ann L. W. Stodder, and Charles A. Bollong (editors) 2010 *Animas–La Plata Project: Bioarchaeology*. SWCA Anthropological Research Papers, Vol. XV, No. 10, SWCA Environmental Consultants, Phoenix, Arizona.

Petsche, Jerome E. (compiler) 1968 *Bibliography of Salvage Archaeology in the United States*. Smithsonian Institution, River Basin Surveys, Publications in Salvage Archaeology 10. Lincoln, Nebraska.

Plog, Fred 1978 Cultural Resource Management and the "New Archeology." In *Social Archeology: Beyond Substance and Dating*, edited by Charles L. Redman, Mary Jane Berman, Edward V. Curtin, William T. Langhorne, Jr., Nina M. Versaggi, and Jeffery C. Wanser, pp. 421–429. Academic Press, New York.

Poole, Meredith 2012 A Short History of Archaeology at Colonial Williamsburg. Colonial Williamsburg Foundation. Electronic document. http://research.history.org/Archaeological_Research/Features/History.cfm, accessed April 12, 2012.

Portnoy, Alice W. (editor) 1978 *Scholars as Managers*. U.S. Department of the Interior, Heritage Conservation and Recreation Service, Washington, DC.

——— 1979 *Scholars as Contractors*. U.S. Department of the Interior, Heritage Conservation and Recreation Service, Washington, DC.

Prezzano, Susan C. 2013 Excavations at the Millstone Site (36El204). Report prepared by the Department of AGES, Clarion University for the Allegheny National Forest, United States Forest Service.

Putnam, Fredrick W. 1878 Archaeological Explorations in Tennessee. In *Eleventh Annual Report of the Trustees of the Peabody Museum of American Archaeology and Ethnology* 2(2):305–360, Cambridge.

Pykles, Benjamin C. 2011 A New Archaeology in the New Deal: The Rise of Historical Archaeology in the 1930s. *The SAA Archaeological Record* 11(3):38–41.

Redman, Charles L. 1991 Distinguished Lecture in Archaeology: In Defense of the Seventies—The Adolescence of New Archaeology. *American Anthropologist* 93(2):295–303.

Redman, Charles L., Mary Jane Berman, Edward V. Curtin, William T. Langhorne, Jr., Nina M. Versaggi, and Jeffery C. Wanser (editors) 1978 *Social Archeology: Beyond Substance and Dating*. Academic Press, New York.

Reid, Kenneth C. and James C. Gallison 1996 Lower Snake River Basin. In *An Overview of Cultural Resources in the Snake River Basin: Prehistory and Paleoenvironments*, edited by Kenneth B. Reid, pp. 15–104. Northwest Anthropological Research Notes, 30(1–2).

Relethford, John H. 2010 Race and the Conflicts within the Profession of Physical Anthropology during the 1950s and 1960s. In *Histories of American Physical Anthropology in the Twentieth Century*, edited by Michael A. Little and Kenneth A. R. Kennedy, pp. 207–219. Lexington Books, Lanham, Maryland.

Resek, Carl 1960 *Lewis Henry Morgan: American Scholar*. University of Chicago Press, Chicago.

Reyman, Jonathan E. 1992 Women in American Archaeology: Some Historical Notes and Comments. In *Rediscovering Our Past: Essays on the History of American Archaeology*, edited by Jonathan E. Reyman, pp. 69–80. Ashgate Publishing Company, Burlington, Vermont.

Rice, David G. 1969 Preliminary Report, Marmes Rockshelter Archaeological Site, Southern Columbia Plateau. Unpublished report submitted to the National Park Service. Manuscript on file, Washington State University, Laboratory of Anthropology, Pullman.

——— 1972 *The Windust Phase in Lower Snake River Region Prehistory*. Reports of Investigations No. 50, Washington State University, Laboratory of Anthropology, Pullman.

Rice, Harvey S. 1965 The Cultural Sequence at Windust Caves. Unpublished Master's thesis, Department of Anthropology, Washington State University, Pullman.

Richards, Horace G. 1939 Reconsideration of the Dating of the Abbott Farm Site at Trenton, New Jersey. *American Journal of Science* 237(5):345–354.

Riddell, Francis A. 1949 Appraisal of the Archeological Resources of Farmington Reservoir, Littlejohns Creek, San Joaquin and Stanislaus Counties, Calif. Smithsonian Institution, River Basin Surveys, Pacific Coast Division, Appraisals. Manuscript on file, U.S. Army Corps of Engineers, Sacramento District, Sacramento, California.

Ritchie, William A. 1944 *The Pre-Iroquoian Occupation of New York State*. Museum Memoir No. Rochester Museum of Arts and Sciences. Rochester, New York.

——— 1954 Dutch Hollow, an Early Historic Period Seneca Site in Livingston County, New York. *Researches and Transactions of the New York State Archeological Association* 13(1).

Roberts, Frank. H. H., Jr. 1947 The "Head Office" in the Smithsonian Institution and Cooperation with State and Local Institutions. *American Antiquity* 12:217–220.

——— 1948 A Crisis in U.S. Archaeology. *Scientific American* December 1948:12–17.

——— 1951 River Basin Surveys. In "Report on the Bureau of American Ethnology" by M. W. Stirling. *Annual Report of the Board of Regents of the Smithsonian Institution for the Year Ended June 30, 1950*, pp. 48–72. Washington, DC.

——— 1952a River Basin Surveys. In *Annual Report of the Board of Regents of the Smithsonian Institution: Operations, Expenditures, and Condition of the Institution for the Year Ended June 30, 1951*, pp. 59–87. Washington, DC.

——— 1952b River Basin Surveys: The First Five Years of the Inter-agency Archeological and Paleontological Salvage Program. In *Annual Report of the Board of Regents of the Smithsonian Institution: Operations, Expenditures, and Condition of the Institution for the Year Ended June 30, 1951*, pp. 351–383. Smithsonian Institution, Washington, DC.

——— 1953 River Basin Surveys. In *Annual Report of the Board of Regents of the Smithsonian Institution: Operations, Expenditures, and Condition of the Institution for the Year Ended June 30, 1952*, pp. 58–79. Washington, DC.

——— 1954 River Basin Surveys. In *Annual Report of the Board of Regents of the Smithsonian Institution: Operations, Expenditures, and Condition of the Institution for the Year Ended June 30, 1953*, pp. 64–88. Washington, DC.

——— 1955 River Basin Surveys. In *Annual Report of the Board of Regents of the Smithsonian Institution: Operations, Expenditures, and Condition of the Institution for the Year Ended June 30, 1954*, pp. 32–41. Washington, DC.

——— 1956 River Basin Surveys. In *Seventy-Second Annual Report of the Bureau of American Ethnology to the Secretary of the Smithsonian Institution: 1954–1955*, pp. 5–18. Washington, DC.

——— 1957 River Basin Surveys. In *Annual Report of the Board of Regents of the Smithsonian Institution: Operations, Expenditures, and Condition of the Institution for the Year Ended June 30, 1956*, pp. 45–64. Washington, DC.

——— 1958 River Basin Surveys. In *Annual Report of the Board of Regents of the Smithsonian Institution: Operations, Expenditures, and Condition of the Institution for the Year Ended June 30, 1957*, pp. 44–62. Washington, DC.

——— 1959 River Basin Surveys. In *Annual Report of the Board of Regents of the Smithsonian Institution: Operations, Expenditures, and Condition of the Institution for the Year Ended June 30, 1958*, pp. 50–75. Washington, DC.

——— 1960 River Basin Surveys. In *Annual Report of the Board of Regents of the Smithsonian Institution: Operations, Expenditures, and Condition of the Institution for the Year Ended June 30, 1959*, pp. 56–91. Washington, DC.

——— 1961a River Basin Surveys. In *Annual Report of the Board of Regents of the Smithsonian Institution: Operations, Expenditures, and Condition of the Institution for the Year Ended June 30, 1960*, pp. 52–77. Washington, DC.

——— 1961b River Basin Salvage Program: After 15 Years. *In Annual Report of the Board of Regents of the Smithsonian Institution: Operations, Expenditures, and Condition of the Institution for the Year Ended June 30, 1960*, pp. 523–549. Washington, DC.

Ritchie, William A. 1963 River Basin Surveys. In *Annual Report of the Board of Regents of the Smithsonian Institution: Operations, Expenditures, and Condition of the Institution for the Year Ended June 30, 1962*, pp. 46–65. Washington, DC.

——— 1964 River Basin Surveys. In "Report on the Bureau of American Ethnology." *Annual Report of the Board of Regents of the Smithsonian Institution for the Year Ended June 30, 1963*, pp. 74–106. Washington, DC.

Rogers, Jerry L. 2006 The Antiquities Act and Historic Preservation. In *The Antiquities Act: A Century of American Archaeology, Historic Preservation, and Nature Conservation*, edited by David Harmon, Francis P. McManamon, and Dwight T. Pitcaithley, pp. 176–186. The University of Arizona Press, Tucson.

Rolingson, Martha A. 1963 An Archeological Survey of the Grayson Reservoir. Manuscript on file, University of Kentucky Museum of Anthropology, Lexington.

——— 1967 Temporal Perspective on the Archaic Cultures of the Middle Green River Region, Kentucky. Unpublished Ph.D. dissertation, University of Michigan, Ann Arbor.

——— 1968 Preliminary Excavations in the Eagle Creek Reservoir, Grant and Owen Counties, Kentucky. Manuscript on file, University of Kentucky Museum of Anthropology, Lexington.

Rolingson, Martha A. and Michael J. Rodeffer 1968a Archeological Excavation in Cave Run Reservoir, Kentucky: Progress Report, 1968. Manuscript on file, University of Kentucky Museum of Anthropology, Lexington.

——— 1968b The Zilpo Site, BY37, Preliminary Excavations in the Cave Run Reservoir, Kentucky: 1968. Manuscript on file, University of Kentucky Museum of Anthropology, Lexington.

Rolingson, Martha A. and Douglas W. Schwartz 1963 Archeological Survey of the Laurel River Reservoir. Manuscript on file, University of Kentucky Museum of Anthropology, Lexington.

Rosebrough, Amy L. 1994 Pots, Paints and Polish: Ceramics Variety and Use at Helb. Unpublished Master's thesis, University of Arkansas, Fayetteville.

——— 1995 A Functional Analysis of Ceramics from the Helb Site, 39CA208, Campbell County, South Dakota. *Plains Anthropologist* 40(152):103–124.

Ruff, Christopher 1994 Biomechanical Analysis of Northern and Southern Plains Femora: Behavioral Implications. In *Skeletal Biology in the Great Plains: Migration, Warfare, Health, and Subsistence*, edited by Douglas W. Owsley and Richard L. Jantz, pp. 235–246. Smithsonian Institution, Washington, DC.

Rusco, Mary Lou Kiehl 1960 *The White Rock Aspect*. Notebook No. 4, Laboratory of Anthropology, University of Nebraska, Lincoln.

Salwen, Bert and Barbara Richards 1959 Field Notes, Site 39LM1 39LM227. Lower Brule Reservoir, N. Dakota, June–July 1959. On file, River Basin Surveys, National Anthropological Archives, National Museum of Natural History, Smithsonian Institution, Washington, DC.

Sanger, David 1977 *Cultural Resource Management in the Dickey-Lincoln School Reservoir, Maine*. University of Maine, Orono.

Saxon, Wolfgang 1996 Merrill Mattes, 85; Historian Traced the Trail West. *New York Times*, 12 May. New York.

Schaafsma, Polly Dix 1962 Rock Art of the Navajo Reservoir. *El Palacio* 69:193–212.

——— 1963 *Rock Art in the Navajo Reservoir District*. Museum of New Mexico Papers in Anthropology 7. Santa Fe.

——— 1965 Kiva Murals from Pueblo del Encierro (LA 70). *El Palacio* 72:6–16.

——— 1975 *Rock Art in the Cochiti Reservoir District*. Museum of New Mexico Papers in Anthropology 16. Santa Fe.

Schoenwetter, James, Frank W. Eddy, and Eleanor Jane Nettle 1964 *Alluvial and Palynological Reconstruction of Environments, Navajo Reservoir District*. Museum of New Mexico Papers in Anthropology 13, Navajo Project Studies 11. Santa Fe.

Schiffer, Michael B. and John H. House 1975 *The Cache River Archeological Project: An Experiment in Contract Archeology*. Arkansas Archeological Survey Research Series 8. Fayetteville.

Schiffer, Michael B. and John H. House 1977 Cultural Resource Management and Archeological Research: The Cache Project. *Current Archeology* 18(1):43–53.

Schiffer, Michael B. and George J. Gumerman (editors) 1977 *Conservation Archaeology: A Guide for Cultural Resource Management Studies*. Academic Press, New York.

Schulyer, Robert L. 1970 Historical and Historic Sites Archaeology as Anthropology: Basic Definitions and Relationships. *Historical Archaeology* 4:83–89.

Scott, Douglas D., Richard A. Fox, Jr., Melissa A. Connor, and Dick Harmon 1989 *Archaeological Perspectives on the Battle of the Little Bighorn*. University of Oklahoma Press, Norman.

Scovill, Douglas H. 1974a History of Archaeological Conservation Policy and the Moss-Bennett Bill. In *Proceedings of the 1974 Cultural Resource Management Conference, Federal Center, Denver, Colorado*, edited by William D. Lipe and Alexander J. Lindsay, Jr., pp. 1–11. Museum of Northern Arizona, Technical Series 14. Flagstaff, Arizona.

——— 1974b Gearing Up for Contract Archaeology. In *Proceedings of the 1974 Cultural Resource Management Conference, Federal Center, Denver, Colorado*, edited by William D. Lipe and Alexander J. Lindsay, Jr., pp. 57–67. Museum of Northern Arizona, Technical Series 14. Flagstaff, Arizona.

Sebastian, Lynne 2006 The Conservation Model Today and Historic Preservation. In *Tracking Ancient Footsteps: William D. Lipe's Contributions to Southwestern Prehistory and Public Archaeology*, edited by R. G. Matson and Timothy A. Kohler, pp. 109–126. Washington State University Press, Pullman,

——— 2009 Deciding What Matters: Archaeology, Eligibility, and Significance. In *Archaeology and Cultural Resource Management: Visions for the Future*, edited by Lynne Sebastian and William D. Lipe, pp. 91–114. School for Advanced Research Press, Santa Fe.

Shafer, Harry 2011 Dee Ann Story. *Texas Archeology, Newsletter of the Texas Archeological Society* 55(2):12–15.

Shiner, Joel L. 1951 An Appraisal of the Archeological Resources of the Priest Rapids Reservoir on the Columbia River, Washington. Columbia Basin Project, River Basin Surveys, Smithsonian Institution, Eugene, Oregon. Manuscript on file, The University of Washington Libraries, Seattle.

——— 1954 The McNary Reservoir, A Study in Plateau Archeology. Unpublished Ph.D. dissertation, Department of Anthropology, University of Arizona, Tucson.

——— 1961 *The McNary Reservoir, a Study in Plateau Archeology*. Smithsonian Institution, Bureau of American Ethnology Bulletin 179, River Basin Surveys Papers 23. Washington, DC.

Shutler, Richard, Jr. 1961 *Lost City. Pueblo Grande de Nevada*. Nevada State Museum Anthropological Papers No. 5. Carson City.

Silverberg, Robert 1967 *Men Against Time: Salvage Archaeology in the United States*. Macmillan, New York.

——— 1986 *Mound Builders of Ancient America: The Archaeology of a Myth*. New York Graphic Society, New York.

Simonds, Wm. Joe 1998 The Columbia Basin Project. Electronic document, www.usbr.gov/projects/Project.jsp?proj_Name=Columbia Basin Project, accessed May 18, 2011.

Smith, Bruce D. 1985 The Role of Chenopodium as a Domesticate in Pre-Maize Garden Systems of the Eastern United States. *Southeastern Archeology* 4:51–72.

Smith, Carlyle S. 2006 From Long Island to the Great Plains. In *Plains Archaeology's Past: A Collection of Personal Narratives*, edited by M. F. Hawley and V. A. Wulfkuhle. *Plains Anthropologist* 51(200):527–552.

Smith, Carlyle S. (editor) 1977 *The Talking Crow Site*. Publications in Salvage Archeology 9. University of Kansas Museum of Anthropology, Lawrence.

Smith, Carlyle S. and Roger T. Grange, Jr. 1958 *The Spain Site (39LM301), a Winter Village in Fort Randall Reservoir, South Dakota*. Smithsonian Institution, Bureau of American Ethnology Bulletin 169, River Basin Surveys Papers 11. Washington, DC.

Smith, Carlyle S. and Alfred E. Johnson 1968 *The Two Teeth Site*. Smithsonian Institution Publications in Salvage Archeology 8. Washington, DC.

Smith, Clarence E. ca. late 1940s Archeological Reconnaissance of Wilson Valley Reservoir, Lake County, California. Manuscript on file, Phoebe A. Hearst Museum of Anthropology, Berkeley.

——— 1947 Preliminary Appraisal of the Archeological Resources of Kelsey Creek Reservoir, Lake County, California. NW Information Center, Anthropology Department, Sonoma State University, Sonoma, California.

Smith, Harlan I. 1904 Archeological Survey of the Interior of the State of Washington, During the Summer of 1903. *Science* 19:579–580.

——— 1906a Preliminary Notes on the Archeology of the Yakima Valley, Washington. *Science* 23:551–555.

——— 1906b Recent Archaeological Discoveries in Northwestern North America. *Bulletin of the American Geographical Society* 38:1–9.

——— 1910 The Archaeology of the Yakima Valley. *American Museum of Natural History Anthropological Papers* 6(1):1–151.

Smith, G. Hubert 1939 Archaeological Report, Fort Laramie National Monument, Summer 1939. Manuscript on file, National Park Service, Midwest Archaeological Center. Lincoln, Nebraska.

——— 1957 Archeological Investigations at Old Stone House, Georgetown (Washington, DC). Manuscript on file, National Park Service, Harpers Ferry Center, Virginia.

——— 1960a Fort Pierre II (39ST217), a Historic Trading Post in the Oahe Dam Area, South Dakota. In *Smithsonian Institution, Bureau of American Ethnology Bulletin* 176, *River Basin Surveys Paper* 18, pp. 83–158. Washington, DC.

——— 1960b Archeological Investigations at the Site of Fort Stevenson (32MLI), Garrison Reservoir, North Dakota. In *Smithsonian Institution, Bureau of American Ethnology Bulletin* 176, *River Basin Surveys Paper* 19, pp. 159–238. Washington, DC.

——— 1968 *Big Bend Historic Sites*. Smithsonian Institution, River Basin Surveys, Publications in Salvage Archaeology 9. Lincoln, Nebraska

——— 1972 *Like-a-Fishhook Village and Fort Berthold, Garrison Reservoir, North Dakota*. Anthropological Papers 2, National Park Service, Washington, DC.

——— 1980 *The Exploration of the La Verendryes in the Northern Plains, 1738–43*, edited by W. Raymond Wood. University of Nebraska Press, Lincoln.

Smith, G. Hubert and John Ludwickson 1981 *Fort Manuel: The Archeology of an Upper Missouri Trading Post of 1812–1813*. Special Publication Number 7, South Dakota Archeological Society. University of South Dakota Archeological Laboratory. Vermillion.

Smithsonian Institution 1946–1947 *Sixty-Fourth Annual Report of the Bureau of American Ethnology to the Secretary of the Smithsonian Institution*, pp. 21–22. Washington DC.

——— 1948 *Annual Report of the Board of Regents of the Smithsonian Institution*, pp. 61–65. Washington, DC.

——— 1949 *Annual Report of the Board of Regents of the Smithsonian Institution*, pp. 84–85. Washington, DC.

——— 1950 *Annual Report of the Board of Regents of the Smithsonian Institution*, pp. 54–59. Washington, DC.

——— 1951 *Annual Report of the Board of Regents of the Smithsonian Institution*, pp. 61–87. Washington, DC.

Smithsonian Institution 1952 *Annual Report of the Board of Regents of the Smithsonian Institution*, pp. 62–63. Washington, DC.

——— 1953 *Annual Report of the Board of Regents of the Smithsonian Institution*, pp. 9–11. Washington, DC.

——— 1954 *Annual Report of the Board of Regents of the Smithsonian Institution*, pp. 7–12. Washington, DC.

——— 1956–1957 *Sixty-Fourth Annual Report of the Bureau of American Ethnology*, pp. 5–29. Smithsonian Institution, Washington, DC.

——— 1956–1957 *Seventy-Fourth Annual Report of the Bureau of American Ethnology*, pp. 76–111. Smithsonian Institution, Washington, DC.

——— 1957–1958 *Seventy-Fifth Annual Report of the Bureau of American Ethnology*, pp. 5–29. Smithsonian Institution, Washington, DC.

——— 1958–1959 *Seventy-Sixth Annual Report of the Bureau of American Ethnology*, pp. 10–15. Smithsonian Institution, Washington, DC.

——— 1960 *Annual Report of the Board of Regents of the Smithsonian Institution*, pp. 8–11. Washington, DC.

——— 1961 *Annual Report of the Board of Regents of the Smithsonian Institution*, pp. 56–61. Washington, DC.

——— 1961–1962 *Seventieth Annual Report of the Bureau of American Ethnology*, pp. 8–11. Smithsonian Institution, Washington, DC.

——— 1967–1968 *Eightieth Annual Report of the Bureau of American Ethnology*, p. 8. Smithsonian Institution, Washington, DC.

——— 2012a Archaeologists and Scholars, Waldo R. Wedel, 1908–1996. Smithsonian Olmec Legacy. Electronic document, anthropology.si.edu/olmec/english/archaeologists/wedel.htm, accessed April 12, 2012.

——— 2012b Collections Search Center. Electronic document, http://collections.si.edu/search/, accessed April 1, 2012.

Smithsonian Institution Archives 1952 Record Unit 89, Box 5, Folder: Research Program, 1952 (River Basin Surveys). Washington, DC.

Snow, Dean R., Mark Gahegan, C. Lee Giles, Kenneth G. Hirth, George R. Milner, Prasenjit Mitra, and James Z. Wang 2006 Cybertools and Archaeology. *Science* 311:958–959.

Snyder, Lynn M. 1995 *Assessing the Role of the Domestic Dog as a Native American Food Resource in the Middle Missouri Subarea A.D. 1000–1840*. Unpublished Ph.D. dissertation, University of Tennessee, Knoxville, 1995.

Snyder, Lynn M., Deborah Hull-Walski, Thomas D. Thiessen, and Myra J. Giesen 2000 Postwar Partners in Archeology: The Bureau of Reclamation, the National Park Service, and the River Basin Surveys in the Missouri River Basin (1945–1969). *CRM* 23(1): 17–20.

Society for Historical Archaeology 1967a Beginnings. *Historical Archaeology* 1:1–12.

——— 1967b Founding Membership. *Historical Archaeology* 1:12–22.

Solecki, Ralph 1948 Appraisal of the Archeological Resources of the Bluestone Reservoir, West Virginia. Manuscript on file, Smithsonian Institution, Washington, DC.

——— 1949a Archaeological Survey of Two River Basins in West Virginia. *Eastern States Archeological Federation* 8:5–6.

——— 1949b An Archaeological Survey of Two Rivers in West Virginia: The Bluestone Reservation. *West Virginia History* 10(3–4):319–432.

——— 1949c Appraisal of the Archaeological Resources of West Fork Reservoir, West Virginia: The Surveys. Manuscript on file, Smithsonian Institution, Washington, DC.

——— 1951 An Appraisal of the Archeological Resources of the Morgantown Locks in West Virginia, and the Conemaugh and Clarion East Branch Reservoirs in Pennsylvania. River Basin Survey, Smithsonian Institution, Washington, DC. On file Tozzer Library, Harvard University, Cambridge, Massachusetts.

Solecki, Ralph 1953 Exploration of an Adena Mound at Natrium, West Virginia. *Anthropological Papers 40, Bureau of American Ethnology Bulletin* 151:313–396. Washington, DC.

Solland, Sonja 1974 A Study of Conflict in a Multi-Ethnic Community. Unpublished Ph.D. dissertation, University of Washington, Seattle.

South, Stanley 1971 Evolution and Horizon as Revealed in Ceramic Analysis in Historical Archeology. *The Conference on Historic Site Archeology Papers* 6(2):71–10.

Sprague, Roderick 1959 A Comparative Cultural Analysis of an Indian Burial Site in Southeast Washington. Unpublished Master's thesis, Department of Anthropology, Washington State University, Pullman.

——— 1965 *The Descriptive Archaeology of the Palus Burial Site, Lyons Ferry, Washington.* Washington State University, Laboratory of Anthropology, Report of Investigations No. 32. Pullman.

——— 1967 Aboriginal Burial Practices in the Plateau Region of North America. Ph.D. dissertation, University of Arizona. University Microfilms, Ann Arbor.

——— 1973 The Pacific Northwest. In *The Development of North American Archaeology*, edited by James E. Fitting, pp. 250–285. Anchor Press/Doubleday, Garden City, New York.

——— 1984 A Check List of Columbia Basin Project Papers. *Northwest Anthropological Research Notes* 18:256–259.

Stein, Mary Anne 1966 An Archaeological Survey of Paiute Mesa, Arizona. Unpublished Master's thesis, University of Oklahoma, Norman.

——— 1984 Pottery Pueblo: A Tsegi Phase Village on Paiute Mesa, Utah. Unpublished Ph.D. dissertation, Southern Methodist University, Dallas.

Stephenson, Robert L. 1952 The Hogge Bridge Site and the Wylie Focus. *American Antiquity* 7:299–312.

——— 1954 Taxonomy and Chronology in the Central Plains-Middle Missouri River Area. *Plains Anthropologist* 1:15–21.

——— 1958 Statement of How Evelyn B. Stewart's Performance during the Rating Period Was Outstanding and of How It Has Compared with the Requirements of the Position. On file, River Basin Surveys, National Anthropological Archives, National Museum of Natural History, Smithsonian Institution, Washington, DC.

——— 1964a River Basin Surveys. In *Annual Report of the Board of Regents of the Smithsonian Institution: Operations, Expenditures, and Condition of the Institution for the Year Ended June 30, 1963*, pp. 83–106. Washington, DC.

——— 1964b Letter to Miss Helen Phillips, March 17, 1964. Copy on file, National Anthropological Archives, Smithsonian Institution, River Basin Survey Papers Box 6. Washington, DC.

——— 1967 Frank H. H. Roberts, Jr., 1897–1966. *American Antiquity* 32:84–94.

Stewart, R. Michael. 1983 Soils and the Prehistoric Archaeology of the Abbott Farm. *North American Archaeologist* 4(1):27–49.

Stirling, Matthew W. 1949 River Basin Surveys. In *Annual Report of the Board of Regents of the Smithsonian Institution: Operations, Expenditures, and Condition of the Institution for the Year Ended June 30, 1948*, pp. 60–84. Washington, DC.

——— 1950 River Basin Surveys. In *Annual Report of the Board of Regents of the Smithsonian Institution: Operations, Expenditures, and Condition of the Institution for the Year Ended June 30, 1949*, pp. 61–84. Washington, DC.

Stini, William A. 2010 Sherwood L. Washburn and the "The New Physical Anthropology." In *Histories of American Physical Anthropology in the Twentieth Century*, edited by Michael A. Little and Kenneth A. R. Kennedy, pp. 173–185. Lexington Books, Lanham, Maryland.

Story, Dee Ann Suhm 1963 Polyisopleths: A New Approach to Distribution Studies in Archaeology. Unpublished Ph.D. dissertation, University of California, Los Angeles.

Stout, Carol 1972 Zuni Transitivity: A Generative Approach. Unpublished Ph.D. dissertation, University of New Mexico, Albuquerque.

Strong, William D. 1933 The Plains Culture Area in the Light of Archaeology. *American Anthropologist* 35:271–287.

——— 1935 An Introduction to Nebraska Archaeology. *Smithsonian Miscellaneous Collections,* Vol. 93, No. 10. Washington, DC.

——— 1940 From History to Prehistory in the Northern Great Plains. In *Smithsonian Miscellaneous Collections,* Vol. 100, pp. 353–394. Smithsonian Institution, Washington, DC.

Strong, William Duncan and W. Egbert Schenck 1925 Petroglyphs Near The Dalles of the Columbia River. *American Anthropologist* 27:76–90.

Strong, William Duncan, W. Egbert Schenck, and Julian H. Steward 1930 *Archaeology of The Dalles-Deschutes Region.* University of California Publications in American Archaeology and Ethnology 29(1). Berkeley.

Suhm, Dee Ann 1959a Extended Survey of the Right Bank of the Glen Canyon. In *The Glen Canyon Archeological Survey, Part I*, by Don D. Fowler, James H. Gunnerson, Jesse D. Jennings, Robert H. Lister, Dee Ann Suhm, and Ted Weller, pp. 163–284. University of Utah Anthropological Papers 39, Glen Canyon Series 6. Salt Lake City.

——— 1959b A Report on Investigations at Two Archeological Sites in the Flaming Gorge Reservoir Area, Daggett County, Utah. Manuscript on file, University of Utah, Anthropology Department, Salt Lake City.

——— 1960a Additional Artifacts from the 1957 Excavations in the Glen Canyon. In *1959 Excavations, Glen Canyon Area*, by William D. Lipe, Floyd W. Sharrock, David S. Dibble, and Keith M. Anderson, Addendum, pp. i–iv, 1–19. University of Utah Anthropological Papers 49, Glen Canyon Series 13. Salt Lake City.

——— 1960b Cataloguing Archeological Collections. *Utah Archaeology: A Newsletter* 6(2):5–10.

Suhm, Dee Ann, Alex D. Krieger, and Edward B. Jelks 1954a An Introductory Handbook of Texas Archeology Part II, Selected Pottery and Projectile-Point Types: Descriptions and Illustrations—Spear Point Types. In An Introductory Handbook of Texas Archeology, edited by Alex D. Kreiger. *Bulletin of the Texas Archaeological Society* 25:396–399.

——— 1954b An Introductory Handbook of Texas Archeology Part II, Selected Pottery and Projectile-Point Types: Descriptions and Illustrations—Dart Point Types. In An Introductory Handbook of Texas Archeology, edited by Alex D. Kreiger. *Bulletin of the Texas Archaeological Society* 25:400–493.

——— 1954c An Introductory Handbook of Texas Archeology Part II, Selected Pottery and Projectile-Point Types: Descriptions and Illustrations—Arrow-Point Types. In An Introductory Handbook of Texas Archeology, edited by Alex D. Kreiger. *Bulletin of the Texas Archaeological Society* 25:494–511.

Swanson, Earl H., Jr. 1956 Archeological Studies in the Vantage Region of the Columbia Plateau. Unpublished Ph.D. dissertation, University of Washington, Seattle.

——— 1962 *The Emergence of Plateau Culture.* Occasional Papers of the Idaho College Museum No. 8. Pocatello.

Sweeney, Catherine L. 1961 Ethnohistoric Study in the Glen Canyon. *Utah Archaeology: A Newsletter* 9(3):9–13.

Sweeney, Catherine L. and Robert C. Euler 1963 *Southern Paiute Archaeology in the Glen Canyon Drainage: A Preliminary Report.* Nevada State Museum Anthropological Papers 9. Carson City.

Talmage, Valerie and Olga Chesler 1977 *The Importance of Small, Surface, and Disturbed Sites as Sources of Significant Archeological Data.* Cultural Resource Management Studies, Office of Archeology and Historic Preservation, National Park Service, Department of the Interior, Washington, DC.

Teague, Lynn S. and Patricia L. Crown 1982 *Hohokam Archaeology along the Salt-Gila Aqueduct Central Arizona Project, Volume I: Research Design*. Arizona State Museum, The University of Arizona, Tucson. (available at tDAR ID: 376375; doi:10.6067/XCV8G73FKG)

Teufel, Hugo, III 2003 Letter to Senator Chuck Hagel Regarding Legal Ownership of River Basin Survey Archeological Collections Housed in the University of Nebraska State Museum, June 10, 2003. U.S. Department of the Interior, Office of the Solicitor, Washington, DC.

Thiessen, Thomas D. 1994 The National Park Service and the Interagency Archeological Salvage Program in the Missouri River Mainstem Reservoirs. In "40 Something: The Smithsonian Institution River Basin Surveys." edited by Kimball Banks. *North Dakota Archaeology: Journal of the North Dakota Archaeological Association* 5:15–26.

——— 1999 *Emergency Archeology in the Missouri River Basin: The Role of the Missouri River Basin Project and the Midwest Archeological Center in the Interagency Archeological Salvage Program, 1946–1975*. National Park Service, Midwest Archeological Center, Special Report No. 2. Lincoln, Nebraska.

——— 2012 A Glance Back in Time. National Park Service, Midwest Archeological Center. Electronic document. www.nps.gov/mwac/history/index.htm, accessed April 12, 2012.

Thiessen, Thomas D. and Karin M. Roberts 2009 The River Basin Surveys Collections: A Legacy for American Archaeology. *Plains Anthropologist* 54(210):121–136.

Thomas, Cyrus F. 1888 Burial Mounds of the Northern Sections of the United States. *Fifth Annual Report of the Bureau of American Ethnology to the Secretary of the Smithsonian Institution, 1883–1884*, pp. 3–119. Washington DC.

Thomas, David Hurst 2000 *Skull Wars: Kennewick Man, Archaeology, and the Battle for Native American Identity*. Basic Books, New York.

Thompson, Raymond H. 1974 Institutional Responsibilities in Conservation Archaeology. In *Proceedings of the 1974 Cultural Resource Management Conference, Federal Center, Denver, Colorado*, edited by William D. Lipe and Alexander J. Lindsay, Jr., pp. 13–25. Museum of Northern Arizona Technical Series 14. Flagstaff.

——— 2000 Edgar Lee Hewett and the Political Process. *Journal of the Southwest* 42(2):271–318.

Thompson, Robert S. 1985 *Paleoenvironmental Investigations at Seed Cave (Windust Cave H–45FR46), Franklin County, Washington*. Eastern Washington University Reports in Archaeology and History 100–41. Archaeological and Historical Services, Cheney, Washington.

Towne, Arlean 1997 Shovels, Stones, and Old Bones: A First for the Sacramento Junior College. *Golden Notes* 43(4). Sacramento County Historical Society Sacramento.

Townsend, Jan 1994 Archeology and the National Register. *CRM* 17(2):10–12.

Treganza, Adán E. ca. 1950 Salvage Archaeology at Sites NAP-74 and NAP-93 in the Monticello Reservoir Area, California. Manuscript on file, NW Information Center, Sonoma State University, Sonoma, California.

——— 1952 *Archaeological Investigations in the Farmington Reservoir Area, Stanislaus County, California*. University of California Archaeological Survey Reports 14. Berkeley.

——— 1957 Archaeological Excavations in the Coyote Valley Reservoir Area, Mendocino County, California. Manuscript on file, Phoebe A. Hearst Museum of Anthropology, Berkeley.

——— 1964 Warm Springs Dam Project Archaeological Survey. Manuscript on file, National Park Service, Pacific West Regional Library, San Francisco.

Treganza, Adán E. and Martin Heiksen 1969 Salvage Archaeology in the Black Butte Area, Glenn County, California. In *The Archaeology of the Black Butte Reservoir Region,*

Glenn and Tehama Counties, California, Part 1, pp. 1–59. San Francisco State College Anthropology Museum, Occasional Papers 2. San Francisco.

Trigger Bruce G. 1984 Alternative Archaeologies: Nationalist, Colonialist, Imperialist. *Man* 19:355–370.

——— 1986 Prehistoric Archaeology and American Society. In *American Archaeology Past and Future: A Celebration of the Society for American Archaeology 1935–1985*, edited by David Meltzer, Don Fowler, and Jeremy Sabloff, pp. 87–215. Smithsonian Institution Press, Washington, DC.

Turner, Christy G. II and Jacqueline Turner 1998 *Man Corn: Cannibalism and Violence in the Prehistoric American Southwest*. University of Utah Press, Salt Lake City.

Tuross, Noreen and Marilyn L. Fogel 1994 Stable Isotope Analysis and Subsistence Patterns at the Sully Site. In *Skeletal Biology in the Great Plains: Migration, Warfare, Health, and Subsistence*, edited by Douglas W. Owsley and Richard L. Jantz, pp. 283–290. Smithsonian Institution, Washington, DC.

Ubelaker, Douglas H. 1994 An Overview of Great Plains Human Skeletal Biology. In *Skeletal Biology in the Great Plains: Migration, Warfare, Health, and Subsistence*, edited by Douglas W. Owsley and Richard L. Jantz, pp. 391–395. Smithsonian Institution, Washington, DC.

Ubelaker, Douglas H. and David R. Hunt 1995 The Influence of William M. Bass, III on the Development of American Forensic Anthropology. *Journal of Forensic Sciences* 40(5):729–734.

United States Bureau of Reclamation 2012 The Inception of the Pick-Sloan Missouri Basin Program. Electronic document, http://www.usbr.gov/gp/multimedia/archive/glimpse/picksloan.html, accessed April 12, 2012.

United States Army Corp of Engineers 2012 Conemaugh River Lake. Electronic document, http://www.lrp.usace.army.mil/Missions/Recreation/Lakes/ConemaughRiverLake.aspx, accessed July 10, 2012.

United States Conference of Mayors, Special Committee on Historic Preservation 1966 *With Heritage So Rich*. Random House, New York.

Vaughan, Alden T. 1982 From White Man to Redskin: Changing Anglo-American Perceptions of the American Indian. *American Historical Review* 87:917–935.

Wagner, Holly C. 1955 Projectile Point Analysis Talking Crow, South Dakota. Manuscript on file, Missouri River Basin Survey Collections, Archaeological Research Center, Biodiversity Institute, University of Kansas, Lawrence.

Walker, Danny N. 2011 New Deal and River Basin Survey Archaeology at Fort Laramie National Historic Site. *The SAA Archaeological Record* 11(5):21–23.

Walthall, John A. 1980 *Prehistoric Indians of the Southeast: Archaeology of Alabama and the Middle South*. University of Alabama Press, Tuscaloosa.

Washburn, Sherwood L. 1951 The New Physical Anthropology. *Transactions of the New York Academy of Science* 213:298–304. New York.

Watson, Patty Jo, Steven A. LeBlanc, and Charles L. Redman 1971 *Explanation in Archeology: An Explicitly Scientific Approach*. Columbia University Press, New York.

Webb, William S. 1938 *An Archaeological Survey of the Norris Basin in Eastern Tennessee*. Smithsonian Institution, Bureau of American Ethnology Bulletin 118. Washington, DC.

Webb, William S. and Charles E. Snow 1945 *The Adena People*. University of Kentucky Reports in Anthropology and Archaeology 6. Lexington.

Webb, William S. and William G. Haag 1947 *Archaic Sites in McLean County, Kentucky*. University of Kentucky, Reports in Anthropology 7. Lexington.

Wedel, Mildred Mott 1959 Oneota Sites on the Upper Iowa River. *Missouri Archaeologist* 21:2–4.

——— 1961 Indian Villages on the Upper Iowa River. *The Palimpsest* 42:561–592.

Wedel, Mildred Mott 1976 Ethnohistory: Its Payoffs and Pitfalls for Iowa Archeologists. *Journal of the Iowa Archeological Society* 23:1–44.

Wedel, Waldo R. 1936 *An Introduction to Pawnee Archeology*. Smithsonian Institution, Bureau of American Ethnology Bulletin 112. Washington, DC.

——— 1938 *The Direct-Historical Approach in Pawnee Archeology*. Smithsonian Miscellaneous Collections, 97(7):1-21. Washington, DC.

——— 1941 *Archeological Investigations at Buena Vista Lake, Kern County, California*. Smithsonian Institution, Bureau of American Ethnology Bulletin 130. Washington, DC.

——— 1953 Prehistory and the Missouri Valley Development Program: Summary Report on the Missouri River Basin Archeological Survey in 1948. In *Smithsonian Institution, Bureau of American Ethnology Bulletin* 154, *River Basin Surveys Paper* 1, pp. 1–59. Washington, DC.

——— 1961 *Prehistoric Man on the Great Plains*. University of Oklahoma Press, Norman.

——— 1977 The Education of a Plains Anthropologist. *Plains Anthropologist* 22(75):1–11.

——— 2006 William. Duncan Strong: Misadventures in Nebraska Archaeology in the 1930s. In *Plains Archaeology's Past: A Collection of Personal Narratives*, edited by M. F. Hawley and V. A. Wulfkuhle. *Plains Anthropologist* 51(200):521-526.

Wendorf, Fred 1962 *A Guide for Salvage Archaeology*. Museum of New Mexico Press, Santa Fe.

——— 2008 *Desert Days. My Life as a Field Archaeologist*. Southern Methodist University Press, Dallas.

Wendorf, Fred D. and Raymond H. Thompson 2002 The Committee for the Recovery of Archaeological Remains: Three Decades of Service to the Archaeological Profession. *American Antiquity* 67:317–330.

Werner, David J. 1972 The Zimmerman Site (36 Pi 14). In *Archeology in the Upper Delaware Valley: A Study of the Cultural Chronology of the Tocks Island Reservoir*, edited by W. Fred Kinsey, pp. 55–130. Anthropological Series 2. Pennsylvania Historical and Museum Commission, Harrisburg.

West Virginia Archeological Society 2010 West Virginia Archeological Society Website. Electronic document, http://wvarch.org/, accessed July 1, 2012.

Weston, Timothy, Stanley A. Ahler, David E. Griffin, and John M. Kjos 1980 Cultural Resource Survey of the Cross Ranch, Oliver County, North Dakota. Department of Anthropology and Archaeology, University of North Dakota, Grand Forks.

Wetmore, Alexander 1947 River Basin Surveys. In *Annual Report of the Board of Regents of the Smithsonian Institution: Operations, Expenditures, and Condition of the Institution for the Year Ended June 30, 1946*. Washington, DC.

——— 1948 River Basin Surveys. In *Annual Report of the Board of Regents of the Smithsonian Institution: Operations, Expenditures, and Condition of the Institution for the Year Ended June 30, 1947*, pp. 64–79. Washington, DC.

Wheat, Joe Ben 1953 An Archaeological Survey of the Addicks Dam Basin, Southeast Texas. *Smithsonian Institution, Bureau of American Ethnology Bulletin* 154, *River Basin Surveys Papers* 4, pp. 143–252. Washington, DC.

Wheeler, Richard Page 1953 Appraisal of the Archeological and Paleontological Resources of the Jamestown Reservoir, North Dakota: Supplement. Smithsonian Institution, Missouri Basin Project, Lincoln, Nebraska. Manuscript on file, Colorado State University Libraries, Fort Collins.

——— 1963 *The Stutsman Focus: An Aboriginal Culture Complex in the Jamestown Reservoir Area*. Smithsonian Institution, Bureau of American Ethnology Bulletin 185, Inter-Agency Archeological Salvage Program, River Basin Surveys Paper 30. Washington, DC.

——— 1995–1997 *Archeological Investigations in Three Reservoir Areas in South Dakota and Wyoming, Parts I–III*. Reprints in Anthropology, Vol. 46. J&L Reprints. Lincoln, Nebraska.

White, Nancy Marie 1999a Hester A. Davis. In *Grit Tempered. Early Women Archaeologists in the Southeastern United States*, edited by Nancy Marie White, Lynne P. Sullivan, and Rochelle A. Marrinan, pp. 206–229. University Press of Florida, Gainesville.

——— 1999b Martha A. Rollingson. In *Grit Tempered. Early Women Archaeologists in the Southeastern United States*, edited by Nancy Marie White, Lynne P. Sullivan, and Rochelle A. Marrinan, pp. 230–255. University Press of Florida, Gainesville.

White, Nancy Marie, Rochelle A. Marrinan, and Hester A. Davis 1994 Early Women in Southeastern Archaeology: A Preliminary Report on Ongoing Research. In *Women in Archaeology*, edited by Cheryl Classen, pp. 96–109. University of Pennsylvania Press, Philadelphia.

White, Theodore E. 1947a Preliminary Appraisal of the Paleontological Resources of the Republican River Sub-basin, Nebraska, Kansas, and Colorado. Manuscript on file, Tozzer Library, Harvard University, Cambridge, Massachusetts.

——— 1947b Preliminary Appraisal of the Paleontological Resources of the Smoky Hill River Sub-basin, Kansas, and Colorado. Manuscript on file, Morgan Library, Colorado State University, Fort Collins.

——— 1948 A Preliminary Appraisal of the Paleontological Resources of Certain Basins and Reservoirs in the Missouri River Basin. Manuscript on file, Morgan Library, Colorado State University, Fort Collins.

——— 1952a Preliminary Analysis of the Vertebrate Fossil Fauna of the Boysen Reservoir Area. *Proceedings of the U.S. National Museum* 102(3296):185–208. Washington, DC.

——— 1952b Observations on the Butchering Techniques of Some Aboriginal Peoples: No. 1. *American Antiquity* 17:337–338.

——— 1953 Appraisal of Paleontological Resources of Nine Reservoirs in the Missouri River Basin: Supplement. Manuscript on file, State Historical Society of North Dakota State Archives, Bismarck.

——— 1954 Preliminary Analysis of the Fossil Vertebrates of the Canyon Ferry Reservoir Area. *Proceedings of the U.S. National Museum.* 103(3326):395–438. Washington, DC.

White, Theodore E. and Paul L. Cooper 1951 Appraisal of the Archeological and Paleontological Resources of the Niobrara River Basin, Nebraska. Missouri River Basin Project, River Basin Surveys, Smithsonian Institution, Washington, DC. Manuscript on file, Morgan Library, Colorado State University, Fort Collins.

Will, George F. and Thad C. Hecker 1944 The Upper Missouri River Valley Aboriginal Culture in North Dakota. *North Dakota Historical Quarterly* 11(1–2).

Willey, Gordon R. 1949 *Excavations in Southeast Florida.* Yale University Publications in Anthropology 42. New Haven.

Willey, Gordon R. and Philip Phillips 1958 *Method and Theory in American Archaeology.* The University of Chicago, Chicago.

Willey, Gordon R. and Jeremy A. Sabloff 1974 *A History of American Archaeology.* Thames and Hudson, London.

Wilshusen, Richard H., Karin Burd, Johnathan Till, Christine G. Ward, and Brian Yunker 1999 The Dolores Legacy: a User's Guide to the Archaeological Program Data Data, Anasazi Heritage Center. Electronic document, www.blm.gov/pgdata/etc/medialib/blm/co/field_offices/ahc/documents.Par.24258.File.dat/doloresguide.pdf, accessed July 1, 2012.

Wilson, Rex 1978 Changing Directions in the Federal Archeology Programs. In *Social Archeology: Beyond Substance and Dating*, edited by Charles L. Redman, Mary Jane Berman, Edward V. Curtin, William T. Langhorne, Jr., Nina M. Versaggi, and J. C. Wanser, pp. 439–448. Academic Press, New York.

Witty, Thomas A. 1961a Appraisal of the Archeological Resources of the Council Grove Reservoir, Morris County, Kansas, November, 1961. Manuscript on file, Kansas State Historical Society, Topeka.

Witty, Thomas A. 1961b Appraisal of the Archeological Resources of the John Redmond Reservoir, Coffey and Lyon Counties, Kansas, February 1961. Manuscript on file, Kansas State Historical Society, Topeka.

——— 1962a *Archeological Investigations of the Hell Creek Valley in the Wilson Reservoir, Russell and Lincoln Counties, Kansas*. Kansas State Historical Society Anthropological Series 1. Topeka.

——— 1962b Archeological Investigations in the Wilson Reservoir, Russell and Lincoln Counties, Kansas, May 1962. Manuscript on file, Kansas State Historical Society, Topeka.

——— 1962c Appraisal of the Archeological Resources of the Cheney Reservoir, Reno, Kingman and Sedgwick Counties, Kansas. Manuscript on file, Kansas State Historical Society, Topeka.

——— 1962d Appraisal of the Archeological Resources of the Elk City Reservoir, Montgomery County, Kansas January, 1962. Manuscript on file, Kansas State Historical Society, Topeka.

——— 1963a Appraisal of the Archeological Resources of the Marion Reservoir, Marion County, Kansas, June 1963. Manuscript on file, Kansas State Historical Society, Topeka.

——— 1963b Archeological Investigations on the Milford Reservoir, Clay County, Kansas. Manuscript on file, Kansas State Historical Society, Topeka.

——— 1964 Appraisal of the Archeological Resources of the Perry Reservoir, Jefferson County, Kansas, April 1964. Manuscript on file, Kansas State Historical Society, Topeka.

——— 1965 Appraisal of the Archeological Resources of the Upper Verdigris Watershed, Chase, Greenwood and Lyon Counties, Kansas, October 1965. Manuscript on file, Kansas State Historical Society, Topeka.

Wood, W. Raymond 1960 The Boundary Mound Group (32SI-1): An Eastern Woodland Mound Complex in North Dakota. *Plains Anthropologist* 5(10):71–78.

——— 1967 *An Interpretation of Mandan Culture History*. Smithsonian Institution, Bureau of American Ethnology, Bulletin 198, River Basin Surveys Papers 39. Washington, DC.

——— 1993 Integrating Ethnohistory and Archaeology at Fort Clark State Historic Site, North Dakota. *American Antiquity* 58:544–559.

——— 1996 The Missouri River Basin on the 1795 Soulard Map: A Cartographic Landmark. *Great Plains Quarterly* 16(3):183–198.

——— 2006 The Missouri River Basin Surveys: Archeology Without the Middle "A". In *Plains Archaeology's Past: A Collection of Personal Narratives*, edited by M. F. Hawley and V. A. Wulfkuhle. *Plains Anthropologist* 51(200):671–682.

——— 2011 *A White-bearded Plainsman: The Memoirs of Archaeologist W. Raymond Wood*. University of Utah Press, Salt Lake City.

Wood, W. Raymond (editor) 1976 *Fay Tolton and the Initial Middle Missouri Variant*. Missouri Archaeological Society Research Series 13. Columbia.

——— 2008 *Twilight of the Upper Missouri River Fur Trade: The Journals of Henry Boller,* with an introduction by W. Raymond Wood. State Historical Society of North Dakota, Bismarck.

Wood, W. Raymond, William J. Hunt, Jr., and Randy H. Williams 2011 *Fort Clark and its Indian Neighbors: An Upper Missouri River Trading Post*. University of Oklahoma Press, Norman.

Wood, W. Raymond, Robert K. Nickel, and David E. Griffin 1984 *Remote Sensing the American Great Plains*. Supplement No. 9 to Remote Sensing: A Handbook for Archeologists and Cultural Resource Managers. National Park Service, Washington, DC.

Woodard, Robert S. Jr. 2006 The Appalachian Power Company along the New River: The Defeat of the Blue Ridge Project in Historical Perspective. Unpublished Master's thesis, Department of History, Virginia Polytechnic Institute, Blacksburg.

Woolfenden, Wallace B. 1969 A Study of 4–Glenn-10: The Brownell Indian Cemetery. In *The Archaeology of the Black Butte Reservoir Region, Glenn and Tehama Counties, California*, Part 2, pp. 1–100. San Francisco State College Anthropology Museum, Occasional Papers 2. San Francisco.

Woolworth, Alan R. 1956 Archeology of Fort Berthold I and Explorations at Like-a-Fishhook Village (32ML2) in 1954. Unpublished manuscript on file, State Historical Society of North Dakota State archives, Bismarck.

——— 2006 Adventures of a Plains Anthropologist-Historian. In *Plains Archaeology's Past: A Collection of Personal Narratives*, edited by M. F. Hawley and V. A. Wulfkuhle. *Plains Anthropologist* 51(200):649–670.

Woolworth, Alan R. and W. Raymond Wood 1960 The Archeology of a Small Trading Post (Kipp's Post, 32MNI) in the Garrison Reservoir, North Dakota. In *Smithsonian Institution, Bureau of American Ethnology Bulletin* 176, *River Basin Survey Papers* 20, pp. 239–306. Washington, DC.

Wormington, H. M. 1939 *Ancient Man in North America.* Popular Series 4. The Colorado Museum of Natural History, Denver.

——— 1944 *Ancient Man in North America*. Popular Series 4, 2nd (revised) ed. The Colorado Museum of Natural History, Denver.

——— 1948 Report of a Survey of the Denver Museum of Natural History in the Cherry Creek Reservoir Area, Arapahoe County, Colorado. Manuscript on file, National Park Service, Midwest Archeological Center, Lincoln, Nebraska.

——— 1957 *Ancient Man in North America*. Popular Series 4, 4th (revised) ed. The Colorado Museum of Natural History, Denver.

Wright John R. and Lori Rohrer 2002 Between a Rock and a Hard Place: Archeology in a National Recreation Area. *CRM* 3:18–20.

Wyoming Authors Wiki 2012 Merrill J. Mattes, 1910–1996. Electronic document, http://wiki.wyomingauthors.org/w/page/28426863/Merrill%20J%20Mattes, accessed April 12, 2012.

Index

About the Authors

Gerard Baker—Yellow Wolf is the retired Assistant Director, NPS, Tribal Relations. He is an enrolled member of the Mandan—Hidatsa-Arikara Nation of the Fort Berthold Indian Reservation, Mandaree, North Dakota. He grew up on his parent's cattle ranch on the West side of the Reservation. He is a 1979 graduate of Southern Oregon State University with degrees in Criminology and Sociology. He worked and retired from the National Park Service after 35 years. In his career he has worked in various appointments in many states, always working with local American Indian Tribes for inclusion in the NPS. He and Mary Kay live in the Black Hills of South Dakota.

Kimball M. Banks earned his Ph.D. from Southern Methodist University. He has spent most of his career in Federal service, first with the Bureau of Indian Affairs and then the Bureau of Reclamation. He has conducted field work in the Southwest, the Southeast, Texas, California, the Northern Plains, and Egypt. While with the Bureau of Reclamation, he administered investigations at the Angostura Site, the McKean Site, and other sites that the River Basin Surveys and Interagency Archeological Survey Program investigated, which stimulated his interest in the history of the River Basin Surveys. He currently manages the Regional Office of Metcalf Archaeological Consultants, Inc. in Bismarck, North Dakota.

Jon S. Czaplicki has been an archaeologist in the Bureau of Reclamation's Phoenix Area Office for 24 years; prior to coming to Reclamation he was a Project Director at the Arizona State Museum's Cultural Resource Management Division for 12 years. He received his BA with Honors (1970) and MA (1971) from the University of Arizona. He's active in the Society for American Archaeology serving on the Public Education Committee from 1995 to 2001 during which time he was Chair for the Subcommittee for Native American Education. He was on the SAA Board from 2001–2003 and chaired Media Relations Committee from 2006–2013.

Don D. Fowler is the Mamie Kleberg Distinguished Professor of Anthropology, Emeritus, University of Nevada, Reno. He was a staff archaeologist for the Glen Canyon Project for six years beginning in 1957, his junior year of the University of Utah. He served as President of the Society for American Archaeology (SAA), 1985–87, and has received many honors, including Lifetime Achievement awards from the SAA and the RPA and a Distinguished Doctoral Graduate Medal from the University of Pittsburgh. He has completed 80 plus grants and contracts in CRM and related fields, and authored over 100 publications, most recently *The Glen Canyon Country, a Personnal Memoir* (2011).

David Mayer Gradwohl is a Professor Emeritus of Anthropology at Iowa State University and the Founding Director of the Iowa State University Archaeological Laboratory. He obtained his BA at the University of Nebraska with majors in anthropology and geology. While an undergraduate, he worked on archaeological crews fielded by the Nebraska State Historical Society and the University of Nebraska State Museum. After a graduate year at the University of Edinburgh, Scotland, under the auspices of a Fulbright Fellowship, he returned to the U.S. and obtained a Ph.D. in anthropology at Harvard University. His field work has centered on prehistoric and historic archaeology in the Prairies and Plains and, most recently, ethnoarchaeology dealing with the relationship between material culture and ethnicity as manifested in cemeteries and mortuary behavior.

John W. Hoganson is the State Paleontologist of North Dakota with the North Dakota Geological Survey and curator of the North Dakota State Fossil Collection housed at the North Dakota Heritage Center in Bismarck. John graduated from North Dakota State University with a Bachelor's Degree in Earth Science. He then obtained a Master's Degree in geology from the University of Florida and a Doctorate in geology from the University of North Dakota. John has authored numerous scientific and popular articles and chapters in books about North Dakota geology and paleontology. In 2003, he co-authored *Geology of the Lewis and Clark Trail in North Dakota*. His 2007 book, *Dinosaurs, Sharks, and Woolly Mammoths: Glimpses of Life in North Dakota's Prehistoric Past*, was the winner of the State Historical Society of North Dakota's 2007 editor's award and was recognized as a 2007 Notable Government Document by the American Library Association. In 1993, he received the Governor's Award for Excellence in Public Service and in 2001 was honored as a master alumnus by the College of Science and Mathematics at North Dakota State University. In 2011 he was presented with the Leonard Award for outstanding scholarship from the Department of Geology and Geological Engineering at the University of North Dakota.

Deborah Hull-Walski is a Collections Manager at the Smithsonian Institution's National Museum of Natural History (NMNH) and a part-time faculty member for The George Washington University's Museum Studies Program. She has a BA in Anthropology and History from the University of Colorado, and a MS in Historical Archaeology from Illinois State University. At the NMNH, she has provided oversight and coordination of collections management activities for the Anthropology Department. Since 2011, Deb has assembled and led a team to upgrade, rehouse, and digitize the Office of Education and Outreach's 20,000 object teaching collection for use in Q?rius, the NMNH's interactive learning center. Her professional interests include museum collections management, genealogical research, and improving researcher access to River Basin Surveys collections. Most recently, Deb has been going quietly insane in the attics of the Natural History building while reviewing chapters and looking for the consistent use of the word "fieldwork."

Edward B. Jelks, in a career spanning more than 60 years, excavated scores of archaeological sites in Texas, Virginia, Illinois, New York, Wyoming, Newfoundland, and Micronesia, published prolifically on his fieldwork and on other archaeological subjects, and taught archaeology at The University of Texas, Southern Methodist University, and Illinois State University. He was president of the Society for Historical Archaeology and of the Society of Professional Archaeologists, and served on numerous committees of those and other scholarly societies. Awards include the J. C. Harrington Medal (historical archaeology), the Curtis Tunnell Award (Texas archaeology), and the Clarence Webb Award (Caddoan archaeology).

Ruthann Knudson began working as a archaeologist in 1963, and has analyzed several Missouri River Project collections that originated within the River Basin Surveys (especially Anton Rygh, Red Smoke). She has had the privilege of knowing well most of the women about whom she writes in this volume; they have been colleagues and role models throughout her career. She has also been acquainted with many of the men involved in these projects across the United States—leaders, crew members, and researchers. It has been a privilege to document some of the critical contributions made by RBS women.

Richard A. Krause received a BA from The University of Nebraska in 1961, an MA from The University of Nebraska in 1962, and a Ph.D. from Yale University in 1967. He was formerly a Chairman and is now a Professor Emeritus of Anthropology at The University of Alabama. He is also a Senior Archaeological Fellow at Tennessee Valley Archaeological

Research, a Senior Consultant for Soltec International, and is currently affiliated with Tennessee Valley Archaeological Research.

William B. Lees is Executive Director of the University of West Florida's Florida Public Archaeology Network. He holds the MA and Ph.D. in Anthropology from Michigan State University and is a Registered Professional Archaeologist. He specializes in historical archaeology of the Great Plains and Southeastern US, the Antebellum South, and Civil War archaeology and memory. He is past president of the Plains Anthropological Society, the Register of Professional Archaeologists, and the Society for Historical Archaeology.

R. Lee Lyman is Professor of Anthropology at the University of Missouri-Columbia. His research interests include zooarchaeology, conservation paleozoology, prehistory of the interior Pacific Northwest, and the history of North American archaeology.

Francis P. McManamon is the Executive Director of the Center for Digital Antiquity in the School for Human Evolution and Social Change at Arizona State University. The Center devoted to improving access to archaeological data and documents and ensuring their long-term preservation and availability for current and future uses. Until he joined Digital Antiquity in November, 2009, he was the Chief Archeologist of the National Park Service and Departmental Consulting Archeologist for the Department of the Interior. He is coeditor of *The Antiquities Act: A Century of American Archaeology, Historic Preservation, and Nature Conservation*, which received a New Mexico Heritage Preservation Book Award in 2006. He is the general editor of the four-volume encyclopedia, *Archaeology in America*, named by the American Library Association an outstanding reference publication in 2010.

Vanessa A. Mirro has served in various capacities on cultural resources projects throughout the United States, Canada, and Belize. She holds a Masters' degree in Physical Anthropology from Colorado State University. Her areas of expertise include the prehistory of California, cave archaeology of the ancient Maya, human osteology, and mortuary practices. She is currently a Regional Manager and Senior Archaeologist in southern California with Applied EarthWorks, Inc. Ms. Mirro is certified by the Register of Professional Archaeologist (RPA).

Michael J. Moratto (Ph.D., RPA) is Principal Archaeologist with Applied EarthWorks, Inc. in California. A retired university professor, he has directed archaeological projects throughout the American West and has served as president of local, statewide, and national professional societies. Notable among his many scholarly works is the textbook, *California Archaeology* (1984, 2004).

Susan C. Prezzano is a professor of anthropology at Clarion University of Pennsylvania where she has taught since 1995. She specializes in Late Woodland community patterns in the Northeast and ceramic analysis. Current research includes the analysis of the multicomponent Millstone site situated along the Clarion River in northwest Pennsylvania, a site with a Middle Archaic occupation dating to 8,000 years ago. She is a Co-PI on a recently awarded National Science Foundation Major Instrumentation grant to obtain a scanning electron microscope.

Francis "Fritz" A. Riddell (MA), after serving as a Marine during WWII, "Fritz" attended the University of California, Berkeley, where he participated in SI-RBS projects in the Pacific states as well as in field research in Peru. Later, he directed the California State Indian Museum and in 1960 became California's first State Archaeologist, a position held until his retirement in 1983. In 2002, he lost a long and courageous battle with cancer.

Lynn M. Snyder is a semi-retired contractor and researcher, living on a small farm in Adams County, Illinois. She has a doctorate in anthropology from the University of Tennessee, Knoxville, with a concentration in zooarchaeology, and a Masters from the University of Nebraska, Lincoln. She worked as a contractor/researcher at the Smithsonian Natural History Museum, in Washington, DC for many years. Her doctoral research explored the role of dogs in Native American culture and economy, utilizing the extensive candid collections produced by River Basin Surveys projects in Nebraska and the Dakotas. She has worked on North American zooarchaeological materials from Archaic through Historic, as well as Minoan through Classical Greece and Crete.

Thomas D. Thiessen, a native of Omaha, Nebraska, received bachelor's and master's degrees in anthropology from the University of Nebraska-Lincoln. After more than 33 years as an archeologist with the National Park Service, he retired in 2006 and resides near Pleasant Dale, Nebraska. He worked at the Midwest Archeological Center in Lincoln; the Interagency Archeological Services office in Denver; and the Knife River Indian Villages National Historic Site in North Dakota. His professional interests include the history of Plains archeology. In 1999, the Midwest Archeological Center published his *Emergency Archeology in the Missouri River Basin: The Role of the Missouri Basin Project and the*

Midwest Archeological Center in the Interagency Archeological Salvage Program, 1946–1975.

W. Raymond Wood was trained at the University of Nebraska (BA, MA) and at the University of Oregon (Ph.D.) and taught anthropology at the University of Missouri from 1963 to 2002. During that time he was involved in archaeological and historical work in Missouri and in each of the states that the Missouri River touches. His work has been largely historical for the past two decades, having specialized in Lewis and Clark and their predecessors, the explorations of Prince Maximilian, the early cartography of the Missouri River, and the Native Americans that lived along the Missouri, especially the Mandans and Hidatsas in present North Dakota. Among his latest books are *Prologue to Lewis and Clark* (University of Oklahoma Press, 2002), *Karl Bodmer's Studio Art* (University of Illinois Press, 2003), and *Fort Clark and its Indian Neighbors* (University of Oklahoma Press, 2011). He has been professor emeritus at the University of Missouri, Columbia since 2002.